COMPUTERS, ETHICS, AND SOCIETY

M. David Ermann
University of Delaware

Mary B. Williams
University of Delaware

Claudio Gutierrez
University of Delaware

New York   Oxford
OXFORD UNIVERSITY PRESS
1990

For Natalie, Mike, and Marlene

Oxford University Press

Oxford   New York   Toronto
Delhi Bombay Calcutta Madras Karachi
Petaling Jaya Singapore Hong Kong Tokyo
Nairobi Dar es Salaam Cape Town
Melbourne   Auckland

and associated companies in
Berlin   Ibadan

Copyright © 1990 by Oxford University Press, Inc.

Published by Oxford University Press, Inc.,
200 Madison Avenue, New York, New York 10016

Oxford is a registered trademark of Oxford University Press

Library of Congress Cataloging-in-Publication Data
Computers, ethics, and society / [edited by] M. David Ermann,
Mary B. Williams, Claudio Gutiérrez.
p. cm.   Bibliography: p.
ISBN 0-19-505850-X
1. Computers and civilization. 2. Computers—Access control.
3. Human-Computer interaction. I. Ermann, M. David.
II. Williams, Mary B. III. Gutiérrez Carranza, Claudio, 1930–
QA76.9.C66C6575   1990
303.48'34—adc20   89-3410   CIP

Art in reading 16 from "The Mechanization of Office Work" by
Vincent E. Guiliano, *Scientific American,* Volume 247, Number 3.
Copyright © 1982 by Scientific American, Inc. All rights reserved.

2 3 4 5 6 7 8 9

Printed in the United States of America

Preface

Faculty at many universities now recognize a need to sensitize students (before they begin their careers) to ethical and social implications of the growing use of computers. In this collection of essays we have tried to provide stimulating analytical selections to meet that need. The readings are intended for junior- and senior-level students with some computer-related interests or coursework, making the book appropriate as a supplement for courses addressing the question of how computers affect society and examining the ethical issues raised by their use.

Selections in this book are a subset of readings we considered while teaching interdisciplinary courses on "Computers, Ethics, and Society" at the University of Delaware, with initial funding by a grant from the National Endowment for the Humanities. We used our classes to identify promising readings and to decide whether each selection accomplished its intended goals. Anonymous questionnaires, requiring students to comment on individual selections, indicated whether they shared our evaluations. Among the areas these readings cover are the ethical obligations of computer specialists, the effects of computers on the work experiences of nonspecialists, and the loss of privacy possible with increased computerization. Specific topics within these issue areas include property rights to computer technologies, codes of ethics for computer professionals, and computer-related unemployment.

We prepared this book because we believe formal and focused discussions of such issues are important. On the negative side, for instance, as computers become increasingly important to business and government, and as these machines continue to gain power, their potential for stripping us of our privacy increases exponentially. Currently this is only a possibility; there is nothing inherent in computer technology that results in a loss of privacy. But unless the technologists who are designing and working with computers understand how their work might reduce privacy, this right may not be protected.

A positive potential also exists. Computers can enhance the quality of human life. Computer technologists, if they strive to make that potential a reality, can go forward with an awareness of how their work can be socially beneficial. To do so, however, they must recognize that the benefits (as well as the risks) of computers are only possibilities. Nothing inherent in computers will ensure that they will be used to improve the quality of life.

Students generally understand the positive potential of computers, but they have a stronger faith than we consider warranted that it will become a reality without conscious effort. We believe these readings will provide them with a greater desire to discover what they will have to do if their work in the computer field is to contribute to creating a better society.

The University of Delaware's Center for Science and Culture has been providing the base from which we have tried to bridge the gulf separating the humanities and social sciences from technical fields. When we approached the computer science department with a proposal for this course, the department responded enthusiastically and made it a requirement for all undergraduate majors. The sociology department showed consistent and strong support for our efforts. The National Endowment for the Humanities (Grant #EL-20142-86) provided financial support to the Center for Science and Culture that enabled us to continue to develop this course.

Among our colleagues we particularly want to thank Ron Martin for his encouragement and his contributions as a coinstructor in the most recent offering of the course. Finally, we thank Constance Weber, our secretary, who kept us coordinated with good cheer and perfect thoroughness through endless revisions.

Newark, Del.   M.D.E.
February 1989   M.B.W.
                                                           C.G.

Contents

Introduction

Like any powerful tool, computers can be a force for good or harm. They can give physicians instant access to all of the information available on a patient in crisis, but they can also give unscrupulous charlatans the names of all cancer patients. They can give a business important new control over its inventory, but they can also give it privacy-invading control over its employees. Computerized robots might make work obsolete while producing abundance for everyone, but they can also throw millions of people into unemployment and poverty. To maximize the benefits and minimize the harms, those making decisions about the implementation of computer systems (e.g., programmers, systems designers, computer scientists, managers, legislators) must be sensitive to the potential problems as well as to the potential advantages of computers.

Because the computer gives us fundamentally new power, we are faced with decisions for which our experiences may give little guidance. The danger of applying old standards to a fundamentally new situation is well illustrated by the law, passed soon after the production of the first automobiles, which required cars traveling the roads to be preceded by a man on foot carrying a red flag. This law reduced danger, but robbed the auto of its intrinsic power. Similarly we could stop one type of computer crime by outlawing electronic fund transfer, or prevent a potentially dangerous accumulation of governmental power by outlawing the interconnection of computers storing different sets of information about individuals (see Burnham, reading 10), or prevent robots from taking workers' jobs by outlawing robotization. It is possible to respond to every danger by cutting off the power that leads to that danger. But it is more productive to respond by analyzing each situation as it occurs. This way, we may conclude that our fundamental values are better served by changing our expectations or rules rather than by denying ourselves opportunities to take advantage of what the computer can do. Such an analysis requires some understanding both of the social problems that computers may cause and of the nature of our moral system. The first part of this introduction focuses on how computer-related moral problems are approached; the second part focuses on potential effects of computers and the moral problems generated by these powerful machines.

Approaching Moral Problems

Decisions About the Morality of Particular Actions

We will begin by sketching an analysis of the morality of actions in a particular case. This case is based on an actual occurrence, though we have changed some details.

> Dr. Pierre Leveque, a French computer scientist visiting an American computer science department for one year, discovered a bug in the mainframe computer operating system which allowed him to borrow the identity of other users, bypassing their passwords. If a user who was logged on left her terminal unattended briefly while Dr. Leveque was in the room, he could enter a few keystrokes and forever gain access to her files or have her billed for the time he used the computer. Having discovered this bug, Dr. Leveque proceeded to collect as many identities as he could. He never abused the power he had; he neither read others' files nor billed his computing time to them.
>
> When Dr. Leveque's colleagues discovered what he had been doing and discussed the case among themselves, they disagreed about how to characterize his behavior. Some believed it was morally wrong; others, although they felt uneasy about the behavior, could not justify condemning Dr. Leveque when it seemed he had not hurt anyone. All agreed, however, that he should be forced to destroy his access to those identities he had collected.

We can partially resolve this case by recognizing that we can condemn the act without condemning the actor. Since all agreed that Dr. Leveque should destroy his access to the collected identities, the consensus was that there was something wrong about his having them. The disagreement was between one group that felt the act was wrong but Dr. Leveque was not blameworthy, and a second group that felt the computer scientist should be blamed for committing an immoral act. This kind of disagreement is to be expected in a new field, with new types of situations for which the relevant ethics have not yet been the subject of much discussion or experience. Once the moral aspects of a situation have been the subject of considerable analysis and discussion, we can expect more agreement about which actions are ethical and which are not. Hence if another faculty member is caught next year collecting identities, we can predict that he or she is more likely to be judged blameworthy, since precedent has been set.

For the purposes of the last paragraph we have simply assumed the correctness of the premise that the act was immoral. Let us now test this assumption with the help of moral theory. First we will use utilitarian theory, since some of the disagreement in the computer science department apparently came from those who felt that the only morally relevant feature of the act was whether it hurt anyone. Utilitarians argue that we should judge all actions solely by their consequences for the well-being of those affected by the action (see Hospers, reading 1). Thus, since Dr. Leveque's identity collection did not actually harm anyone, it seems that the utilitarians would conclude that the act was morally neutral.

However, utilitarians count the long-term effects of an act as well as its immediate effects. It is probable that some people who hear of Dr. Leveque's actions will look for and discover his method of collecting identities, and that some of these people will abuse that knowledge by invading others' privacy or by charging computer time to others' accounts. (The bug cannot be removed from the operating system, though the system manager can install checks to minimize the harm.) Utilitarians would hold Dr. Leveque accountable for the damage caused by people who used his method to gain access to files. On the other hand, the moral discussion within the computer science department that was generated by Dr. Leveque's deed expanded the department's understanding of the novel moral problems presented by computers, and this must be counted as a beneficial consequence of his actions. The benefits and damage to Dr. Leveque himself also must be counted in the utilitarian analysis; in this case, the pleasure that Dr. Leveque gained from collecting identities was probably far outweighed by the distress he suffered as the object of angry debate within the department.

The utilitarian would conclude that Dr. Leveque's actions were moral if, compared to the alternatives available to him, his behavior produced more total benefit (or less total harm) than any other course would have produced. Dr. Leveque's obvious alternative action upon discovering the bug in the computer operating system was to inform the system manager so that the manager could take steps to prevent anyone from collecting identities. While this alternative clearly seems to be the morally right thing to do, it is at least arguable that, because his actions significantly upgraded the department's moral understanding, Dr. Leveque made a beneficial contribution that outweighed the harmful effects of his actions and was therefore morally right.

Many people believe that the utilitarian analysis, because it relies solely on the consequences of the action, misses the important point that there is something about the nature of some acts that make them morally wrong. Critics of the utilitarian method would say, for example, that lying is wrong even when, in a particular case, it produces a favorable result. Immanuel Kant (see Rachels, reading 3) is a principal proponent of this view, so those who object to the utilitarian conclusion given above might find that a Kantian analysis better captures their moral intuitions.

For the Kantian, the important ethical principle is that a person should always treat others as ends in themselves, never merely as means to one's own ends. In collecting the computer identities of others, Dr. Leveque was treating others as means to demonstrate (to himself) his ability to gain power over them. By ignoring the desire of others to determine for themselves whether any particular person has power over them, he failed to treat them as people with the right to have their goals respected. Thus, for the Kantian Dr. Leveque's actions would have been wrong even if they had no harmful consequences.

An analysis such as this is useful when a new situation appears that doesn't seem to be covered by the intuitive rules that we use every day in making moral decisions. But once the members of a community agree that a particular type of action is immoral, that agreement is encoded in a rule of behavior. A profession may formally gather these rules into a code of professional conduct (see reading 32), but a far more powerful way of ensuring that members of a professional community follow ethical precepts is to use a cautionary tale.

Implementing Decisions About Professional Morality: Cautionary Tales

When one of us was taking a medical school anatomy class many years ago, the professor told the class about a student who had taken home the calvarium (top of the skull) of his cadaver and used it as an ashtray. When this was discovered, the student was expelled from medical school.

In another incident, a friend who had been a British army officer during World War II repeated to one of us a tale he had heard of an officer who refused a very dangerous assignment (to try to sneak through enemy lines to carry a message from his surrounded battalion) and had been dismissed from the service for cowardice. In repeating this story, the friend commented that he never understood how the officer could refuse; he himself would have faced certain death rather than the contempt of his fellow officers.

These stories are cautionary tales; by means of memorable narratives, neophyte professionals learn what types of behavior are considered contemptible by members of the group to which they belong. Cautionary tales are a normal part of socialization into a profession; scholars hear cautionary tales about plagiarists, scientists hear cautionary tales about data-fakers, and so forth. These tales are most important when the behavior is prohibited by virtue of a special obligation of the particular profession, and would be considered only a misdemeanor if exhibited by someone outside the profession. Strong prohibitions against such behavior are not learned as a part of one's general acculturation into society.

The computer profession is very young and has not yet developed strong cautionary tales. At this time, the profession is still defining the special obligations of computer professionals, and the cautionary tales that are being told have not yet been stripped of irrelevant and distracting elements. Consider the following two illustrations. The first is a tale told by a computer science professor leading a class discussion on the propriety of looking at other people's files.

> Two years earlier, a graduate student had written a program so that a person logging on at an apparently vacant terminal was actually logging on through the student's program, exposing his or her password in the process. When the professor discovered this, he discussed with the graduate student why it was wrong to use such programs.

The professor told this story in order to emphasize that it was wrong to do this, but his intention was undermined by his tone (which disclosed his admiration for the cleverness of the method), by his failure to specify how the graduate student had been punished, and by his casual mention of subsequent professional triumphs of the student involved.

The second illustration involves a tale told by John Dvorak in *PC Magazine* (1988). After describing the Lehigh University virus, which can infect personal computers and then thoroughly erase every disk (including any hard disks), nulling the boot tracks, the FAT tables, and so forth, Dvorak writes:

> The mainstream computer magazines seldom discuss these destructive little gags, even though there are plenty of them. PC users must make themselves aware of these things. If a virus program got into a corporation and started eating hard disks, you can be sure that the next time someone brought in some software from home, it would be quickly confiscated. . . . Remember, the most talented of the hackers love to design programs like this just to harrass the average PC user.

Dvorak's article is intended to warn PC users that these viruses are dangerous. But he unconsciously weakens his point by his choice of the words "little" and "gag." And notice how he shows his respect for the talent shown in developing viruses, and how he unintentionally encourages virus-writing behavior by telling his readers (many of them hackers) that the most talented hackers love to write such programs. In fact, his admiration has prompted a statement which is simply false; if the most talented hackers loved to design such programs to harass average users, every PC in the country would already be infected. A person who wanted to warn against computer viruses in language proper for a cautionary tale—that is, in language loaded with contempt for the virus programmers—could have written: "The creators of computer viruses are either vicious creeps who love to destroy other people's work, or moral morons who are unable to understand that such destruction is wrong."

While the emotional tone of the tales told in medical school and in the military was one of unadulterated contempt for the transgressors, the emotional tone of the tales told to the computer science students was one of anger adulterated with admiration for the skill of the transgressors. The military tale does not distract the listener with the possibility that it took great courage to refuse the order, because according to military ethics, refusing to accept a dangerous assignment represents a contemptible misuse of courage. Similarly, the computer tales should not distract the listener with a mention of the skill needed for the transgression; even the greatest skill, if contemptibly used, is contemptible.

Decisions About the Morality of a Life-style

In addition to raising questions about how computer professionals should act, computerization also raises questions about the kind of life we all will lead. In fact, computerization may initiate changes so radical that we will need to rethink the fundamental values of our society. For example, by causing the disappearance of work as we know it for most people, computerization may require reexamination of the contemporary American belief that a person's work is fundamental to his or her identity as a worthwhile person. Would such a computerized society deprive the majority of people of the possibility of a worthwhile life? (Kurt Vonnegut suggests in *Player Piano* that it would.) Or is the belief that productive labor is essential for a fully-rounded human existence merely a myth fostered by a culture deeply influenced by the Protestant work ethic?

To grapple with such issues we must ask even more fundamental questions: What is the ideal life for a person? What set of opportunities for work, play, intellectual enrichment, spiritual enrichment, physical activity, creativity, achievement, contribution to society, power, luxury, and so forth, enables a person to fulfill his or her potential? Looking at the way thinkers in other societies in other centuries answered these questions gives us a glimpse of the range of possible answers, and may help to give us the perspective needed to seriously consider what is really important for a full life.

Aristotle (see reading 7) is so far from subscribing to the belief that work is fundamental to a worthwhile life that he does not even mention the possibility in his discussion. He assumes that the work necessary to provide food, clothing, and other necessities is done by slaves (as in our hypothesized computerized society it would be done by computers), and he is thus specifically concerned with the ideal life for those who do not have to work for a living. After rejecting the possibility that enjoyment, fame, virtue, or wealth is of paramount importance, he concludes that the active exercise of one's reasoning powers is the fundamental good making life worthwhile. If this is correct, the fully computerized society would free people to lead more fulfilling lives than is now possible for them. Similarly, if St. Augustine (see reading 8) is right in concluding that the fundamental good making life worthwhile is the perfecting of one's soul by following God, then the disappearance of work would be irrelevant to the achievement of the good life.

Many observers would claim that questions about whether a worthwhile life could be achieved in any ideal computerized society are irrelevant, since such an ideal society is not achievable. To implement even an approximation to this ideal we must grapple not only with the nature of the good life but also with the concrete problems that may be caused by the computerization of a society; therefore let us consider the impact of computers on work, economic justice, privacy, power, and sense of self.

The Potential Impact of Computers

On Work

Computers can be used in many ways by business: to automate work processes, to monitor employees' work and efficiency, to maintain massive amounts of personnel data, and even to reduce building and transportation costs by having employees work at home using the telecommunication features of computers. Each of these uses has potential risks and benefits.

Thanks to computers, for instance, managers now can monitor employees' business calls, their minute-by-minute work patterns, and the time they spend in contact with customers. Already today, about 5 million workers, mostly in clerical or repetitive jobs, have some or all of their work evaluated on the basis of computer-generated data, and many more have computer-generated data collected but not currently used in evaluation. This information, according to critics, can invade employees' privacy, reduce their personal dignity, and even affect their health (Office of Technological Assessment, 1987). Consider the case of Patricia Johnson, a post office employee in Washington, D.C. She sorts fifty letters per minute, remembering thousands of addresses in two zip codes in order to assign each letter correctly to one of seventy letter carriers. And now computerization allows her supervisor to watch her like a hawk, creating greater stress than she previously experienced: "The mail is running by me and running by me and the machine kind of hypnotizes you. And this computer is looking over your shoulder, watching you. It gets very stressful. . . . The supervisor knows everything about you, right in that machine" (Perl, 1984).

Of course, Patricia Johnson's experience may reflect our current inability to use computers to full advantage. In the future, computers may become less tools for external supervision and more integrated aids to workers. For instance, computers may revolutionize the conditions of office work. Technologists are already experimenting on artificial intelligence–inspired systems that actually create a worker/machine partnership and coordinate the interactions of different workstations. These integrated systems keep track of relationships among administrative tasks and subtasks, of peoples' promises to complete tasks, and of actual fulfillment of tasks (Gutierrez and Hidalgo, 1988). The result is an environment in which the worker has up-to-date information about which of his unfinished tasks is most crucial to the system. Such a partnership between worker and computer might well increase the worker's job satisfaction and decrease the need for external supervision.

Large organizations are ravenous information consumers. Because computers can collect, process, and exchange massive quantities of information, they can help corporations violate an individual's rights. The potential for misuse of some information about employees is inescapable. Many employers have tried to counter the possibility of misuse of personnel records by initiating policies that limit the amount of computerized (and other) data in employee records. Control Data Corporation's policy, for instance, states that individually identifiable employee records should be collected only if justified by the needs of (1) a specific business decision, (2) payroll, benefits, or other administrative procedures, or (3) requirements of government reporting. The policy addresses four ways employee records could be used unfairly: asking for unneeded information; gathering information in unfair ways (e.g., polygraphs, secret investigations); failing to keep accurate and complete records; and failing to properly restrict access to records. Among policies adopted to regulate data collection are requirements that employees be informed of data acquired "through interfaces with external organizations (insurance carriers, credit card companies, etc.)," and that the company not collect or store information regarding "political opinions, religious or other beliefs and sex life" of employees (Control Data Corporation, 1984). Doubters point out that such prohibitions are nice, but they are not always obeyed.

The introduction of computers for work in the home, as a total or partial substitute for work at the office or factory, creates additional concerns. On the one hand, there are specialized professionals, like computer programmers, for whom work at home has many advantages and few disadvantages. Conversely, there are the legions of office workers for whom working at home will result in lower salaries and benefits. Unions systematically oppose such arrangements, both because the dispersion of workers is inimical to the esprit de corps so important for mobilizing members to fight for better conditions, and because the history of home–work arrangements is replete with worker exploitation. (Such fears might cause a union to oppose a proffered arrangement that is actually beneficial to the workers.) Additional disadvantages for at-home workers are less visibility for promotion, problems with supervision and security of sensitive materials, and diminished interaction among co-workers. But there are also advantages: more availability of jobs for parents with small children, for the handicapped, and for the aged; better integration between personal and work life; more and better time for recreation (avoidance of weekend crowds); and savings on fuel and clothes.

Finally, consider the worker experience with wholesale computerization within a workplace. Computers can be responsible for deskilling workers; for fragmenting complex jobs into small, meaningless pieces, each done by a different person; for reducing the skill and initiative and hence the psychic rewards of a job; and for making work machine-paced and hence out of the control of workers. Past technologies, particularly the assembly line, have done this. Computers can also reduce the total number of jobs available. Virtually all studies suggest that computers eliminate more jobs than they create, though (predictably) studies sponsored by unions and liberal groups show greater reductions than studies sponsored by managements and conservative groups.

We would suggest that the pessimists are correct on the reduction of total employment, while the optimists are correct about the nature of work. Computers have eliminated more jobs than they have created, and will continue to do so. But, with important exceptions such as the use of computer controls for machine tools, they have not deskilled many jobs. Deskilling has already been accomplished by the assembly line and other methods of speeding production; the computer is putting important skills, responsibilities, and autonomy back into jobs. This was clearly illustrated by past research, and by interviews our students conducted with people who recently had computers introduced into their work (see box). Secretaries overwhelmingly liked the improved quality and quantity of work they produced, and the fact that they knew more about the computer than their bosses did. Factory workers being retrained at a newly computerized automobile plant said that in the past they were hired from their neck down, but now they could use their heads as well (Ermann, Gutierrez, and Williams; unpublished interviews at General Motors' Boxwood Plant, October 16, 1986). The computerization of the plant caused great anxiety and the loss of 1,300 jobs (held by those with the least social power and the fewest skills). But it is hard to imagine how the computer could have further deskilled the jobs of auto assemblers, and it seems instead to have enhanced skills used in this type of work.

Computerization of a factory appears to bring with it a pattern of improved working conditions—for those who are not laid off—as illustrated in the General Motors assembly plant near our university in Delaware, and the General Electric Appliance Park in Louisville, Kentucky (Swaboda, 1987). In both cases, employment dropped drastically. At G.E., it dropped from 19,000 to 10,000 in one decade. On the positive side, however, jobs of the remaining workers were less narrowly prescribed, leaving more room for individual initiative; and the company made concerted and somewhat successful efforts to improve the skills of workers, to listen to their suggestions, and to give them more control of the pace of the assembly line and the ability to stop it when problems arise. In sum, in the United States and other economically developed societies, computers cost more jobs than they create, but improve many of the jobs that remain.

On Economic Justice

On an international level, advances in communications through the use of cable, telephone lines, or satellite links make possible the transmission of data across national boundaries. But even though we might expect that the mobility of information would accelerate technical progress all over the world, facilitate world commerce, and help to solve the special needs of underdeveloped countries, in fact it seems that this is not happening (see reading 22). There are good reasons to believe that computers help the dominant classes of third-world nations form alliances with their counterparts in industrialized countries, while at the same time putting distance between themselves and the impoverished majorities in their own countries. The new technology thus seems to be increasing rather than decreasing the gap between rich and poor in the

Experiences with Computers

From 1986 to 1989, each of our students interviewed two people who experienced significant computerization in a work setting that previously had not been computerized. Based on the 200-plus interview reports submitted, our students' research suggested that people tended to approve of computerization of their workplaces when:

1. their jobs were not threatened by the computer. White collar workers were least likely to feel that computers might cause them to lose their jobs.

2. their skills and responsibilities were upgraded, or they received supervisory approval for their new skills and outputs. Though the computer tended to enhance skills in both blue-collar and white-collar settings, it was particularly job-enhancing in office work. Prople with office jobs also appreciated the feeling of being at the cutting edge of their occupations.

3. their job productivity increased, or their tedious and unvalued job routines or clerical work declined. This satisfaction increase was greatest for people who placed relatively more emphasis on efficiency than on interpersonal relations.

4. their personal control over their work increased. Those in assembly work and typing pools tended not to get this increased control.

5. they were young, or they were older and felt proud of overcoming their initial anxieties.

Computers also had negative side effects for a minority of our interviewees. Some respondents reported that:

1. interpersonal contact declined, as accountants, travel agents, supervisors, analysts, and others had fewer reasons to consult in person with colleagues at nearby desks or down the hall.

2. privacy declined, as others could monitor employees' work, send junk-mail messages, and in one case access private information.

3. tension increased when the computers they had learned to depend on were down for more than a few seconds.

Our conclusion is that computers tend to give the greatest rewards to young, white-collar, technologically comfortable, and organizationally secure people who can use computers to gain increased satisfaction from their work.

third world. The basic reasons for this are that advanced technology requires high levels of education which the underprivileged of the world lack, and that the desired goods of the information age are "knowledge-intensive" rather than "labor-intensive." Since the underprivileged have only their labor to sell, their position seems likely to become worse as computerization increases.

Reactions to these worrisome facts are varied. Some think this is only a temporary phenomenon which will soon be overcome by global transformations of a more positive character. They mention the precedent of the industrial revolution, when prophesies of doom about explosive unemployment did not come true because the very energy of the revolution created new and better jobs for displaced workers. But, as Nilsson points out in his essay (see reading 21), there is something new and eminently different about the information revolution. The displaced labor force of the industrial revolution retreated from dangerous jobs, or jobs that required great physical effort, or were repetitive and unintellectual; most of these people then found more interesting and intellectually satisfying jobs in the area of services and state or private bureaucracies. But the new technological unemployment is eliminating the types of jobs that the industrial revolution brought into being. With the exception of the robotization of industry, computerization is occurring and will continue to occur mainly in areas (like banking, commerce, and administration) where displaced workers do not have the alternatives that their predecessors had during the industrial revolution.

For the first time in history, therefore, many of the unemployed may not be able to find a new job, interesting or otherwise. It is clear that this qualitatively new problem has to be faced in ways which will be profoundly distinct and for which there is little precedent. With the elimination of work opportunities by the progressive automation of the intellectual functions in factories, stores, and offices, large masses of the population will lose their means of support. If they join the present unemployed as marginal citizens, they will become a sign of our moral failure to create a humane computerized world.

The question then becomes how to provide incomes to people when there is not enough work to go around. Is it sufficient to keep diminishing the weekly hours of work, to use early retirement, and to protract the years of formal education? Louis Kelso and Mortimer Adler (1975) anticipated this situation and proposed drastic measures of economic reform in the *Capitalist Manifesto*. These authors explored the possibility of having a society of "owners," in which every citizen receives his/her income from ownership of machines (intelligent robots) that produce all the necessary goods of society without the need for human labor (apart from governance and the administration of property). In such a society, people would end up occupying positions similar to the owners of slaves; being owner of the means of production, they would avoid servile toil and concentrate on governance and administration, the practice of the sciences and the arts, the cultivation of friendship, sports, and the like.

But how can we make a transition to such a utopia? History teaches us that the powerful are unlikely to altruistically relinquish much of their economic power in order to benefit the weak. The industrial revolution resulted in a considerable transfer of power to the masses, but in that case the powerful could not realize enormous profits unless they paid the masses to operate the industrial machinery. Will the masses find a way to use the power of the computer to strengthen their position even though they are no longer economically necessary? Will the powerful use the computer to prevent an increasingly disaffected population from exploding into violence? Much as we would like to believe that the intrinsic dynamics of the computer revolution will inevitably lead to economic justice, it seems more likely that those involved in computerization decisions (management and labor, government and citizens, computer professionals and computer users) will have to consciously and innovatively create a form of computerization that will result in a just distribution of the benefits of the computer revolution.

On Power

Some observers (e.g., Weizenbaum, 1976) have concluded that if the computer had not come when it did, the social and political institutions of the United States would have had to be transformed radically, in the direction of decentralization. The computer appeared just in time to permit the system to deal with massive files that otherwise would have decreed the death of centralized government. It made possible the continuation of the status quo, with an enhanced level of efficiency. Conclusions like this have contributed to a fear that computers inevitably strengthen current institutions (with their defects and inequalities), existing power relationships, and political and economic centralization.

Those who fear that computers increase economic centralization have in mind organizations in which a few people make decisions that affect many subordinates whom they control. Since computers permit those higher up to deal with more information and be more independent of consultation with the lower echelons, computers seem to favor centralization. But computers could also permit more initiative to those at the lower levels of decentralization. Thus, computers could allow Social Security headquarters to assume all responsibility for decisions, using massive data bases. On the other hand, computers could allow local offices to resolve local cases more flexibly and responsively, using tele-communications to access centralized data bases. Perhaps the best illustration of this type of decentralization is the airline reservation system, which has given local travel agents great latitude to serve individual needs.

There are some reasons to hope computers will aid political decentralization. For instance, with widespread computerization, people will be better able to let their preferences be known to decision makers. Computer bulletin boards might evolve into vehicles for a computerized direct democracy. Citizens might even legislate on public matters from their homes, by means of their personal computers and the informational networks that would become available for that purpose. On the other hand, experience to date indicates that computer users apply their technologies to their own special purposes: to electronically send out hundreds of thousands of "personalized" partisan political messages, to lobby for keeping down the costs of computerized communication, and so forth. And the powerless may be less able to use computer technology to influence the government than they are to use simpler and cheaper technologies like the mail, the phone, and the ballot.

On Privacy

With the help of "mobile data access terminals," police now can discover immediately whether an illegally parked car has unpaid parking tickets and therefore should be towed. Universities can determine credit histories before granting student loans, and students who fail to repay loans can be traced, their employers informed, and their credit ratings harmed. The federal government can ascertain if an applicant for a student loan has registered for the draft. Landlords can find out if a prospective tenant has had a dispute with a former landlord. The list of types of information about you that is available from some computer data base increases daily. Without computers, large bureaucratic organizations would be unable to use and exchange records easily. Are these data bases a serious threat to individual privacy?

Let us begin by asking if it is important that privacy be respected. Different societies have different attitudes about privacy: in a primitive or rural society privacy may be almost nonexistent, since the smallness of the population allows each person to know the intimate details of everyone else's life; in a large industrial state, the citizen may expect almost complete control over who gains access to information about his personal life, although the resulting anonymity may deprive him of the benefits of neighborliness. Similarly, in different circumstances in the same society, different degrees of respect for privacy may be considered proper: during wartime, or the early period of a newly installed revolutionary regime, there may be widespread popular support for actions that, in other circumstances, would be considered insupportable invasions of privacy. Privacy, then, is not an intrinsic good, to be valued in all circumstances, but an instrumental good, to be valued when it is useful and proper.

In a society that esteems personal growth, creativity, and progress, privacy is important. It allows us an opportunity to accept help from trusted friends without exposing to the public intermediate stages of our transitions from an old self to a new self, or from a battered self to a recovered self. Similarly it allows great thinkers to experiment with unorthodox ideas—testing, discarding, and refining theories without exposing them to public ridicule until they are strong enough to withstand the test of time. It allows individuals to adapt to the dislocations inevitable as some occupations become obsolete and new ones are created. As Fried (see reading 4) argues, "Privacy provides the rational context for a number of our most significant ends, such as love, trust and friendship, respect and self-respect." Thus, privacy allows adolescents to grow into adulthood, scientists to produce revolutionary breakthroughs, and everyone to adapt to life in an ever-changing society.

Clearly privacy is important for our society, and for the computerized society in our future. But does the unlimited proliferation of computer-accessible data about individuals undermine the kind of privacy we need? Can the scientist discuss her ideas with a trusted colleague using electronic mail without risking a destructive early release of the ideas? Can the political dissident fight the draft if his actions will haunt him in every subsequent encounter with government or industry? Can a wild teenager mature into a productive citizen if the misdeeds of her youth are enshrined in a computer file and prejudice everyone against her? Computerization has a tremendous potential for destructive invasion of privacy, but this potential can be kept in check with a combination of legislative and technical devices (e.g., legislation to prevent the retention of certain types of data longer than five years; computer security systems which prevent electronic eavesdropping). Computerization is neither necessarily beneficial nor necessarily harmful; the actions of decisionmakers will determine the extent of benefit and harm.

On Our Sense of Self

The discussion of privacy gives us a glimpse of the possibility that computers might change in a very basic way the kind of people we are. Even if this potential is not realized, computerization might join other important milestone events in deeply changing our perception of ourselves. The first such milestone, a great (and greatly humbling) challenge to our sense of human beings as uniquely important, occurred when the Copernican revolution established that Earth, the human home, was not the center of the universe. The second milestone was Charles Darwin's conclusion that the emergence of Homo sapiens was not the result of a special act of God, but the result of evolution from lower species by the process of natural selection. The third milestone resulted from the work of Karl Marx and Sigmund Freud, which showed intellectual, social, and individual creativity to be the result of nonrational (unconscious) libidinal or economic forces—not, as had been believed, the products of the almost godlike powers of the human mind. As a consequence of these three key events we have a much humbler view of our place in the universe than did our ancestors.

Computers may provide a fourth major blow to our self-esteem. For millennia the human ability to think rationally has been considered to be our most important and uniquely distinguishing feature. Work in artificial intelligence may lead to two different challenges to this belief: If, as Minsky suggests (reading 14), computers can think, then humans are not unique in this ability. And even if computers cannot think, research in artificial intelligence may lead cognitive scientists to conclude that rational thought is a mundane process of inputs, internal states, and outputs. Either discovery would challenge our belief in the transcendent importance of rational thought, and we would have to reevaluate our place in the universe.

Such a reappraisal may force changes in our answers to fundamental moral questions, particularly the question of what is the ideal life for a person. Indeed, if human thought were shown (against the expectation of most philosophers) to be wholly deterministic, then the entire concept of moral blame would have to be rethought.

We have discussed several different types of potential moral problems generated by the possible impacts of computers. Our focus on the problems computers may generate should not be taken to indicate that we are pessimistic about the computer's effects on society. Computerization will be neither the utopia promised by its most fervent adherents nor the unadulterated calamity foreseen by its most gloomy opponents. But the important lesson to learn from a serious consideration of the problems is that our chances for reaping the maximum benefits of computers and avoiding the hazards will depend on our willingness to seriously consider both kinds of outcome.

Appendix

Preparing Ourselves to Be Part of the Solution, Not Part of the Problem:
Assumptions Underlying the Different Positions

News stories, television talk shows, books, and magazines continually bombard us with claims about what is going on in the world. If a claim is critical of our favorite belief or institution, we tend to discount it as biased; if it is supportive we tend to accept it. All too often this natural tendency causes us to disregard problems during the stage when they are easiest to correct. A useful way to immunize ourselves against this tendency is to recognize the assumptions underlying our own positions and the opposing positions in a controversy.

In the case of computers, we will label the people who have thought about the social implications of computers to be either "defenders" or "doubters." According to the defenders, computer technology is a tool that will enhance the efficiency and effectiveness of our economy, possibly even creating a "computopia" of abundance and the absence of exploitation (Masuda, 1983). The people who adopt these arguments tend to emphasize the positive contributions to society made by technologies, whether computerized or otherwise, and tend to believe that the problems computers might cause are amenable to technical fixes. They emphasize the benefits that could come from the intelligent and humane spread of computerization. Defenders believe new technologies are adopted because they satisfy existing needs. They suggest that computers are being blamed for evils that have existed without computers—for example, loss of privacy (witness the McCarthy era), the deskilling of jobs (the assembly line), and unemployment (the Great Depression). Computers, the defenders say, enable hospitals to cure more patients, auto assembly plants to make better cars, and governments to serve the people more effectively.

Defenders make a number of assumptions about how people and organizations make computerization decisions. They tend to assume: (1) socially shared criteria for deciding what is good for a society or an organization; (2) mutual trust among the various groups that might want to control the implementation of computerization; and (3) large-scale dissemination of information on which to base choices. For instance, they assume that people working in hospitals want only to cure illness (not gain power, prestige, income). Likewise, they assume it is desirable that computerization enhance factory managers' information about what is happening on plant floors, draft boards' knowledge about the identities of draft-aged men who have not registered, and Internal Revenue Service agents' awareness of the sources of each person's income. In general, they tend to accept the legitimacy and motives of those in positions of political and economic power.

Doubters have a very different perspective. They fear standardization of behavior. Though the roots of this position go back far in history, the introduction of computers has radicalized this malaise. According to doubters, new technologies should be viewed skeptically because increasingly complex systems unavoidably bring with them hazards which even the most benevolent and bright technologists could not have foreseen. Doubters see far less consensus in society than advocates see, and they believe that people inside various organizations and societies have multiple, changing, and sometimes conflicting goals. They emphasize the problems that could come from haphazard and malevolent spread of computerization, rather than the benefits that could come from planned and humane use of computers. And they believe that the exponents of computers are people who have the skills and data to use this technology to expand their own power and influence.

Thus, doubters emphasize that hospitals have conflicting goals between administration and doctors (with administrators wanting costs to be kept down and doctors wanting the latest equipment and services), and that improved management information systems will enhance the control of hospital administrators at the expense of doctors. Doubters prefer limiting the power of groups like factory managers, draft boards, and tax collectors. In support of their views, they cite studies, such as one focusing on the insurance industry, that show that decision making becomes more centralized as a result of computerization (Whistler, 1970). In general, doubters tend to fear that computerization will undermine the rights and preferences of those not in positions of political and economic power.

At a societal level, doubters are more concerned with losses of privacy, deskilling of jobs, and increases in unemployment than with gains in economic efficiency and effectiveness. They are less enthused by the availability of new consumption items than about the loss of what they consider basic human rights. They are worried about what evil may come from the computer in some future and less benevolent time. And they tend to see the decision about what technologies to develop and deploy as being made by those in power to serve their own, rather than the collective, good. In the view of most doubters, people in power do not authorize new technologies that would limit their influence or make their positions obsolete.

Doubters also note that we live mobile lives in a complex society where the key events in our lives (birth, educational attainment, marriage, criminal convictions, lawsuits, eviction, death, and so forth) are recorded by a variety of organizations in a variety of locales. Many mundane events (phone calls, plane trips, bank withdrawals) also are recorded. In the past, these records could not be collected and used effectively, even if they were publicly available. The federal government could not readily check if you had registered for the military draft before granting you a student loan, or check if the interest you had received from a bank appeared on your tax forms. Your landlord had no practical way to determine if you had previously been evicted, a prospective employer would have considered it too expensive to try to find out if you had a criminal record in another jurisdiction, and your new doctor had no reasonable way to ascertain if you had sued another doctor in the past. The computer has changed the economics of these situations; in the process, it has reduced your privacy.

The debate between doubters and defenders cannot be resolved in any objective way. What separates the two groups are their assumptions and predictions (Kling, 1980). However, it is still too soon to tell what current computer technologies (let alone future ones) will bring to our way of life. The computer age is young, and technology and its uses are still evolving. Computers have definite potential to harm or enhance the quality of our lives. If it is true that faith is strongest where knowledge is uncertain, then the tenacious convictions of both the doubters and the defenders suggest that we have much to learn about the uses and abuses of computers.

References

Control Data Corporation. 1984. Approved policy and procedure on data collection of employee information. *Personnel Manager's Manual,* January.

Dvorak, John C. 1988. Virus wars: A serious warning. *PC Magazine,* February 29: 71.

Gutierrez, Claudio and J. S. Hidalgo. 1988. Suggesting what to do next. Proceedings of 1988 ACM Symposium on Personal and Small Computers. Paris: INRIA.

Kelso, Louis O. and Mortimer Adler. 1975. *Capitalist manifesto.* Westport, CT: Greenwood Press.

Kling, Rob. 1980. Social analyses of computing: Theoretical perspectives in recent empirical research. *Computing Surveys* 12 (March): 62.

Masuda, Yoneji. 1983. *The information society as post-industrial society.* Bethesda, MD: World Future Society.

Office of Technological Assessment. 1987. *The electronic supervisor.* Office of Technological Assessment, U.S. Congress. Washington, DC: U.S. Government Printing Office, Sept.

Perl, Peter. 1984. "Monitoring by computer sparks employee concerns." *Washington Post,* September 2: 1.

Swaboda, Frank. 1987. "A good thing is brought to life in Louisville." *Washington Post Weekly Edition,* Nov. 2, p. 20.

Vonnegut, Kurt. 1952. *Player piano.* New York: Dell.

Weizenbaum, Joseph. 1976. *Computer power and human reason: From judgement to calculation.* San Francisco: W. H. Freeman.

Whistler, T. 1970. *The impact of computers on organizations.* New York: Praeger.

I

Ethical Frameworks for Discussing Computer-Related Issues

General Ethical Theory

General ethical theories attempt to capture the most fundamental ethical insights of our culture; these insights can then be applied to determine the morality of particular actions, or types of actions. The readings in this section present and explain the most important of these fundamental ethical insights.

Utilitarians (see reading 1) claim that the most fundamental ethical insight is: of those actions available to you, you are morally obliged to choose that action which maximizes total happiness (summed over all affected persons). This is particularly useful in illuminating those problems in which many people are affected in different ways by an action; for example, a utilitarian analysis would be especially useful in the look-and-feel issue (readings 24 and 25), the liability issue (reading 28), and the whistle-blowing issue (readings 6, 32, and 33).

Many philosophers feel that although the utilitarian insight is indeed fundamental, it does not give sufficient importance to the view that the benefits and burdens of society ought to be justly distributed. Hospers (reading 2) examines the concept of justice, showing that a fair distribution need not be strictly equal since it should reflect what different individuals deserve. Is the distribution of the benefits and burdens of the computer revolution just? If the computer revolution, as it is presently being implemented, leads to the widening of the economic gap between developed and less developed countries (reading 22), reinforces the patterns of an elite-dominated society (reading 19), and exacerbates the exploitation of women (reading 18), then this implementation could be judged to be unjust and therefore immoral.

Kantians (reading 3) claim that the fundamental ethical principle is:

people should always be treated as ends, never as simply a means. That is, it is wrong to ignore another person's legitimate desires and to use him/her just to attain what you want. For example, Bereano's essay about the use of computers to further control workers (reading 23) centers on the extent to which computers enable employers to exploit workers while ignoring the workers' legitimate desires. In another use of Kantian reasoning, Stallman (reading 27) argues that programmers should share software freely instead of selling it. Although Stallman buttresses his Kantian argument with a utilitarian argument (that society would benefit from the system he advocates), his stress on treating other computer users as persons whose desires should be respected, rather than as economic units, is clearly primary.

— 1

Utilitarian Theory

John Hospers

Once one admits that one's own personal good is not the only consideration, how can one stop short of the good of everyone—"the general good"? This conclusion, at any rate, is the thesis of the ethical theory known as *utilitarianism.* The thesis is simply stated, though its application to actual situations is often extremely complex: whatever is intrinsically good should be promoted, and, accordingly, our obligation (or duty) is always to act so as to promote the greatest possible intrinsic good. It is never our duty to promote a lesser good when we could, by our action, promote a greater one; and the act which we should perform in any given situation is, therefore, the one which produces more intrinsic good than any other act we could have performed in its stead. In brief, the main tenet of utilitarianism is the maximization of intrinsic good.

The description just given is so brief that it will almost inevitably be misleading when one attempts to apply it in actual situations unless it is spelled out in greater detail. Let us proceed at once, then, to the necessary explanations and qualifications.

From John Hospers, *Human Conduct, Problems of Ethics*, pp. 199–208. Harcourt Brace Jovanovich, Inc., 1972. Reprinted by permission.

1. When utilitarians talk about right or wrong acts, they mean—and this point is shared by the proponents of all ethical theories—*voluntary* acts. Involuntary acts like the knee jerk are not included since we have no control over them: once the stimulus has occurred the act results quite irrespective of our own will. The most usual way in which the term "voluntary act" is defined is as follows:[1] an act is voluntary if the person *could* have acted differently *if* he had so chosen. For example, I went shopping yesterday, but if I had chosen (for one reason or another) to remain at home, I would have done so. My choosing made the difference. Making this condition is not the same as saying that an act, to be voluntary, must be *premeditated* or that it must be the outcome of *deliberation,* though voluntary acts often are planned. If you see a victim of a car accident lying in the street, you may rush to help him at once, without going through a process of deliberation; nevertheless your act is voluntary in that if you had chosen to ignore him you would have acted differently. Though not premeditated, the action *was* within your control. "Ought implies can," and there is no ought when there is no can. To be right or wrong, an act must be within your power to perform: it must be performable as the result of your choice, and a different choice must have led to a different act or to no act at all.

2. There is no preference for immediate, as opposed to remote, happiness. If Act A will produce a certain amount of happiness today and Act B will produce twice as much one year hence, I should do B, even though its effects are more remote. Remoteness does not affect the principle at all: happiness is as intrinsically good tomorrow or next year as it is today, and one should forgo a smaller total intrinsic good now in favor of a larger one in the future. (Of course, a remote happiness is often less certain to occur. But in that case we should choose A not because it is more immediate but because it is more nearly certain to occur.) . . .

3. Unhappiness must be considered as well as happiness. Suppose that Act A will produce five units of happiness and none of unhappiness and Act B will produce ten units of happiness and ten of unhappiness. Then A is to be preferred because the *net* happiness—the resulting total after the unhappiness has been subtracted from it—is greater in A than in B: it is five in A and zero in B. Thus the formula "You should do what will produce the greatest total happiness" is not quite accurate; you should do what will produce the most *net* happiness. This modification is what we shall henceforth mean in talking about "producing the greatest happiness"—we shall assume that the unhappiness has already been figured into the total.

4. It is not even accurate to say that you should always do what leads to the greatest *balance* of happiness over unhappiness, for there may be no such balance in any alternative open to the agent: he may have to choose between "the lesser of two evils." If Act A leads to five units of happiness

and ten of unhappiness and Act B leads to five units of happiness and fifteen of unhappiness, you should choose A, not because it produces the most happiness (they both produce an equal amount) and not because there is a greater balance of happiness over unhappiness in A (there is a balance of unhappiness over happiness in both), but because, although both A and B produce a balance of unhappiness over happiness, A leads to a *smaller* balance of unhappiness over happiness than B does. Thus we should say, "Do that act which produces the greatest balance of happiness over unhappiness, or, if no act possible under the circumstances does this, do the one which produces the smallest balance of unhappiness over happiness." This qualification also we shall assume to be included in the utilitarian formula from now on in speaking of "producing the greatest happiness" or "maximizing happiness."

5. One should not assume that an act is right according to utilitarianism simply because it produces more happiness than unhappiness in its total consequences. If one did make this assumption, it would be right for ten men collectively to torture a victim, provided that the total pleasure enjoyed by the sadists exceeded the pain endured by the victim (assuming that pain is here equated with unhappiness and that all the persons died immediately thereafter and there were no further consequences). The requirement is not that the happiness exceed the unhappiness but that it do so *more* than any other act that could have been performed instead. This requirement is hardly fulfilled here: it is very probable indeed that the torturers could think of something better to do with their time.

6. When there is a choice between a greater happiness for yourself at the expense of others, and a greater happiness for others at the expense of your own, which should you choose? You choose, according to the utilitarian formula, whatever alternative results in the greater total amount of *net* happiness, precisely as we have described. If the net happiness is greater in the alternative favorable to yourself, you adopt this alternative; otherwise not. Mill says, "The happiness which forms the utilitarian standard of what is right in conduct, is not the agent's own happiness, but that of all concerned. As between his own happiness and that of others, utilitarianism requires him to be as strictly impartial as a disinterested and benevolent spectator."[2] To state this in different language, you are not to ignore your own happiness in your calculations, but neither are you to consider it more important than anyone else's; you count as one, and only as one, along with everyone else. Thus if Act A produces a total net happiness of one hundred, and Act B produces seventy-five, A is the right act even if you personally would be happier in consequence of B. Your choice should not be an "interested" one; you are not to be prejudiced in favor of your own happiness nor, for that matter, against it; your choice should be strictly *dis*interested as in the

case of an impartial judge. Your choice should be dictated by the greatest-total-happiness principle, not by a *your*-greatest-happiness principle. If you imagine yourself as a judge having to make a decision designed to produce the most happiness for all concerned *without* knowing which of the people affected would be *you,* you have the best idea of the impartiality of judgment required by the utilitarian morality.

In egoistic ethics . . . your sole duty is to promote your own interests as much as possible, making quite sure, of course, that what you do will make you really happy (or whatever else you include in "your own interest") and that you do not choose merely what you *think* at the moment will do so; we have called this policy the policy of "*enlightened* self-interest." In an *altruistic* ethics, on the other hand, you sacrifice your own interests completely to those of others: you ignore your own welfare and become a doormat for the fulfillment of the interests of others. . . . But the utilitarian ethics is neither egoistic nor altruistic: it is a *universalistic* ethics, since it considers your interests equally with everyone else's. You are not the slave of others, nor are they your slaves. Indeed, there are countless instances in which the act required of you by ethical egoism and the act required by utilitarianism will be the same: for very often indeed the act that makes you happy will also make those around you happy, and by promoting your own welfare you will also be promoting theirs. (As support for this position, consider capitalistic society: the producer of wealth, by being free to amass profits, will have more incentive to produce and, by increasing production, will be able to create more work and more wealth. By increasing production, he will be increasing the welfare of his employees and the wealth of the nation.) Moreover, it is much more likely that you can effectively produce good by concentrating on your immediate environment than by "spreading yourself thin" and trying to help everyone in the world: "do-gooders" often succeed in achieving no good at all. (But, of course, sometimes they do.) You are in a much better position to produce good among those people whose needs and interests you already know than among strangers; and, of course, the person whose needs and interests you probably know best of all (though not always) is yourself. Utilitarianism is very far, then, from recommending that you ignore your own interests.

It is only when your interests cannot be achieved except at the cost of sacrificing the *greater* interests of others that utilitarianism recommends self-sacrifice. When interests conflict, you have to weigh your own interest against the general interest. If, on the one hand, you are spending all your valuable study time (and thus sacrificing your grades and perhaps your college degree) visiting your sick aunt because she wants you to, you would probably produce more good by spending your time studying. But on the other hand, if an undeniably greater good will result from your sacrifice, if, for instance, your mother is seriously ill and no one else is available to care for her, you might have to drop out of school for a semester to care for her. It might even, on occasion, be your utilitarian duty to sacrifice your very life for a cause, when the cause is extremely worthy and requires your sacrifice for its fulfillment. But you must first make quite sure that your sacrifice will indeed produce the great good intended; otherwise you would be throwing your life away uselessly. You must act with your eyes open, not under the spell of a martyr complex.

. . . . .

7. The general temper of the utilitarian ethics can perhaps best be seen in its attitude toward moral rules, the traditional dos and don'ts. What is the utilitarian's attitude toward rules such as "Don't kill," "Don't tell lies," "Don't steal"?

According to utilitarianism, such rules are *on the whole* good, useful, and worthwhile, but they *may* have exceptions. None of them is sacrosanct. If killing is wrong, it is not because there is something intrinsically bad about killing itself, but because killing leads to a diminution of human happiness. This undesirable consequence almost always occurs: when a man takes another human life, he not only extinguishes in his victim all chances of future happiness, but he causes grief, bereavement, and perhaps years of misery for the victim's family and loved ones; moreover, for weeks or months countless people who know of his act may walk the streets in fear, wondering who will be the next victim—the amount of insecurity caused by even one act of murder is almost incalculable; and in addition to all this unhappiness, every violation of a law has a tendency to weaken the whole fabric of the law itself and tends to make other violations easier and more likely to occur. If the guilty man is caught, he himself hardly gains much happiness from lifelong imprisonment, nor are other people usually much happier for long because of his incarceration; and if he is not caught, many people will live in fear and dread, and he himself will probably repeat his act sooner or later, having escaped capture this time. The good consequences, if any, are few and far between and are overwhelmingly outweighed by the bad ones. Because of these prevailingly bad consequences, killing is condemned by the utilitarian, and thus he agrees with the traditional moral rule prohibiting it.

He would nevertheless admit the possibility of exceptions: if you had had the opportunity to assassinate Hitler in 1943 and did not, the utilitarian would probably say that you were doing wrong in *not* killing him. By not killing him, you would be sealing the death of thousands, if not millions, of other people: political prisoners and Jews whom he tortured and killed in concentration camps and thousands of soldiers (both Axis and Allied) whose lives would have been saved by an earlier cessation of the war. If you had refrained from killing him when you had the chance, saying "It is my duty never to take a life, therefore I shall not take his," the man whose life you saved would then turn around and have a thousand others killed, and for his act the victims would have you to thank. Your conscience, guided by the traditional moral rules, would have helped to bring about the torture and death of countless other people.

Does the utilitarian's willingness to adopt violence upon occasion mean that a utilitarian could never be a pacifist? Not necessarily. He *might* say that *all* taking of human life is wrong, but if he took this stand, he would do so because he believed that killing *always* leads to worse consequences (or greater unhappiness) than not killing and *not* because there is anything intrinsically bad about killing. He might even be able to make out a plausible argument for saying that killing Hitler would have been wrong: perhaps even worse men would have taken over and the slaughter wouldn't have been prevented (but then wouldn't it have been right to kill *all* of them if one had the chance?); perhaps Hitler's "intuitions" led to an earlier defeat for Germany than if stabler men had made more rationally self-seeking decisions on behalf of Nazi Germany; perhaps the assassination of a bad leader would help lead to the assassination of a good one later on. With regard to some Latin American nations, at any rate, one might argue that killing one dictator would only lead to a revolution and another dictator just as bad as the first, with the consequent assassination of the second one, thus leading to revolution and social chaos and a third dictator. There are countless empirical facts that must be taken into consideration and carefully weighed before any such decision can safely be made. The utilitarian is not committed to saying that any one policy or line of action is the best in any particular situation, for what is best depends on empirical facts which may be extremely difficult to ascertain. All he is committed to is the statement that when the action is one that does not promote human happiness as much as another action that he could have performed instead, then the action is wrong; and that when it does promote more happiness, it is right. Which particular action will maximize happiness more than any other, in a particular situation, can be determined only by empirical investigation. Thus, it is possible that killing is always wrong—at least the utilitarian could consistently say so and thus be a pacifist; but *if* killing is always wrong, it is wrong not because killing is wrong per se but because it always and without exception leads to worse consequences than any other actions that could have been performed instead. Then the pacifist, if he is a consistent utilitarian, would have to go on to show in each instance that each and every act of killing is worse (leads to worse consequences) than any act of refraining from doing so—even when the man is a trigger-happy gunman who will kill dozens of people in a crowded street if he is not killed first. That killing is worse in every instance would be extremely difficult—most people would say impossible—to prove.

Consider the syllogism:

> *The action which promotes the maximum happiness is right.*
>
> *This action is the one which promotes the maximum happiness.*

Therefore, *This action is right.*

The utilitarian gives undeviating assent only to the *first* of these three statements (the major premise); this statement is the chief article of his utilitarian creed, and he cannot abandon it without being inconsistent with his own doctrine. But this first premise is not enough to yield the third statement, which is the conclusion of the argument. To know that the conclusion is true, even granting that the major premise is, one must also know whether the second statement (the minor premise) is true; and the second statement is an empirical one, which cannot be verified by the philosopher sitting in his study but only by a thorough investigation of the empirical facts of the situation. Many people would accept the major premise (and thus be utilitarians) and yet disagree among themselves on the conclusion because they would disagree on the minor premise. They would agree that an act is right if it leads to maximum happiness, but they would not agree on whether this action or that one is the one which *will* in fact lead to the most happiness. They disagree about the empirical facts of the case, not in their utilitarian ethics. The disagreement could be resolved if both parties had a complete grasp of all the relevant empirical facts, for then they would know *which* action *would* lead to the most happiness. In many situations, of course, such agreement will never be reached because the consequences of people's actions (especially when they affect thousands of other people over a long period of time, as happens when war is declared) are so numerous and so complex that nobody will ever know them all. Such a disagreement will not be the fault of ethics, or of philosophy in general, but of the empirical world for being so complicated and subtle in its workings that the full consequences of our actions often can not be determined. Frequently it would take an omniscient deity to know which action in a particular situation was right. Finite human beings have to be content with basing their actions on estimates of probability.

According to utilitarianism, then, the traditional moral rules are justified for the most part because following them will lead to the best consequences far more often than violating them will; and that is why they are useful rules of thumb in human action. But, for the utilitarian, this is *all* they are—rules of thumb. They should never be used blindly, as a pat formula or inviolable rule subject to no exceptions, without an eye to the detailed consequences in each particular situation. The judge who condemned a man to death in the electric chair for stealing $1.95 (as in a case in Alabama in 1959) was probably not contributing to human happiness by inflicting this extreme penalty, even though he acted in accordance with the law of that state. The utilitarian would say that if a starving man steals a loaf of bread, as in Victor Hugo's *Les Miserables,* he should not be condemned for violating the rule "Do not steal"; in fact he probably did nothing morally wrong by stealing in this instance because the effects of not stealing would . . . have meant starvation and preserving a life (the utilitarian would say) is more important to human happiness than refraining from stealing a loaf of bread—especially since the man stole from one who was far from starving himself (the "victim" would never have missed it). He is probably blameless furthermore because the whole episode was made possible in the first place by a system of laws and a social structure which, by any utilitarian standard, were vicious in the extreme. (But see the effects of lawbreaking, below.)

Moral rules are especially useful when we have to act at once without being able adequately to weigh the consequences; for *usually* (as experience shows) better—i.e., more-happiness-producing—consequences are obtained by following moral rules than by not following them. If there is a drowning person whom you could rescue, you should do so without further investigation; for if you stopped to investigate his record, he would already have drowned. True, he might turn out to be a Hitler, but unless we have such evidence, we have to go by the probability that the world is better off for his being alive than his being dead. Again, there may be situations in which telling a lie will have better effects than telling the truth. But since, on the whole, lying has bad effects, we have to have special evidence that this situation is different before we are justified in violating the rule. If we have no time to gather such evidence, we should act on what is most probable, namely that telling a lie in this situation will produce consequences less good than telling the truth.

The utilitarian attitude toward moral rules is more favorable than might first appear because of the hidden, or subtle, or not frequently thought of, consequences of actions which at first sight would seem to justify a violation of the rules. One must consider *all* the consequences of the action and not just the immediate ones or the ones that happen to be the most conspicuous. For example: the utilitarian would not hold that it is *always* wrong to break a law, unless he had good grounds for saying that breaking the law *always* leads to worse consequences than observing it. But if the law is a bad law to begin with or even if it is a good law on the whole but observing the law in this particular case would be deleterious to human happiness, then the law should be broken in this case. You would be morally justified, for example, in breaking the speed law in order to rush a badly wounded person to a hospital. But in many situations (probably in most) in which the utilitarian criterion at first *seems* to justify the violation of a law, it does not really do so after careful consideration because of the far-flung consequences. For example, in a more typical instance of breaking the speed law, you might argue as follows: "It would make me happier if I were not arrested for the violation, and it wouldn't make the arresting officer any the less happy, in fact it would save him the trouble of writing out the ticket, so—why not? By letting me go, wouldn't the arresting officer be increasing the total happiness of the world by just a little bit, both his and mine, whereas by giving me a ticket he might actually decrease the world's happiness slightly?"

But happiness would be slightly increased only if one considers only the immediate situation. For one thing, by breaking the speed limit you are endangering the lives of others—you are less able to stop or to swerve out of the way in an emergency. Also those who see you speeding and escaping the penalty may decide to do the same thing themselves; even though you don't cause any accidents by your violation, *they* may do so after taking their cue from you. Moreover, lawbreaking may reduce respect for law itself; although there may well be unjust laws and many laws could be improved, it is usually better (has better consequences) to work for their repeal than to break them while they are still in effect. Every violation decreases the effectiveness of law, and we are surely better off having law than not having it at all—even the man who violently objects to a law and complains bitterly when he's arrested will invoke the law to protect *himself* against the violations of others. In spite of these cautions, utilitarianism does not say that one should *never* break a law but only that the consequences of doing so are far more often bad than good; a closer look at the consequences will show how true their reasoning is.

Notes

1. This term is most precisely defined by G. E. Moore in chapter 1, "Utilitarianism," of his book *Ethics*. [New York: Henry Holt and Company, 1912]. For the clearest and most rigorous statement of utilitarianism in its hedonistic form, see chapters 1 and 2 of [Moore's] book.
2. J. S. Mill, *Utilitarianism*. [ed. Oskar Piest (New York: Bobbs-Merrill, 1957; originally published 1863.], chap. 2.

2

Justice as Part
of an Ethical Theory

John Hospers

Justice and Equality

There is no single unambiguous meaning for the word "justice." The
word may seem at times to be a hopeless tangle; but we shall try to work
our way into the complexities gradually, step by step. (Mill, for example,
in Chapter 5 of his *Utilitarianism,* lists five separate senses of "justice,"
and other writers have isolated even more.) We shall attempt to handle
every problem concerning justice under one or the other of two main
headings: equality and desert.

There is no doubt that in daily life we associate the idea of justice with
that of equality. If a parent is kind to one child and cruel to another, we
call this treatment unjust. If a judge is severe to one prisoner and lenient
to another for the same offense, this treatment too is called unjust. In
both cases we attribute the injustice to "unequal treatment."

A. *Equal Treatment*

If the judge fines you one hundred dollars and lets your neighbor go,
and if the offense was the same, without mitigating circumstances in
either case, we accuse him of injustice. "If the offense was the same, the
treatment should be the same," we say. Or if the judge lets the second
person go because that person happens to be his personal friend or
relative, or because the judge happened to feel good on the one occasion but not on the other, again we cry "Injustice!" If the judge had a
quarrel with his wife at breakfast that morning, felt a need to vent his
aggression on somebody, and consequently imposed extra heavy fines
that day, we consider this situation an example of injustice. We cannot,
of course, escape entirely this human frailty: when we feel satisfied, we
are inclined to be more benevolent toward the world; and if we are
angry or resentful or envious, we take it out on innocent people, if

From John Hospers, *Human Conduct, Problems of Ethics,* pp. 344–75. Harcourt Brace
Jovanovich, Inc., 1972. Reprinted by permission.

circumstances or guilt feelings prevent us from taking it out on the guilty ones. No one is perfectly just, and people in positions of power are often less just than others; therefore we hear the popular complaint, "There ain't no justice."

But suppose that the judge sentences Smith heavily because he killed someone deliberately (homicide) but sentences Jones less heavily because he killed someone accidentally (manslaughter). Here we do not consider the sentence to be unjust. Why? We may be tempted to say that it is not unjust bcause Smith deserved a heavier punishment than Jones. This answer introduces the tangled concept of deserts, which we shall discuss later in the chapter. At the moment we are concerned with justice as equal treatment; and we do not consider the differential sentence in this instance to be an example of unequal treatment. Unequal treatment for the same offense we consider unjust; but unequal treatment for unequal offenses we do not. We could put the matter still otherwise in terms of impartiality: if the roles had been reversed and it had been Jones who was guilty of homicide and Smith of manslaughter, then the sentences too would presumably have been reversed; and so, thus far, there would have been no injustice.

But now a problem arises: if the judge lets Jones off more easily than Smith because Jones is a personal friend of his or Smith an enemy, or because Jones was a member of his national fraternity, or because he feels benign toward the world that day, we consider the sentences unjust; but if he lets off Jones more easily because his act was accidental or because someone forced him at the point of a gun, we do not consider the difference in sentences to be unjust. In other words, we consider the first group of reasons to be *irrelevant* ones (for a differential decision) and the second group to be *relevant*. But why? What is there about the first group of reasons that there is not about the second? Why does "He sentenced Jones more lightly because Jones is his personal friend" invoke a charge of unequal treatment, whereas "He sentenced Jones more lightly because Jones's act was accidental" does not? What is the criterion for making such a distinction?

. . . . .

B. Equal Distribution

"There are some things that all men are equally entitled to, regardless of their rank or station in life, which it would be unjust for everyone not to have." "It is unjust for some to have caviar before everybody has bread." These and many similar sentiments are voiced every day, and they indicate that, however confusedly, the notion of equality is somehow tied up with that of justice.

But equality of what? Would it be desirable if everyone in the world had an equal supply of insulin, an equal number of electric fans, an equal number of potatoes, an equal number of books? Such equality would be about as absurd as anything could be: diabetics need insulin, while to nondiabetics it is not of the slightest use; some people desire many books, whereas others are unable or disinclined to read; some live in arctic climates and need no electric fans, and some live where no electricity is available and the fans would be useless. Nothing could be more ridiculous than to provide everyone with an equal number of each kind of material thing.

.   .   .   .   .

In any possible scheme for equal distribution, money is about the only thing we can distribute, because people can satisfy so many, though not all, of their desires with it. Economists often assume that an equal distribution of money will come nearer to bringing about an equal distribution of happiness (or welfare) than would anything else. Therefore we shall be using the distribution of money in many of our examples, assuming for the moment the principle "If equal money, then equal happiness," even though this principle is not true.

At once, however, a possible conflict arises between the utilitarian ideal . . . and the requirements of justice. For [utilitarianism has] one end in mind: the production of the *largest possible amount (quantity) of intrinsic good*. But does not justice require something additional that is not taken into account thus far at all, namely an equal *distribution* of that intrinsic good? This requirement can be illustrated as follows: Suppose you have a choice between two acts, A and B. If you do A, eleven units of intrinsic good will result, all going to Mr. Smith. If you do B, only ten units of intrinsic good will result, but instead of all going to Mr. Smith, they will be divided equally among ten people. Which is preferable? The follower of . . . utilitarianism, it seems, will have to say that A is preferable, since the act of doing A . . . will result in more intrinsic good than doing B. Yet many people would say that you should do B—not because B produces more good, for it doesn't, but because in B there is an equal *distribution* of good, and we tend to believe that there is some value in equality of distribution too, even *over and above* the total *amount* of good to be distributed.

Of course, we may try to beat the game by such maneuvers as this: if Smith gets all the intrinsic good, the other nine will be envious of him, and this unhappiness is sufficient to tip the scales in favor of doing B after all. But this solution won't work; for we have changed the original situation. If that were the situation, there would be no problem. The problem arises when the amount of good resulting from A *really is* greater. Suppose that the other nine will never know that Smith gets everything; or suppose that even if they do know, the total intrinsic good (happiness) to Smith is still greater, even when the envy is included in the total calculations as an already deducted minus quantity. Then our problem remains.

Perhaps it seems strange that this problem should arise: doesn't the utilitarian already provide for distribution of good—the greatest happiness (quantity) for the greatest number (distribution)? Aren't they both already included in the original Bentham-Mill principle of utility?

Let us try to unravel this matter. Bentham struck the egalitarian keynote in utilitarianism when he said that each person was to count for one, and only for one. Each person was to count equally; that is to say, in calculating the total good to be produced by an act, the good was to be counted no matter *to whom* it went—whether to your father or to the stranger in the burning building. . . .

However, in another way utilitarianism is not specifically egalitarian. We are never told that we should try to bring about an equal amount of happiness (or anything else) in all people; we are told only that we should aim at the *greatest possible* amount of happiness, *counting everybody's happiness into the total*. Perhaps Bentham and Mill meant to include equal distribution as well as maximum quantity into their principle; but if they did, they never said so. In fact, the motto "greatest happiness for the greatest number" is not very clear. It does not tell us what we are supposed to do in situations where the greatest total *amount* of happiness is achieved when not everyone has it or when it is unequally distributed; the motto does not tell us whether we should then decide for the largest amount or the most nearly equal distribution of what happiness there is.

The twentieth-century utilitarians, at any rate, have always interpreted the classical utilitarians as meaning that one should aim at the largest total quantity of intrinsic good, with no qualifications or additions saying that quantity of good is to be sacrificed when a more nearly equal distribution can thereby be achieved. (Why then did Bentham and Mill include the phrase "for the greatest number"? Probably to ensure that every person was included in the calculations of the greatest total quantity.) Our problem, then, is this: does this classical utilitarian account of the matter (largest total quantity of good, with everybody being figured into the total) need revision in the light of the principle of equal distribution which we have said is included in our idea of justice?

Most thoughtful people, it seems, desire both ideals to be achieved: they would like to have a society in which the largest total *amount* of good is present, and if they had to choose between a society containing more good and a society containing less, they would unhesitatingly choose the first. Similarly, however, they would like to have a society in which good is, as nearly as possible, *equally distributed* (with exceptions we shall take up in the next section); and if they had to choose between a society in which good was equally distributed and one in which there were glaring inequalities, they would choose the first. The question is, what is to be done when the two ideals conflict? Are we—as the classical utilitarians would say, or at any rate as we are taking them to mean— always to select the alternative that contains the maximum total quantity of good, irrespective of its distribution? Or are we, as the supporters of justice would say, to select the alternative that contains the most nearly equal distribution of good, regardless of the amount? Or are we some-how to mediate between the two views by considering *both* principles and by believing that the right act should embody them both—the greatest total possible good that is compatible with the most nearly equal distribution thereof? It is probably fair to say that most people, once they have thought of it, would consider the third alternative—the one bringing in both principles—to be the best.

The possible clash between these two principles when each is taken alone is obscured because of the empirical fact that equal distribution is usually a *means* of achieving the greatest possible quantity. That is why classical utilitarians have usually been egalitarians. Mill worked all his life to break down the glaring inequalities between rich and poor; but, if one is to interpret his formula in the way that we have, he did so *in order* thereby to secure the largest total *quantity* of good in the nation as a whole. Whether equal distribution would or would not secure the largest total quantity of good is an empirical problem which is not ours at the moment. Our problem is, Should we aim at two ideals or one? Should we aim at maximum quantity of good (equal distribution being merely a means to it—perhaps a good means, perhaps not), or should we aim at maximum quantity of good *plus* equal distribution of good?

. . . . .

Most people . . . try to mediate, when there is a conflict, between achieving the highest total quantity of good (of whatever kind) and the most nearly equal possible distribution thereof. How does one mediate between these principles? In the opinion of many people, the principle of utility—interpreted as maximum total *quantity* of intrinsic good—will have to be revised. "For," they will allege, "it is not only the quantity of intrinsic good (happiness, knowledge, or whatever) that counts but also its *distribution:* it is just as important that there be a just distribution of intrinsic good as that there be a large quantity of it." If we accept this addition, we need not abandon . . . utilitarianism, but we will have to abandon it in the form in which we have thus far considered it. We can still retain it *if* we say that not only happiness (and according to the pluralist, certain other things as well) is intrinsically good but the *distribution* of happiness (or whatever else is intrinsically good) as well. In other words, justice is added to the list of things intrinsically worth attaining or producing.

.   .   .   .   .

But are there not occasions on which equality is not just at all? If one man works hard and long while another is lazy, is it just that they should receive the same wage? Does not justice sometimes require *in*equality?

.   .   .   .   .

Justice and Desert: Reward

Why should people have unequal rewards and punishments? Why should A receive more money than B, and why should C move about freely among his fellow men while D is placed behind bars? Because of deserts, we say: A deserves more pay than B because A works harder or produces more, and D has committed a crime as a result of which he deserves punishment whereas C does not. We often speak of justice in terms of desert. We say that a certain individual has received his just reward—that is, the reward that he deserves; or that someone has been unjustly punished—that is, the punishment meted out is more than he deserves.

.   .   .   .   .

Justice is getting what one deserves; what could be simpler? But here the simplicity ends. Often people do not agree on what rewards or punishments are deserved; nor is this disagreement always the result of partiality in their own favor. Even when they themselves are in no way involved, they often disagree on who deserves what and why. But at least they agree that justice has something to do with desert. They only disagree on what, in specific situations, is deserved.

To explore this issue, even in briefest outline, will be a sticky business. But there is no help for it; we cannot ignore it. Many would say that desert is the most important concept contained in the idea of "justice." Indeed, it is often held that the best possible state of society would be one in which each man gets what he deserves: he who deserves most is rewarded most (presumably with happiness, or whatever else is thought to be intrinsically good); he who deserves much is given much; and he who deserves little receives little. In other words, the ideal state of affairs would be one of *perfect apportionment* between one's desert and one's reward.

But when we try to come down to cases, we are baffled at the outset. What *does* a person deserve, and why? What does a plumber deserve? two dollars an hour? five dollars? How is one to say? Shall we say of a man who has committed armed robbery that he *deserves* the ten-year prison sentence that the law prescribes for him—or five, or one, or twenty? and who is to say how much more time the armed robber deserves than the petty thief? It is not merely that cases differ, though they do; but that even in *one* case we seem to be utterly at a loss to say exactly *what* a person deserves. By what criterion are deserts to be estimated?

To guide our steps in this maze, let us restrict our discussion, for the moment, to reward for work performed; and let us consider only monetary rewards, assuming—though it is far from being true—that a greater monetary reward will result in greater intrinsic good to the person who receives it. What criteria, then, can be employed to determine who should receive the most money for his labor?[1]

The most straightforward and seemingly obvious criterion for determining the amount of one's reward is *work*. Surely, we say, what a person should be paid bears a relation to the work that he performs. But unfortunately the word "work" is ambiguous. It may refer either to the *achievement*—what one actually produces, whether it be crops, manufactured articles, or trained minds—or to the *effort* leading to that achievement. Let us consider these in order.

1. *Achievement*—what one has actually produced, accomplished or brought about through the expenditure of one's effort, regardless of how easy or difficult it was to produce or how long or short was the time required to produce it. Many people believe that actual achievement should be the only criterion for estimating reward. It *is* the only criterion employed (at least officially) in many areas, for example the assignment of grades to students in courses. The student who has accomplished the most in the course gets the A, regardless of whether he had to work hard or spend much time to get it. If he worked hard but, through lack of aptitude or stupidity, still did not accomplish much in the course, we may praise him for his effort but we still withhold the high grade. The winner of a hundred-yard dash may not have tried a tenth as hard as the competitor with a stiff knee, but he wins the trophy anyway. We admire the man with a stiff knee for his effort, but we reward the winner for his achievement. Normally, but not always, wages in industry tend to be in proportion to the achievement of the worker. And, some would say, this is the way it ought to be. Not only is reward according to achievement the only practicable system (it is easier to determine actual production than effort), but it has the highest utility; for in this way the people with the

greatest capacity for achievement will be more likely to have the incentive to attain it.

. . . . .

Practical difficulties, however, arise in attempting to estimate achievement. Can one measure one type of achievement against another? Should a good teacher receive less pay than a good lawyer, a good lawyer less than a good surgeon? In our society they are rewarded most unequally. But if we concede that this inequality is unjust, how *would* we propose to reward them? Perhaps we would do well to examine other possible criteria.

2. *Effort*—the other sense of the word "work." Some would say that the more effort you put forth, the higher the wage you deserve, regardless of the actual quality or quantity of what you produce. Effort, of course, is not a function only of the amount of time expended; it is also a function of the energy expended during that time. Two men can each spend eight hours at a certain job, but if one of them concentrates fully on it and gives it all he can during that time while the other works in a slovenly or half-hearted manner, clearly the amount of effort expended is not the same, even though the time spent is. If the two men work the same amount of time and at 100 percent of their capacity, they may be said to have put forth equal effort.

. . . . .

Achievement and effort are the two principal competitors for criteria of distributive justice. Others, however, should be mentioned:

3. *Ability.* Perhaps rewards should be dispensed in proportion to one's ability.

. . . . .

4. *Need.* It may be suggested that one's reward should depend on how much one needs; the one who needs the most should receive the most.

. . . . .

Thus far, all our criteria have had to do with the worker or what he can produce. Further suggestions, however, may come from another quarter: criteria may be found having nothing to do with the worker but with the society in which he lives.

5. *The open market.* Some have suggested that what a worker deserves depends entirely on the wage his work can command in an open market. When there is a great demand for engineers—a greater demand than the

market can supply—then the wages of engineers go up, and (it is contended) so they should.

. . . . .

As a criterion of justice, however, this scheme of supply and demand does not run without a hitch. People, once established in a profession for which long training was required, cannot suddenly change their profession no matter how overcrowded it has become. . . . Poets, even great poets, receive very little money from their life's work (in the United States, at any rate)—far less than farm hands and prostitutes. Does this inequality show that the oldest profession is more deserving than the less pecunious ones? Is an engineer less deserving of high wages because a sudden glut on the engineer market or a national depression forces his income down?

6. Another criterion that we can note is *public need* for the product of one's labor. Perhaps not the need of the worker but the need of the public for the fruits of his labor should be a criterion of reward.

. . . . .

7. Yet a further criterion to be considered is *public desire* for the product of one's labor. Perhaps, then, it is not what people need but what they desire that should count. Desire is, of course, different from need: we may need food but not desire it, and we may desire alcohol but not need it.

. . . . .

8. As if this difficulty were not enough, there are still other criteria that are sometimes used in estimating desert; no one suggests that they are sufficient by themselves, but it is often suggested that at least they are relevant to the question of deserts in income. Let us consider them briefly.

a. Anyone who is in a profession that requires *long training* at a considerable expenditure of money and years of his life is said to deserve a larger income, once he is in a position to earn, than one whose job is unskilled.

. . . . .

b. Similarly, a position requiring *expensive equipment* for which the person must pay out of his own pocket (say, that of a physician or dentist in private practice) deserves more compensation; these added expenses are not luxuries but necessities for the professions in question, and they are there primarily for the benefit of the patient rather than the practitioner.

c. If a certain field of work involves great *risk of financial loss,* perhaps success in that field should be more heavily compensated.

. . . . .

d. It would usually be agreed that a person whose job requires him to face considerable *physical danger* deserves a greater compensation than people in safer occupations.

. . . . .

e. If a job is *unpleasant*—dirty, nerve-wracking, or in some way particularly taxing, such as collecting garbage or cleaning boilers or working eight hours a day next to a loudly blowing siren—it is felt that the person who endures the conditions of such a job deserves more than if his job is not unpleasant.

. . . . .

All these factors, and doubtless others besides, have been considered relevant to the estimation of deserts. Many people would exclude some of these factors as irrelevant; many would include the majority of them but would differ with one another in the weight or importance to be attached to each of them. The estimation of the various factors depends, of course, on the ethical position that one accepts: specifically, whether one believes that the criteria of distributive justice are to be judged solely by their relevance to overall utility or whether one must consider more than utility in judging these criteria.

Note

1. The criteria of justice which follow are in no way suggestions that any agency, such as a State Planning Commission, should attempt to regulate people's material rewards so as to bring about a state of justice. The results of such a procedure would doubtless be just as horrible as Mill thought. We are concerned here with what an ideal system of material rewards *would be,* not with how they should be brought about. If they were brought about by force, whim, or the arbitrary fiat of the bureaucrats on a regulatory commission, the results would probably be worse than that of the most flagrant injustices created without such fiat.

3

Kantian Theory: The Idea of Human Dignity

James Rachels

The great German philosopher Immanuel Kant thought that human beings occupy a special place in creation. Of course he was not alone in
thinking this. It is an old idea: from ancient times, humans have considered themselves to be essentially different from all other creatures—and
not just different but *better*. In fact, humans have traditionally thought
themselves to be quite fabulous. Kant certainly did. [I]n his view, human
beings have "an intrinsic worth, i.e., *dignity*," which makes them valuable "above all price." Other animals, by contrast, have value only insofar as they serve human purposes. In his *Lecture on Ethics* (1779), Kant
said:

> But so far as animals are concerned, we have no direct duties. Ani-
> mals . . . are there merely as means to an end. That end is man.

We can, therefore, use animals in any way we please. We do not even
have a "direct duty" to refrain from torturing them. Kant admits that it is
probably wrong to torture them, but the reason is not that *they* would be
hurt; the reason is only that *we* might suffer indirectly as a result of it,
because "he who is cruel to animals becomes hard also in his dealings
with men." Thus [i]n Kant's view, mere animals have no importance at
all. Human beings are, however, another story entirely. According to
Kant, humans may never be "used" as means to an end. He even went so
far as to suggest that this is the ultimate law of morality.

Like many other philosophers, Kant believed that morality can be
summed up in one ultimate principle, from which all our duties and
obligations are derived. He called this principle *The Categorical Imperative*. In the *Groundwork of the Metaphysics of Morals* (1785) he expressed it like this:

> Act only according to that maxim by which you can at the same time will
> that it should become a universal law.

However, Kant also gave *another* formulation of The Categorical Imperative. Later in the same book, he said that the ultimate moral principle
may be understood as saying:

> Act so that you treat humanity, whether in your own person or in that of
> another, always as an end and never as a means only.

Scholars have wondered ever since why Kant thought these two rules
were equivalent. They *seem* to express very different moral conceptions.
Are they, as he apparently believed, two versions of the same basic idea,
or are they really different ideas? We will not pause over this question.
Instead we will concentrate here on Kant's belief that morality requires
us to treat persons "always as an end and never as a means only." What
exactly does this mean, and why did he think it true?

When Kant said that the value of human beings "is above all price," he
did not intend this as mere rhetoric but as an objective judgment about the
place of human beings in the scheme of things. There are two important
facts about people that, in his view, support this judgment.

First, because people have desires and goals, other things have value
for them, in relation to *their* projects. Mere "things" (and this includes
nonhuman animals, whom Kant considered unable to have self-conscious
desires and goals) have value only as means to ends, and it is human ends
that *give* them value. Thus if you want to become a better chess player, a
book of chess instruction will have value for you; but apart from such
ends the book has no value. Or if you want to travel about, a car will have
value for you; but apart from this desire the car will have no value.

Second, and even more important, humans have "an intrinsic worth,
i.e., *dignity,*" because they are *rational agents*—that is, free agents capable of making their own decisions, setting their own goals, and guiding
their conduct by reason. Because the moral law is the law of reason,
rational beings are the embodiment of the moral law itself. The only way
that moral goodness can exist at all in the world is for rational creatures
to apprehend what they should do and, acting from a sense of duty, do it.
This, Kant thought, is the *only* thing that has "moral worth." Thus if
there were no rational beings, the moral dimension of the world would
simply disappear.

It makes no sense, therefore, to regard rational beings merely as one
kind of valuable thing among others. They are the beings *for whom* mere
"things" have value, and they are the beings whose conscientious actions
have moral worth. So Kant concludes that their value must be absolute,
and not comparable to the value of anything else.

If their value is "beyond all price," it follows that rational beings must
be treated "always as an end, and never as a means only." This means, on
the most superficial level, that we have a strict duty of beneficence to-

ward other persons: we must strive to promote their welfare; we must respect their rights, avoid harming them, and generally "endeavor, so far as we can, to further the ends of others."

But Kant's idea also has a somewhat deeper implication. The beings we are talking about are *rational* beings, and "treating them as ends-in-themselves" means *respecting their rationality*. Thus we may never *manipulate* people, or *use* people, to achieve our purposes, no matter how good those purposes may be. Kant gives this example, which is similar to an example he uses to illustrate the first version of his categorical imperative: Suppose you need money, and so you want a "loan," but you know you will not be able to repay it. In desperation, you consider making a false promise (to repay) in order to trick a friend into giving you the money. May you do this? Perhaps you need the money for a good purpose—so good, in fact, that you might convince yourself the lie would be justified. Nevertheless, if you lied to your friend, you would merely be manipulating him and using him "as a means."

On the other hand, what would it be like to treat your friend "as an end"? Suppose you told the truth, that you need the money for a certain purpose but will not be able to repay it. Then your friend could make up his own mind about whether to let you have it. He could exercise his own powers of reason, consulting his own values and wishes, and make a free, autonomous choice. If he did decide to give the money for this purpose, he would be choosing to make that purpose *his own*. Thus you would not merely be using him as a means to achieving *your* goal. This is what Kant meant when he said, "Rational beings . . . must always be esteemed at the same time as ends, i.e., only as beings who must be able to contain in themselves the end of the very same action."

Now Kant's conception of human dignity is not easy to grasp; it is, in fact, probably the most difficult notion discussed [here]. We need to find a way to make the idea clearer. In order to do that, we will consider in some detail one of its most important applications—this may be better than a dry, theoretical discussion. Kant believed that if we take the idea of human dignity seriously, we will be able to understand the practice of criminal punishment in a new and revealing way.

. . . . .

On the face of it, it seems unlikely that we could describe punishing someone as "respecting him as a person" or as "treating him as an end-in-himself." How could taking away someone's freedom, by sending him to prison, be a way of "respecting" him? Yet that is exactly what Kant suggests. Even more paradoxically, he implies that *executing* someone may also be a way of treating him "as an end." How can this be?

Remember that, for Kant, treating someone as an "end-in-himself" means treating him *as a rational being*. Thus we have to ask, What does it mean to treat someone as a rational being? Now a rational being is someone who is capable of reasoning about his conduct and who freely decides what he will do, on the basis of his own rational conception of what is best. Because he has these capacities, a rational being is *responsible* for his actions. We need to bear in mind the difference between:

1. *Treating someone as a responsible being*

and

2. *Treating someone as a being who is not responsible for his conduct.*

Mere animals, who lack reason, are not responsible for their actions; nor are people who are mentally "sick" and not in control of themselves. In such cases it would be absurd to try to "hold them accountable." We could not properly feel gratitude or resentment toward them, for they are not responsible for any good or ill they cause. Moreover, we cannot expect them to understand why we treat them as we do, any more than they understand why they behave as they do. So we have no choice but to deal with them by manipulating them, rather than by addressing them as autonomous individuals. When we spank a dog who has urinated on the rug, for example, we may do so in an attempt to prevent him from doing it again—but we are merely trying to "train" him. We could not reason with him even if we wanted to. The same goes for mentally "sick" humans.

On the other hand, rational beings are responsible for their behavior and so may properly be "held accountable" for what they do. We may feel gratitude when they behave well, and resentment when they behave badly. Reward and punishment—not "training" or other manipulation—are the natural expression of this gratitude and resentment. Thus in punishing people, we are *holding them responsible* for their actions, in a way in which we cannot hold mere animals responsible. We are responding to them not as people who are "sick" or who have no control over themselves, but as people who have freely chosen their evil deeds.

Furthermore, in dealing with responsible agents, we may properly allow *their conduct* to determine, at least in part, how we respond to them. If someone has been kind to you, you may respond by being generous in return; and if someone is nasty to you, you may also take that into account in deciding how to deal with him or her. And why shouldn't you? Why should you treat everyone alike, regardless of how *they* have chosen to behave?

Kant gives this last point a distinctive twist. There is, [i]n his view, a deep logical reason for responding to other people "in kind." The first formulation of The Categorical Imperative comes into play here. When we decide what to do, we in effect proclaim our wish that our conduct be made into a "universal law." Therefore, when a rational being decides to treat people in a certain way, he decrees that in his judgment *this is the way people are to be treated.* Thus if we treat him the same way in return, we are doing nothing more than treating him as *he has decided* people are to be treated. If he treats others badly, and we treat him badly, we are complying with his own decision. (Of course, if he treats others well, and we treat him well in return, we are also complying with the choice he has made.) We are allowing *him* to decide how he is to be treated—and so we are, in a perfectly clear sense, respecting his judgment, by allowing it to control our treatment of him. Thus Kant says of the criminal, "His own evil deed draws the punishment upon himself."

The Ethical Context
of Some Particular Problems

The readings in this section provide general philosophical discussion of privacy, the relationship between law and morals, and whistle-blowing. As is shown in readings 10, 11, and 12, the computer revolution brings massive new potential threats to privacy. These threats exist not only because of the extraordinary power of computers, but also because the concept of privacy is sufficiently ill-defined that computer decision makers may implement a new system without recognizing it as an invasion of privacy, or without having any theory to guide them in balancing the system's benefits against the harm of some sacrifice of privacy. Fried (reading 4) probes the concept of privacy and discusses why it is one of our most basic values.

A new technology may bring difficult new challenges to the legal system, such as the challenge to protect the competing rights of software developers and customers (as discussed in readings 24 and 25) or the challenge to protect the public from harm caused by program malfunction (as discussed in reading 28). In such difficult situations both judges and legislators look for guidance to the moral principles underlying the law. H. L. A. Hart (reading 5) provides a general discussion of the relationship between morality and law which makes explicit the possibility of, and the necessity for, the use of general moral principles in debates about what the law ought to say.

Computer scientists, like other professionals, may find their general obligation to "consider the health, privacy, and general welfare of the public in the performance of [one's] work" (EC5.1 of the ACM code, reading 32) in discord with other loyalties. (See reading 33 for a discussion of such a situation and reading 34 for a different view of the affair.) Bok (reading 6) analyzes the conflicting pressures on the professional in these situations and offers some guidance about when one ought to "blow the whistle."

4

Privacy: A Rational Context

Charles Fried

In this chapter I analyze the concept of privacy and attempt to show why it assumes such high significance in our system of values. There is a puzzle here, since we do not feel comfortable about asserting that privacy is intrinsically valuable, an end in itself—privacy is always for or in relation to something or someone. On the other hand, to view privacy as simply instrumental, as one way of getting other goods, seems unsatisfactory too. For we feel that there is a necessary quality, after all, to the importance we ascribe to privacy. This perplexity is displayed when we ask how privacy might be traded off against other values. We wish to ascribe to privacy more than an ordinary priority. My analysis attempts to show why we value privacy highly and why also we do not treat it as an end in itself. Briefly, my argument is that privacy provides the rational context for a number of our most significant ends, such as love, trust and friendship, respect and self-respect. Since it is a necessary element of those ends, it draws its significance from them. And yet since privacy is only an element of those ends, not the whole, we have not felt inclined to attribute to privacy ultimate significance. In general this analysis of privacy illustrates how the concepts in this essay can provide a rational account for deeply held moral values.

An Immodest Proposal: Electronic Monitoring

There are available today electronic devices to be worn on one's person which emit signals permitting one's exact location to be determined by a monitor some distance away. These devices are so small as to be entirely unobtrusive: other persons cannot tell that a subject is "wired," and even the subject himself—if he could forget the initial installation—need be no more aware of the device than of a small bandage. Moreover, existing technology can produce devices capable of monitoring not only a person's location, but other significant facts about him: his temperature, pulse rate, blood pressure, the alcoholic content of his blood, the sounds in his immediate environment—for example, what he says and what is said to him—and perhaps in the not too distant future even the pattern of his brain waves. The suggestion has been made, and is being actively investigated, that such devices might be employed in the surveillance of persons on probation or parole.

Probation leaves an offender at large in the community as an alternative to imprisonment, and parole is the release of an imprisoned person prior to the time that all justification for supervising him and limiting his liberty has expired. Typically, both probation and parole are granted subject to various restrictions. Most usually the probationer or parolee is not allowed to leave a prescribed area. Also common are restrictions on the kinds of places he may visit—bars, pool halls, brothels, and the like may be forbidden—the persons he may associate with, and the activities he may engage in. The most common restriction on activities is a prohibition on drinking, but sometimes probation and parole have been revoked for "immorality"—that is, intercourse with a person other than a spouse. There are also affirmative conditions, such as a requirement that the subject work regularly in an approved employment, maintain an approved residence, report regularly to correctional, social, or psychiatric personnel. Failure to abide by such conditions is thought to endanger the rehabilitation of the subject and to identify him as a poor risk.

Now the application of personal monitoring to probation and parole is obvious. Violations of any one of the conditions and restrictions could be uncovered immediately by devices using present technology or developments of it; by the same token, a wired subject assured of detection would be much more likely to obey. Although monitoring is admitted to be unusually intrusive, it is argued that this particular use of monitoring is entirely proper, since it justifies the release of persons who would otherwise remain in prison, and since surely there is little that is more intrusive and unprivate than a prison regime. Moreover, no one is obliged to submit to monitoring: an offender may decline and wait in prison until his sentence has expired or until he is judged a proper risk for parole even without monitoring. Proponents of monitoring suggest that seen in this way monitoring of offenders subject to supervision is no more offensive than the monitoring on an entirely voluntary basis of epileptics, diabetics, cardiac patients, and the like.

Much of the discussion about this and similar (though perhaps less futuristic) measures has proceeded in a fragmentary way to catalog the disadvantages they entail: the danger of the information falling into the wrong hands, the opportunity presented for harassment, the inevitable involvement of persons as to whom no basis for supervision exists, the use of the material monitored by the government for unauthorized purposes, the danger to political expression and association, and so on.

Such arguments are often sufficiently compelling, but situations may be envisaged where they are overridden. The monitoring case in some of its aspects is such a situation. And yet one often wants to say the invasion of privacy is wrong, intolerable, although each discrete objection can be met. The reason for this, I submit, is that privacy is much more than just a possible social technique for assuring this or that substantive interest. Such analyses of the value of privacy often lead to the conclusion that the various substantive interests may after all be protected as well by some other means, or that if they cannot be protected quite as well, still those other means will do, given the importance of our reasons for violating privacy. It is just because this instrumental analysis makes privacy so vulnerable that we feel impelled to assign to privacy some intrinsic significance. But to translate privacy to the level of an intrinsic value might seem more a way of cutting off analysis than of carrying it forward.

It is my thesis that privacy is not just one possible means among others to insure some other value, but that it is necessarily related to ends and relations of the most fundamental sort: respect, love, friendship, and trust. Privacy is not merely a good technique for furthering these fundamental relations; rather, without privacy they are simply inconceivable. They require a context of privacy or the possibility of privacy for their existence. To make clear the necessity of privacy as a context for respect, love, friendship, and trust is to bring out also why a threat to privacy seems to threaten our very integrity as persons. To respect, love, trust, or feel affection for others and to regard ourselves as the objects of love, trust, and affection is at the heart of our notion of ourselves as persons among persons, and privacy is the necessary atmosphere for these attitudes and actions, as oxygen is for combustion.

Privacy and Personal Relations

Before going further, it is necessary to sharpen the intuitive concept of privacy. As a first approximation, privacy seems to be related to secrecy, to limiting the knowledge of others about oneself. This notion must be refined. It is not true, for instance, that the less that is known about us the more privacy we have. Privacy is not simply an absence of information about us in the minds of others; rather, it is the control we have over information about ourselves.

To refer, for instance, to the privacy of a lonely man on a desert island would be to engage in irony. The person who enjoys privacy is able to grant or deny access to others. Even when one considers private situations into which outsiders could not possibly intrude, the context implies some alternative situation where the intrusion is possible. A man's house may be private, for instance, but that is because it is constructed—with doors, windows, window shades—to allow it to be made private, and because the law entitles a man to exclude unauthorized persons. And even the remote vacation hideaway is private just because one resorts to it in order—in part—to preclude access to unauthorized persons.

Privacy, thus, is control over knowledge about oneself. But it is not simply control over the quality of information abroad; there are modulations in the quality of the knowledge as well. We may not mind that a person knows a general fact about us, and yet feel our privacy invaded if he knows the details. For instance, a casual acquaintance may comfortably know that I am sick, but it would violate my privacy if he knew the nature of the illness. Or a good friend may know what particular illness I am suffering from, but it would violate my privacy if he were actually to witness my suffering from some symptom which he must know is associated with the disease.

Privacy in its dimension of control over information is an aspect of personal liberty. Acts derive their meaning partly from their social context—from how many people know about them and what the knowledge consists of. For instance, a reproof administered out of the hearing of third persons may be an act of kindness, but if administered in public it becomes cruel and degrading. Thus if a man cannot be sure that third persons are not listening—if his privacy is not secure—he is denied the freedom to do what he regards as an act of kindness.

Besides giving us control over the context in which we act, privacy has a more defensive role in protecting our liberty. We may wish to do or say things not forbidden by the restraints of morality but nevertheless unpopular or unconventional. If we thought that our every word and deed were public, fear of disapproval or more tangible retaliation might keep us from doing or saying things which we would do or say if we could be sure of keeping them to ourselves or within a circle of those who we know approve or tolerate our tastes.

These reasons support the familiar arguments for the right of privacy. Yet they leave privacy with less security than we feel it deserves; they leave it vulnerable to arguments that a particular invasion of privacy will secure to us other kinds of liberty which more than compensate for what is lost. To present privacy, then, only as an aspect of or an aid to general liberty is to miss some of its most significant differentiating features. The value of control over information about ourselves is more nearly absolute than that. For privacy is the necessary context for relationships which we would hardly be human if we had to do without—the relationships of love, friendship, and trust.

Love and friendship . . . involve the initial respect for the rights of others which morality requires of everyone. They further involve the voluntary and spontaneous relinquishment of something between friend and friend, lover and lover. The title to information about oneself conferred by privacy provides the necessary something. To be friends or lovers persons must be intimate to some degree with each other. Intimacy is the sharing of information about one's actions, beliefs, or emotions which one does not share with all, and which one has the right not to share with anyone. By conferring this right, privacy creates the moral capital which we spend in friendship and love.

The entitlements of privacy are not just one kind of entitlement among many which a lover can surrender to show his love. Love or friendship can be partially expressed by the gift of other rights—gifts of property or of service. But these gifts, without the intimacy of shared private information, cannot alone constitute love or friendship. The man who is generous with his possessions, but not with himself, can hardly be a friend, nor—and this more clearly shows the necessity of privacy for love—can the man who, voluntarily or involuntarily, shares everything about himself with the world indiscriminately.

Privacy is essential to friendship and love in another respect besides providing what I call moral capital. The rights of privacy are among those basic entitlements which men must respect in each other; and mutual respect is the minimal precondition for love and friendship.

Privacy also provides the means for modulating those degrees of friendship which fall short of love. Few persons have the emotional resources to be on the most intimate terms with all their friends. Privacy grants the control over information which enables us to maintain degrees of intimacy. Thus even between friends the restraints of privacy apply; since friendship implies a voluntary relinquishment of private information, one will not wish to know what his friend or lover has not chosen to share with him. The rupture of this balance by a third party—the state perhaps—thrusting information concerning one friend upon another might well destroy the limited degree of intimacy the two have achieved.

Finally, there is a more extreme case where privacy serves not to save something which will be "spent" on a friend, but to keep it from all the world. There are thoughts whose expression to a friend or lover would be a hostile act, though the entertaining of them is completely consistent with friendship or love. That is because these thoughts, prior to being given expression, are mere unratified possibilities for action. Only by expressing them do we adopt them, choose them as part of ourselves, and draw them into our relations with others. Now a sophisticated person knows that a friend or lover must entertain thoughts which if expressed would be wounding, and so—it might be objected—why should he attach any significance to their actual expression? In a sense the objection is well taken. If it were possible to give expression to these thoughts and yet make clear to ourselves and to others that we do not thereby ratify them, adopt them as our own, it might be that in some relations, at least, another could be allowed complete access to us. But this possibility is not a very likely one. Thus the most complete form of privacy is perhaps also the most basic, since it is necessary not only to our freedom to define our relations with others but also to our freedom to define ourselves. To be deprived of this control over what we do and who we are is the ultimate assault on liberty, personality, and self-respect.

Trust is the attitude of expectation that another will behave according to the constraints of morality. Insofar as trust is only instrumental to the more convenient conduct of life, its purposes could be as well served by cheap and efficient surveillance of the person upon whom one depends. One does not trust machines or animals; one takes the fullest economically feasible precautions against their going wrong. Often, however, we choose to trust people where it would be safer to take precautions—to watch them or require a bond from them. This must be because, as I have already argued, we value the relation of trust for its own sake. It is one of those relations, less inspiring than love or friendship but also less tiring, through which we express our humanity.

There can be no trust where there is no possibility of error. More specifically, man cannot know that he is trusted unless he has a right to act without constant surveillance so that he knows he can betray the trust. Privacy confers that essential right. And since, as I have argued, trust in its fullest sense is reciprocal, the man who cannot be trusted cannot himself trust or learn to trust. Without privacy and the possibility of error which it protects that aspect of his humanity is denied to him.

The Concrete Recognition of Privacy

In concrete situations and actual societies, control over information about oneself, like control over one's bodily security or property, can only be relative and qualified. As is true for property or bodily security, the control over privacy must be limited by the rights of others. And as in the cases of property and bodily security, so too with privacy: the more one ventures into the outside, the more one pursues one's other interests with the aid of, in competition with, or even in the presence of others, the more one must risk invasions. As with property and personal security, it is the business of legal and social institutions to define and protect the right of privacy which emerges intact from the hurly-burly of social interactions. Now it would be absurd to argue that these concrete definitions and protections, differing as they do from society to society, are or should be strict derivations from general principles, the only legitimate variables being differing empirical circumstances (such as differing technologies or climatic conditions). The delineation of standards must be left to a political and social process the results of which will accord with justice if two conditions are met: (1) the process itself is just, that is, the interests of all are fairly represented; and (2) the outcome of the process protects basic dignity and provides moral capital for personal relations in the form of absolute title to at least some information about oneself.

The particular areas of life which are protected by privacy will be conventional at least in part, not only because they are the products of political processes, but also because of one of the reasons we value privacy. Insofar as privacy is regarded as moral capital for relations of love, friendship, and trust, there are situations where what kinds of information one is entitled to keep to oneself is not of the first importance. The important thing is that there be *some* information which is protected. Convention may quite properly rule in determining the particular areas which are private.

Convention plays another more important role in fostering privacy and the respect and esteem which it protects; it designates certain areas, intrinsically no more private than other areas, as symbolic of the whole institution of privacy, and thus deserving of protection beyond their particular importance. This apparently exaggerated respect for conventionally protected areas compensates for the inevitable fact that privacy is gravely compromised in any concrete social system: it is compromised by the inevitably and utterly just exercise of rights by others, it is compromised by the questionable but politically sanctioned exercise of rights by others, it is compromised by conduct which society does not condone but which it is unable or unwilling to forbid, and it is compromised by plainly wrongful invasions and aggressions. In all this there is a real danger that privacy might be crushed altogether, or, what would be as bad, that any venture outside the most limited area of activity would mean risking an almost total compromise of privacy.

Given these threats to privacy in general, social systems have given symbolic importance to certain conventionally designated areas of privacy. Thus in our culture the excretory functions are so shielded that situations in which this privacy is violated are experienced as extremely distressing, as detracting from one's dignity and self-esteem. Yet there does not seem to be any reason connected with the principles of respect, esteem, and the like why this would have to be so, and one can imagine other cultures in which it was not so, but where the same symbolic privacy was attached to, say, eating and drinking. There are other more subtly modulated symbolic areas of privacy, some of which merge into what I call substantive privacy (that is, areas where privacy does protect substantial interests). The very complex norms of privacy about matters of sex and health are good examples.

An excellent, very different sort of example of a contingent, symbolic recognition of an area of privacy as an expression of respect for personal integrity is the privilege against self-incrimination and the associated doctrines denying officials the power to compel other kinds of information without some explicit warrant. By according the privilege as fully as it does, our society affirms the extreme value of the individual's control over information about himself. To be sure, prying into a man's personal affairs by asking questions of others or by observing him is not prevented. Rather it is the point of the privilege that a man cannot be forced to make public information about himself. Thereby his sense of control over what others know of him is significantly enhanced, even if other sources of the same information exist. Without his cooperation, the other sources are necessarily incomplete, since he himself is the only ineluctable witness to his own present life, public or private, internal or manifest. And information about himself which others have to give out is in one sense information over which he has already relinquished control.

The privilege is contingent and symbolic. It is part of a whole structure of rules by which there is created an institution of privacy sufficient to the sense of respect, trust, and intimacy. It is contingent in that it cannot, I believe, be shown that some particular set of rules is necessary to the existence of such an institution of privacy. It is symbolic because the exercise of the privilege provides a striking expression of society's willingness to accept constraints on the pursuit of valid, perhaps vital, interests in order to recognize the right of privacy and the respect for the individual that privacy entails. Conversely, a proceeding in which compulsion is brought to bear on an individual to force him to make revelations about himself provides a striking and dramatic instance of a denial of title to control information about oneself, to control the picture we would have others have of us. In this sense such a procedure quite rightly seems profoundly humiliating. Nevertheless it is not clear to me that a system is unjust which sometimes allows such an imposition.

In calling attention to the symbolic aspect of some areas of privacy I do not mean to minimize their importance. On the contrary, they are highly significant as expressions of respect for others in a general situation where much of what we do to each other may signify a lack of respect or at least presents no occasion for expressing respect. That this is so is shown not so much on the occasions where these symbolic constraints are observed, for they are part of our system of expectations, but where they are violated. Not only does a person feel his standing is gravely compromised by such symbolic violations, but also those who wish to degrade and humiliate others often choose just such symbolic aggressions and invasions on the assumed though conventional area of privacy.

The Concept of Privacy Applied to the Problem of Monitoring

Let us return now to the concrete problem of electronic monitoring to see whether the foregoing elucidation of the concept of privacy will help to establish on firmer ground the intuitive objection that monitoring is an intolerable violation of privacy. Let us consider the more intrusive forms of monitoring where not only location but conversations and perhaps other data are monitored.

Obviously such a system of monitoring drastically curtails or eliminates altogether the power to control information about oneself. But, it might be said, this is not a significant objection if we assumed the monitored data will go only to authorized persons—probation or parole officers— and cannot be prejudicial so long as the subject of the monitoring is not violating the conditions under which he is allowed to be at liberty. This retort misses the importance of privacy as a context for all kinds of relations, from the most intense to the most casual. For all of these may require a context of some degree of intimacy, and intimacy is made impossible by monitoring.

It is worth being more precise about this notion of intimacy. Monitoring obviously presents vast opportunities for malice and misunderstanding on the part of authorized personnel. For that reason the subject has reason to be constantly apprehensive and inhibited in what he does. There is always an unseen audience, which is the more threatening because of the possibility that one may forget about it and let down his guard, as one would not with a visible audience. Even assuming the benevolence and understanding of the official audience, there are serious consequences to the fact that no degree of true intimacy is possible for the subject. Privacy is not, as we have seen, just a defensive right. It forms the necessary context for the intimate relations of love and friendship which give our lives much of whatever affirmative value they have. In the role of citizen or fellow worker, one need reveal himself to no greater extent than is necessary to display the attributes of competence and morality appropriate to those roles. In order to be a friend or lover one must reveal far more of himself. Yet where any intimate revelation may be heard by monitoring officials, it loses the quality of exclusive intimacy required of a gesture of love or friendship. Thus monitoring, in depriving one of privacy, destroys the possibility of bestowing the gift of intimacy, and makes impossible the essential dimension of love and friendship.

Monitoring similarly undermines the subject's capacity to enter into relations of trust. As I analyzed trust, it required the possibility of error on the part of the person trusted. The negation of trust is constant surveillance—such as monitoring—which minimizes the possibility of undetected default. The monitored parolee is denied the sense of self-respect inherent in being trusted by the government which has released him. More important, monitoring prevents the parolee from entering into true *relations* of trust with persons in the outside world. An employer, unaware of the monitoring, who entrusts a sum of money to the parolee cannot thereby grant him the sense of responsibility and autonomy which an unmonitored person in the same position would have. The parolee in a real—if special and ironical—sense, cannot be trusted.

Now let us consider the argument that however intrusive monitoring may seem, surely prison life is more so. In part, of course, this will be a matter of fact. It may be that a reasonably secure and well-run prison will allow prisoners occasions for conversation among themselves, with guards, or with visitors, which are quite private. Such a prison regime would in this respect be less intrusive than monitoring. Often prison regimes do not allow even this, and go far toward depriving a prisoner of any sense of privacy: if the cells have doors, these may be equipped with peepholes. But there is still an important difference between this kind of prison and monitoring: the prison environment is overtly, even punitively unprivate. The contexts for relations to others are obviously and drastically different from what they are on the "outside." This itself, it seems to me, protects the prisoner's human orientation where monitoring only assails it. If the prisoner has a reasonably developed capacity for love, trust, and friendship and has in fact experienced ties of this sort, he is likely to be strongly aware (at least for a time) that prison life is a drastically different context from the one in which he enjoyed those relations, and this awareness will militate against his confusing the kinds of relations that can obtain in a "total institution" like a prison with those of freer social settings on the outside.

Monitoring, by contrast, alters only in a subtle and unobtrusive way—though a significant one—the context for relations. The subject appears free to perform the same actions as others and to enter the same relations, but in fact an important element of autonomy, of control over one's environment, is missing: he cannot be private. A prisoner can adopt a stance of withdrawal, of hibernation as it were, and thus preserve his sense of privacy intact to a degree. A person subject to monitoring by virtue of being in a free environment, dealing with people who expect him to have certain responses, capacities, and dispositions, is forced to make at least a show of intimacy to the persons he works closely with, those who would be his friends, and so on. They expect these things of him, because he is assumed to have the capacity and disposition to enter into ordinary relations with them. Yet if he does—if, for instance, he enters into light banter with slight sexual overtones with the waitress at the diner where he eats regularly—he has been forced to violate his own integrity by revealing to his official monitors even so small an aspect of his private personality, the personality he wishes to reserve for persons toward whom he will make some gestures of intimacy and friendship. Theoretically, of course, a monitored parolee might adopt the same attitude of withdrawal that a prisoner does, but in fact that too would be a costly and degrading experience. He would be tempted, as in prison he would not be, to "give himself away" and to act like everyone else, since in every outward respect he seems like everyone else. Moreover, by withdrawing, the person subject to monitoring would risk seeming cold, unnatural, odd, inhuman, to the very people whose esteem and affection he craves. In prison the circumstances dictating a reserved and tentative facade are so apparent to all that adopting such a facade is no reflection on the prisoner's humanity.

The insidiousness of a technique which forces a man to betray himself in this humiliating way or else seem inhuman is compounded when one considers that the subject is also forced to betray others who may become intimate with him. Even persons in the overt oppressiveness of a prison do not labor under the burden of this double betrayal.

As against all of these considerations, there remains the argument that so long as monitoring depends on the consent of the subject, who feels it is preferable to prison, to close off this alternative in the name of a morality so intimately concerned with liberty is absurd. This argument may be decisive; I am not at all confident that the alternative of monitored release should be closed off. My analysis does show, I think, that it involves costs to the prisoner which are easily overlooked, that on inspection it is a less desirable alternative than might at first appear. Moreover, monitoring presents systematic dangers to potential subjects as a class. Its availability as a compromise between conditional release and continued imprisonment may lead officials who are in any doubt whether or not to trust a man on parole or probation to assuage their doubts by resorting to monitoring.

The seductions of monitored release disguise not only a cost to the subject but to society as well. The discussion of trust should make clear that unmonitored release is a very different experience from monitored release, and so the educational and rehabilitative effect of unmonitored release is also different. Unmonitored release affirms in a far more significant way the relations of trust between the convicted criminal and the society which he violated by his crime and which we should now be seeking to reestablish. But trust can only arise, as any parent knows, through the experience of being trusted.

Finally, it must be recognized that more limited monitoring—for instance where only the approximate location of the subject is revealed—lacks the offensive features of total monitoring, and is obviously preferable to prison.

The Role of Law

This evaluation of the proposal for electronic monitoring has depended on the general theoretical framework of this whole essay. It is worth noting the kind of evaluation that framework has permitted. Rather than inviting a fragmentation of the proposal into various pleasant and unpleasant elements and comparing the "net utility" of the proposal with its alternatives, we have been able to evaluate the total situation created by the proposal in another way. We have been able to see it as a system in which certain actions and relations, the pursuit of certain ends, are possible or impossible. Certain systems of actions, ends, and relations are possible or impossible in different social contexts. Moreover, the social context itself is a system of actions and relations. The social contexts created by monitoring and its alternatives, liberty or imprisonment, are thus evaluated by their conformity to a model system in which are instantiated the principles of morality, justice, friendship, and love. Such a model, which is used as a standard, is of course partially unspecified in that there is perhaps an infinite number of specific systems which conform to those principles. Now actual systems, as we have seen, may vary in respect to how other ends—for example, beauty, knowledge—may be pursued in them, and they may be extremely deficient in allowing for the pursuit of such ends. But those who design, propose, and administer social systems are first of all bound to make them conform to the model of morality and injustice, for in so doing they express respect and even friendship—what might be called civic friendship—toward those implicated in the system. If designers and administrators fail to conform to this model, they fail to express that aspect of their humanity which makes them in turn fit subjects for the respect, friendship, and love of others.

5

Law and Morals

H. L. A. Hart

Power and Authority

It is often said that a legal system must rest on a sense of moral obligation or on the conviction of the moral value of the system, since it does not and cannot rest on mere power of man over man.

. . . . .

A necessary condition of the existence of coercive power is that some at least must voluntarily cooperate in the system and accept its rules. In this sense it is true that the coercive power of law presupposes its accepted authority. But the dichotomy of "law based merely on power" and "law which is accepted as morally binding" is not exhaustive. Not only may vast numbers be coerced by laws which they do not regard as morally binding, but it is not even true that those who do accept the system voluntarily must conceive of themselves as morally bound to do so, though the system will be most stable when they do so. In fact, their allegiance to the system may be based on many different considerations: calculations of long-term interest; disinterested interest in others; an unreflecting inherited or traditional attitude; or the mere wish to do as others do.

. . . . .

The Influence of Morality on Law

The law of every modern state shows at a thousand points the influence of both the accepted social morality and wider moral ideals. These influences enter into law either abruptly and avowedly through legislation, or silently and piecemeal through the judicial process. In some systems, as in the United States, the ultimate criteria of legal validity explicitly incorporate principles of justice or substantive moral values; in other systems, as in England, where there are no formal restrictions on the competence of the supreme legislature, its legislation may yet no less scrupulously conform to justice or morality. The further ways in which law mirrors morality are myriad, and still insufficiently studied: statutes may be a mere legal shell and demand by their express terms to be filled out with the aid of moral principles; the range of enforceable contracts may be limited by reference to conceptions of morality and fairness; liability for both civil and criminal wrongs may be adjusted to prevailing views of moral responsibility. No "positivist" could deny that these are facts, or that the stability of legal systems depends in part upon such types of correspondence with morals.

. . . . .

Interpretation

Laws require interpretation if they are to be applied to concrete cases, and once the myths which obscure the nature of the judicial processes are dispelled by realistic study, it is patent . . . that the open texture of law leaves a vast field for creative activity which some call legislative. Neither in interpreting statutes nor precedents are judges confined to the alternatives of blind, arbitrary choice, or "mechanical" deduction from rules with predetermined meaning. Very often their choice is guided by an assumption that the purpose of the rules which they are interpreting is a reasonable one, so that the rules are not intended to work injustice or offend settled moral principles. Judicial decision, especially on matters of high constitutional import, often involves a choice between moral values, and not merely the application of some single outstanding moral principle; for it is folly to believe that where the meaning of the law is in doubt, morality always has a clear answer to offer. At this point judges may again make a choice which is neither arbitrary nor mechanical; and here often display characteristic judicial virtues, the special appropriateness of which to legal decision explains why some feel reluctant to call such judicial activity "legislative." These virtues are: impartiality and neutrality in surveying the alternatives; consideration for the interest of all who will be affected; and a concern to deploy some acceptable general principle as a reasoned basis for decision. No doubt because a plurality of such principles is always possible it cannot be *demonstrated* that a decision is uniquely correct: but it may be made acceptable as the reasoned product of informed impartial choice. In all this we have the "weighing" and "balancing" characteristic of the effort to do justice between competing interests.

.   .   .   .   .

The Criticism of Law

. . . A good legal system must conform at certain points . . . to the requirements of justice and morality. Some may regard this as an obvious truism; but is is not a tautology, and in fact, in the criticism of law, there may be disagreement both as to the appropriate moral standards and as to the required points of conformity. Does the morality, with which law must conform if it is to be good, mean the accepted morality of the group whose law it is, even though this may rest on superstition or may withhold its benefits and protection from slaves or subject classes? Or does morality mean standards which are enlightened in the sense that they rest on rational beliefs as to matters of fact, and accept all human beings as entitled to equal consideration and respect?

No doubt the contention that a legal system must treat all human beings within its scope as entitled to certain basic protections and freedoms, is now generally accepted as a statement of an ideal of obvious relevance in the criticism of law. Even where practice departs from it, lip service to this ideal is usually forthcoming. It may even be the case that a morality which does not take this view of the right of all men to equal consideration, can be shown by philosophy to be involved in some inner contradiction, dogmatism, or irrationality. If so, the enlightened morality which recognizes these rights has special credentials as the true morality, and is not just one among many possible moralities. These are claims which cannot be investigated here, but even if they are conceded, they cannot alter, and should not obscure, the fact that municipal legal systems, with their characteristic structure of primary and secondary rules, have long endured though they have flouted these principles of justice. What, if anything, is to be gained from denying that iniquitous rules are law, we consider below.

Principles of Legality and Justice

It may be said that the distinction between a good legal system which conforms at certain points to morality and justice, and a legal system which does not, is a fallacious one, because a minimum of justice is necessarily realized whenever human behavior is controlled by general rules publicly announced and judicially applied. Indeed we have already pointed out, in analyzing the idea of justice, that its simplest form (justice in the application of the law) consists in no more than taking seriously the notion that what is to be applied to a multiplicity of different persons is the same general rule, undeflected by prejudice, interest, or caprice. This impartiality is what the procedural standards known to English and American lawyers as principles of "Natural Justice" are designed to secure. Hence, though the most odious laws may be justly applied, we have, in the bare notion of applying a general rule of law, the germ at least of justice.

Further aspects of this minimum form of justice which might well be called "natural" emerge if we study what is in fact involved in any method of social control—rules of games as well as law—which consists primarily of general standards of conduct communicated to classes of persons, who are then expected to understand and conform to the rules without further official direction. If social control of this sort is to function, the rules must satisfy certain conditions: they must be intelligible and within the capacity of most to obey, and in general they must not be retrospective, though exceptionally they may be. This means that, for the most part, those who are eventually punished for breach of the rules will have had the ability and opportunity to obey. Plainly these features of control by rule are closely related to the requirements of justice which lawyers term principles of legality.

. . . . .

Legal Validity and Resistance to Law

However incautiously they may have formulated their general outlook, few legal theorists classed as positivists would have been concerned to deny the forms of connection between law and morals discussed under the last five headings. What then was the concern of the great battle cries of legal positivism: "The existence of law is one thing; its merit or demerit another."[1] "The law of a State is not an ideal but something which actually exists . . . it is not that which ought to be, but that which is."[2] "Legal norms may have any kind of content."[3]

What these thinkers were, in the main, concerned to promote was clarity and honesty in the formulation of the theoretical and moral issues raised by the existence of particular laws which were morally iniquitous but were enacted in proper form, clear in meaning, and satisfied all the acknowledged criteria of validity of a system. Their view was that, in thinking about such laws, both the theorist and the unfortunate official or private citizen who was called on to apply or obey then, could only be confused by an invitation to refuse the title of "law" or "valid" to them. They thought that, to confront these problems, simpler, more candid resources were available, which would bring into focus far better, every relevant intellectual and moral consideration: we should say, "This is law; but it is too iniquitous to be applied or obeyed."

The opposed point of view is one which appears attractive when, after revolution or major upheavals, the courts of a system have to consider their attitude to the moral iniquities committed in legal form by private citizens or officials under an earlier regime. Their punishment may be felt socially desirable, and yet, to procure it by frankly retrospective legislation, making criminal what was permitted or even required by the law of the earlier regime, may be difficult, itself morally odious, or perhaps not possible. In these circumstances it may seem natural to exploit the moral implications latent in the vocabulary of the law. . . . It may then appear tempting to say that enactments which enjoined or permitted iniquity should not be recognized as valid, or have the quality of law, even if the system in which they were enacted acknowledged no restriction upon the legislative competence of its legislature. It is in this form that Natural Law arguments were revived in Germany after the last war in response to the acute social problems left by the iniquities of Nazi rule and its defeat. Should informers who, for selfish ends, procured the imprisonment of others for offenses against monstrous statutes passed during the Nazi regime, be punished? Was it possible to convict them in the courts of postwar Germany on the footing that such statutes violated the Natural Law and were therefore void so that the victims' imprisonment for breach of such statutes was in fact unlawful, and procuring it was itself an offense?[4] Simple as the issue looks between those who would accept and those who would repudiate the view that morally iniquitous rules cannot be law, the disputants seem often very unclear as to its general character. It is true that we are here concerned with alternative ways of formulating a moral decision not to apply, obey, or allow others to plead in their defense morally iniquitous rules: yet the issue is ill-presented as a verbal one. Neither side to the dispute would be content if they were told, "Yes:

you are right, the correct way in English (or in German) of putting that sort of point is to say what you have said." So, though the positivist might point to a weight of English usage, showing that there is no contradiction in asserting that a rule of law is too iniquitous to be obeyed, and that it does not follow from the proposition that a rule is too iniquitous to obey that it is not a valid rule of law, their opponents would hardly regard this as disposing of the case.

Plainly we cannot grapple adequately with this issue if we see it as one concerning the properties of linguistic usage. For what is really at stake is the comparative merit of a wider and a narrower concept or way of classifying rules, which belong to a system of rules generally effective in social life. If we are to make a reasoned choice between these concepts, it must be because one is superior to the other in the way in which it will assist our theoretical inquiries, or advance and clarify our moral deliberations, or both.

The wider of these two rival concepts of law includes the narrower. If we adopt the wider concept, this will lead us in theoretical inquiries to group and consider together as "law" all rules which are valid by the formal tests of a system of primary and secondary rules, even though some of them offend against a society's own morality or against what we may hold to be an enlightened or true morality. If we adopt the narrower concept we shall exclude from "law" such morally offensive rules. It seems clear that nothing is to be gained in the theoretical or scientific study of law as a social phenomonon by adopting the narrower concept: it would lead us to exclude certain rules even though they exhibit all the other complex characteristics of law. Nothing, surely, but confusion could follow from a proposal to leave the study of such rules to another discipline, and certainly no history or other form of legal study has found it profitable to do this. If we adopt the wider concept of law, we can accommodate within it the study of whatever special features morally iniquitous laws have, and the reaction of society to them. Hence the use of the narrower concept here must inevitably split, in a confusing way, our effort to understand both the development and potentialities of the specific method of social control to be seen in a system of primary and secondary rules. Study of its use involves study of its abuse.

· · · · · ·

But perhaps a stronger reason for preferring the wider concept of law, which will enable us to think and say, "This is law but iniquitous," is that to withhold legal recognition from iniquitous rules may grossly oversimplify the variety of moral issues to which they give rise. Older writers who, like Bentham and Austin, insisted on the distinction between what law is and what it ought to be, did so partly because they thought that unless men kept these separate they might, without counting the cost to society, make hasty judgments that laws were invalid and ought not to be obeyed. But besides this danger of anarchy, which they may well have overrated, there is another form of oversimplification. If we narrow our point of view and think only of the person who is called upon to *obey* evil rules, we may regard it as a matter of indifference whether or not he thinks that he is faced with a valid rule of "law" so long as he sees its moral iniquity and does what morality requires. But besides the moral question of obedience (Am I to do this evil thing?) there is Socrates' question of submission: Am I to submit to punishment for disobedience or make my escape? There is also the question which confronted the postwar German courts, "Are we to punish those who did evil things when they were permitted by evil rules then in force?" These questions raise very different problems of morality and justice, which we need to consider independently of each other: they cannot be solved by a refusal, made once and for all, to recognize evil laws as valid for any purpose. This is too crude a way with delicate and complex moral issues.

A concept of law which allows the invalidity of law to be distinguished from its immorality, enables us to see the complexity and variety of these separate issues; whereas a narrow concept of law which denies legal validity to iniquitous rules may blind us to them. It may be conceded that the German informers, who for selfish ends procured the punishment of others under monstrous laws, did what morality forbade; yet morality may also demand that the state should punish only those who, in doing evil, did what the state at the time forbade. This is the principle of *nulla poena sine lege*. If inroads have to be made on this principle in order to avert something held to be a greater evil than its sacrifice, it is vital that the issues at stake be clearly identified. A case of retroactive punishment should not be made to look like an ordinary case of punishment for an act illegal at the time. At least it can be claimed for the simple positivist doctrine that morally iniquitous rules may still be law, that this offers no disguise for the choice between evils which, in extreme circumstances, may have to be made.

Notes

1. Austin, *The Province of Jurisprudence Defined*, Lecture V, pp. 184–85.
2. Gray, *The Nature and Sources of the Law*, S. 213.
3. Kelsen, *General Theory of Law and State*, p. 113.
4. See the judgment of 27 July 1940, Oberlandesgericht Bamberg, 5 *Süddeutsche Juristen-Zeitung*, 207: discussed at length in H. L. A. Hart,

"Legal Positivism and the Separation of Law and Morals," in *Harvard Law Rev.* lxxi (1958), 598.

6

The Morality of Whistle-blowing

Sissela Bok

"Whistle-blower" is a recent label for those who . . . make revelations meant to call attention to negligence, abuses, or dangers that threaten the public interest. They sound an alarm based on their expertise or inside knowledge, often from within the very organization in which they work. With as much resonance as they can muster, they strive to breach secrecy, or else arouse an apathetic public to dangers everyone knows about but does not fully acknowledge.[1] . . . Most [whistle-blowers know] that their alarms pose a threat to anyone who benefits from the ongoing practice and that their own careers and livelihood may be at risk. The lawyer who breaches confidentiality in reporting bribery by corporate clients knows the risk, as does the nurse who reports on slovenly patient care in a hospital, the engineer who discloses safety defects in the braking systems of a fleet of new rapid-transit vehicles, or the industrial worker who speaks out about hazardous chemicals seeping into a playground near the factory dump.

. . . . .

Would-be whistle-blowers also face conflicting pressures from without. In many professions, the prevailing ethic requires above all else loyalty to colleagues and to clients; yet the formal codes of professional ethics stress responsibility to the public in cases of conflict with such loyalties. Thus the largest professional engineering society asks members to speak out against abuses threatening the safety, health, and welfare of the public.[2] A number of business firms have codes making similar requirements; and the United States Code of Ethics for government servants asks them to

From Sissela Bok, *Secrets: On the Ethics of Concealment and Revelation*, pp. 211–25. Vintage Books, 1984. Reprinted by permission.

"expose corruption wherever uncovered" and to "put loyalty to the highest moral principles and to country above loyalty to persons, party, or Government department."[3] Regardless of such exhortations, would-be whistle-blowers have reason to fear the results of carrying out the duty to reveal corruption and neglect. However strong this duty may seem in principle, they know that in practice, retaliation is likely. They fear for their careers and for their ability to support themselves and their families.

. . . . .

Blowing the Whistle

The alarm of the whistle-blower is meant to disrupt the status quo: to pierce the background noise, perhaps the false harmony, or the imposed silence of "business as usual." Three elements, each jarring, and triply jarring when conjoined, lend acts of whistle-blowing special urgency and bitterness: dissent, breach of loyalty, and accusation.[4]

Like all *dissent*, first of all, whistle-blowing makes public a disagreement with an authority or a majority view. But whereas dissent can arise from all forms of disagreement with, say, religious dogma or government policy or court decisions, whistle-blowing has the narrower aim of casting light on negligence or abuse, of alerting the public to a risk and of assigning responsibility for that risk.

It is important, in this respect, to see the shadings between the revelations of neglect and abuse which are central to whistle-blowing, and dissent on grounds of policy. In practice, however, the two often come together. Coercive regimes or employers may regard dissent of any form as evidence of abuse or of corruption that calls for public exposure. And in all societies, persons may blow the whistle on abuses in order to signal policy dissent. Thus Daniel Ellsberg, in making his revelations about government deceit and manipulation in the Pentagon Papers, obviously aimed not only to expose misconduct and assign responsibility but also to influence the nation's policy toward Southeast Asia.

In the second place, the message of the whistle-blower is seen as a *breach of loyalty* because it comes from within. The whistle-blower, though he is neither referee nor coach, blows the whistle on his own team. His insider's position carries with it certain obligations to colleagues and clients. He may have signed a promise of confidentiality or a loyalty oath. When he steps out of routine channels to level accusations, he is going against these obligations. Loyalty to colleagues and to clients comes to be pitted against concern for the public interest and for those who may be injured unless someone speaks out. Because the whistle-blower criticizes from within, his act differs from muckraking and other forms of exposure by outsiders. Their acts may arouse anger, but not the sense of betrayal that whistle-blowers so often encounter.

The conflict is strongest for those who take their responsibilities to the public seriously, yet have close bonds of collegiality and of duty to clients as well. They know the price of betrayal. They know, too, how organizations protect and enlarge the area of what is concealed, as failures multiply and vested interests encroach. And they are aware that they violate, by speaking out, not only loyalty but usually hierarchy as well.

It is the third element of *accusation,* of calling a "foul" from within, that arouses the strongest reactions on the part of the hierarchy. The charge may be one of unethical or unlawful conduct on the part of colleagues or superiors. Explicitly or implicitly, it singles out specific groups or persons as responsible: as those who knew or should have known what was wrong and what the dangers were, and who had the capacity to make different choices. If no one could be held thus responsible—as in the case of an impending avalanche or a volcanic eruption—the warning would not constitute whistle-blowing.

·   ·   ·   ·   ·

Not only immediacy but also specificity is needed for the whistle-blower to assign responsibility. A concrete risk must be at issue rather than a vague foreboding or a somber prediction. The act of whistle-blowing differs in this respect from the lamentation or the dire prophecy.

Such immediate and specific threats would normally be acted upon by those at risk. But the whistle-blower assumes that his message will alert listeners to a threat of which they are ignorant, or whose significance they have not grasped. It may have been kept secret by members within the organization, or by all who are familiar with it. Or it may be an "open secret," seemingly in need only of being pointed out in order to have its effect. In either case, because of the elements of dissent, breach of loyalty, and accusation, the tension between concealing and revealing is great. It may be intensified by an urge to throw off the sense of complicity that comes from sharing secrets one believes to be unjustly concealed, and to achieve peace of mind by setting the record straight at last. Sometimes a desire for publicity enters in, or a hope for revenge for past slights or injustices. Colleagues of the whistle-blower often suspect just such motives; they may regard him as a crank, publicity-hungry, eager for scandal and discord, or driven to indiscretion by his personal biases and shortcomings.[5]

On the continuum of more or less justifiable acts of whistle-blowing, the whistle-blower tends to see more such acts as justified and even necessary than his colleagues. Bias can affect each side in drawing the line, so that each takes only some of the factors into account—the more so if the action comes at the end of a long buildup of acrimony and suspicion.

The Leak

. . . . .

Both leaking and whistle-blowing can be used to challenge corrupt or cumbersome systems of secrecy—in government as in the professions, the sciences, and business. Both may convey urgently needed warnings, but they may also peddle false information and vicious personal attacks. How, then, can one distinguish the many acts of revelation from within that are genuinely in the public interest from all the petty, biased, or lurid tales that pervade our querulous and gossip-ridden societies? Can we draw distinctions between different messages, different methods and motivations?

We clearly can, in a number of cases. Whistle-blowing and leaks may be starkly inappropriate when used in malice or in error, or when they lay bare legitimately private matters such as those having to do with political belief or sexual life. They may, just as clearly, offer the only way to shed light on an ongoing practice such as fraudulent scientific research or intimidation of political adversaries; and they may be the last resort for alerting the public to a possible disaster. Consider, for example, the action taken by three engineers to alert the public to defects in the braking mechanisms of the Bay Area Rapid Transit System (BART):

> The San Francisco Bay Area Rapid Transit System opened in 1972. It was heralded as the first major breakthrough toward a safe, reliable, and sophisticated method of mass transportation. A public agency had been set up in 1952 to plan and carry out the project; and the task of developing its major new component, a fully automatic train control system, was allocated to Westinghouse.
>
> In 1969, three of the engineers who worked on this system became increasingly concerned over its safety. They spotted problems independently, and spoke to their supervisors, but to no avail. They later said they might well have given up their effort to go farther had they not found out about one another. They made numerous efforts to speak to BART's management. But those in charge were already troubled by costs that had exceeded all projections, and by numerous unforseen delays. They were not disposed to investigate the charges that the control system might be unsafe. Each appeal by the three engineers failed.
>
> Finally, the engineers interested a member of BART's board of trustees, who brought the matter up at a board meeting. Once again, the effort failed. But in March 1973, the three were fired once the complaint

had been traced to them. When they wrote to ask why they had been dismissed, they received no answer.

Meanwhile, the BART system had begun to roll. The control system worked erratically, and at times dangerously. A month after the opening, one train overshot the last station and crashed into a parking lot for commuters. Claiming that some bugs still had to be worked out, BART began to use old-fashioned flagmen in order to avoid collisions.

The three engineers had turned, in 1972, to the California Society of Professional Engineers for support. The society, after investigating the complaint, agreed with their views, and reported to the California State legislature. It too had launched an investigation, and arrived at conclusions quite critical of BART's management.

The engineers filed a damage suit against BART in 1974, but settled out of court in 1975. They had difficulties finding new employment, and suffered considerable financial and emotional hardship in spite of their public vindication.[6]

The three engineers were acting in accordance with the law and with engineering codes of ethics in calling attention to the defects in the train control system. Because of their expertise, they had a special responsibility to alert the company, and if need be its board of directors and the public, to the risks that concerned them. If we take such a clear-cut case of legitimate whistle-blowing as a benchmark, and reflect on what it is about it that weighs so heavily in favor of disclosure, we can then examine more complex cases in which speaking out in public is not so clearly the right choice or the only choice.

Individual Moral Choice

What questions might individuals consider, as they wonder whether to sound an alarm? How might they articulate the problem they see, and weigh its seriousness before deciding whether or not to reveal it? Can they make sure that their choice is the right one? And what about the choices confronting journalists or others asked to serve as intermediaries?

In thinking about these questions, it helps to keep in mind the three elements mentioned earlier: dissent, breach of loyalty, and accusation. They impose certain requirements: of judgment and accuracy in dissent, of exploring alternative ways to cope with improprieties that minimize the breach of loyalty, and of fairness in accusation. The judgment expressed by whistle-blowers concerns a problem that should matter to the public. Certain outrages are so blatant, and certain dangers so great, that all who are in a position to warn of them have a *prima facie* obligation to do so. Conversely, other problems are so minor that to blow the whistle would be a disproportionate response. And still others are so hard to pin down that whistle-blowing is premature. In between lie a great many of the problems troubling whistle-blowers. Consider, for example, the following situation:

> An attorney for a large company manufacturing medical supplies begins to suspect that some of the machinery sold by the company to hospitals for use in kidney dialysis is unsafe, and that management has made attempts to influence federal regulatory personnel to overlook these deficiencies.
>
> The attorney brings these matters up with a junior executive, who assures her that he will look into the matter, and convey them to the chief executive if necessary. When she questions him a few weeks later, however, he tells her that all the problems have been taken care of, but offers no evidence, and seems irritated at her desire to learn exactly where the issues stand. She does not know how much further she can press her concern without jeopardizing her position in the firm.

The lawyer in this case has reason to be troubled, but does not yet possess sufficient evidence to blow the whistle. She is far from being as sure of her case as . . . the engineers in the BART case, whose professional expertise allowed them to evaluate the risks of the faulty braking system . . . The engineers would be justified in assuming that they had an obligation to draw attention to the dangers they saw, and that anyone who shared their knowledge would be wrong to remain silent or to suppress evidence of the danger. But if the attorney blew the whistle about her company's sales of machinery to hospitals merely on the basis of her suspicions, she would be doing so prematurely. At the same time, the risks to hospital patients from the machinery, should she prove correct in her suspicions, are sufficiently great so that she has good reason to seek help in looking into the problem, to feel complicitous if she chooses to do nothing, and to take action if she verifies her suspicions.

Her difficulty is shared by many who suspect, without being sure, that their companies are concealing the defective or dangerous nature of their products—automobiles that are firetraps, for instance, or canned foods with carcinogenic additives. They may sense that merely to acknowledge that they don't know for sure is too often a weak excuse for inaction, but recognize also that the destructive power of adverse publicity can be great. If the warning turns out to have been inaccurate, it may take a long time to undo the damage to individuals and organizations. As a result, potential whistle-blowers must first try to specify the degree to which there is genuine impropriety, and consider how imminent and how serious the threat is which they perceive.

If the facts turn out to warrant disclosure, and if the would-be whistle-

blower has decided to act upon them in spite of the possibilities of repri-
sal, then how can the second element—breach of loyalty—be overcome
or minimized? Here, as in the Pentagon Papers case, the problem is one
of which set of loyalties to uphold. Several professional codes of ethics,
such as those of engineers and public servants, facilitate such a choice at
least in theory, by requiring that loyalty to the public interest should
override allegiance to colleagues, employers, or clients whenever there is
a genuine conflict. Accordingly, those who have assumed a professional
responsibility to serve the public interest—as had . . . the engineers in
the BART case—have a special obligation not to remain silent about
dangers to the public.

Before deciding whether to speak out publicly, however, it is important
for [whistle-blowers] to consider whether the existing avenues for change
within the organization have been sufficiently explored. By turning first
to insiders for help, one can often uphold both sets of loyalties and settle
the problem without going outside the organization. The engineers in the
BART case clearly tried to resolve the problem they saw in this manner,
and only reluctantly allowed it to come to public attention as a last resort.

．　．　．　．　．

It *is* disloyal to colleagues and employers, as well as a waste of time for
the public, to sound the loudest alarm first. Whistle-blowing has to remain a last alternative because of its destructive side effects. It must be
chosen only when other alternatives have been considered and rejected.
They may be rejected if they simply do not apply to the problem at hand,
or when there is not time to go through routine channels, or when the
institution is so corrupt or coercive that steps will be taken to silence the
whistle-blower should he try the regular channels first.

What weight should an oath or a promise of silence have in the conflict
of loyalties? There is no doubt that one sworn to silence is under a
stronger obligation because of the oath he has taken, unless it was obtained under duress or through deceit, or else binds him to something in
itself wrong or unlawful. In taking an oath, one assumes specific obligations beyond those assumed in accepting employment. But even such an
oath can be overridden when the public interest at issue is sufficiently
strong. The fact that one has promised silence is no excuse for complicity
in covering up a crime or violating the public trust.

The third element in whistle-blowing—accusation—is strongest whenever efforts to correct a problem without going outside the organization
have failed, or seem likely to fail. Such an outcome is especially likely
whenever those in charge take part in the questionable practices, or have
too much at stake in maintaining them.

．　．　．　．　．

Given these difficulties, it is especially important to seek more general means of weighing the arguments for and against whistle-blowing; to take them up in public debate and in teaching; and to consider changes in organizations, law, and work practices that could reduce the need for individuals to choose between blowing and "swallowing" the whistle.[7]

Notes

1. I draw, for this chapter, on my earlier essays on whistleblowing: "Whistleblowing and Professional Responsibilities," in Daniel Callahan and Sissela Bok, eds., *Ethics Teaching in Higher Education* (New York: Plenum Press, 1980), pp. 277–95 (reprinted, "Blowing the Whistle," in Joel Fleishman, Lance Liebman, and Mark Moore, eds., *Public Duties: The Moral Obligations of Officials* (Cambridge, Mass.: Harvard University Press, 1981), pp. 204–21.
2. Institute of Electrical and Electronics Engineers, Code of Ethics for Engineers, art. 4, *IEEE Spectrum* 12 (February 1975): 65.
3. Code of Ethics for Government Service, passed by the U.S. House of Representatives in the 85th Congress, 1958, and applying to all government employees and officeholders.
4. Consider the differences and the overlap between whistle-blowing and civil disobedience with respect to these three elements. First, whistleblowing resembles civil disobedience in its openness and its intent to act in the public interest. But the dissent in whistle-blowing, unlike that in civil disobedience, usually does not represent a breach of law; it is, on the contrary, protected by the right of free speech and often encouraged in codes of ethics and other statements of principle. Second, whistle-blowing violates loyalty, since it dissents from within and breaches secrecy, whereas civil disobedience need not and can as easily challenge from without. Whistle-blowing, finally, accuses specific individuals, whereas civil disobedience need not. A combination of the two occurs, for instance, when former CIA agents publish books to alert the public about what they regard as unlawful and dangerous practices, and in so doing openly violate, and thereby test, the oath of secrecy that they have sworn.
5. Judith P. Swazey and Stephen R. Scheer suggest that when whistleblowers expose fraud in clinical research, colleagues respond *more* negatively to the whistle-blowers who report the fraudulent research than to the person whose conduct has been reported. See "The Whistleblower as a Deviant Professional: Professional Norms and Responses to Fraud in Clinical Research," Workshop on Whistleblowing in Biomedical Research, Washington, D.C., September 1981.
6. See Robert J. Baum and Albert Flores, eds., *Ethical Problems in Engi-*

neering (Troy, N.Y.: Center for the Study of the Human Dimension of Science and Technology, 1978), pp. 227–47.

7. Alal Westin discusses "swallowing" the whistle in *Whistle Blowing!*, pp. 10–13. For a discussion of debate concerning whistle-blowing, see Rosemary Chalk, "The Miner's Canary," *Bulletin of the Atomic Scientists* 38 (February 1982): pp. 16–22.

On the Nature of the Good Life

The question of how people ought to live in order to fulfill their deepest potential has been a subject of philosophical discussion for millennia. Americans, by and large, are deeply influenced by the Protestant work ethic, which holds that a person attains personal fulfillment, and his/her life achieves meaning, primarily through productive work. The computer revolution threatens to remove this source of meaning for perhaps the majority of the work force (see readings 20 and 21).

One response to this threat would be to prevent computers from attaining their full potential in taking over jobs now done by people. This response is probably warranted if computerization would prevent a large proportion of the population from attaining personal fulfillment and a meaningful life. But before accepting such a drastic solution, we should reexamine the assumption that a job is necessary for a meaningful life.

The three readings in this section contain quite different answers to the question of how people ought to live in order to fulfill their deepest potential. Aristotle's conclusion (reading 7) is that the good of man (and hence the ideal life for a person) consists in the active exercise of the mind's faculties. Saint Augustine's answer (reading 8) is that the ideal life for a person inheres in following God. Blanshard's answer (reading 9) is that the good life is that which yields the greatest possible fulfillment and satisfaction of humankind's intrinsic drives. (Blanshard calls such a drive an impulse-desire.)

Aristotle, Saint Augustine, and the Protestant work ethic all proffer one-dimensional views of the ideal life, while Blanshard's outlook is multidimensional. Those most involved with computerization, for whom reasoning is a highly valued activity, will probably agree with Aristotle. A deeply religious

person may find Saint Augustine's answer most convincing. Blanshard's multidimensional view might appeal to the pragmatist who is looking for a
general framework to examine whether a computerized society with few
jobs can still provide opportunities for all types of people to achieve a
maximally fulfilling life.

——— 7 ———————————————————————

Nicomachean Ethics

Aristotle

. . . Inasmuch as all study and all deliberate action is aimed at some good
object, let us state what is the good which is in our view the aim of
political science, and what is the highest of the goods obtainable by
action.

Now as far as the name goes there is virtual agreement about this
among the vast majority of mankind. Both ordinary people and persons
of trained mind define the good as happiness. But as to what constitutes
happiness opinions differ; the answer given by ordinary people is not the
same as the verdict of the philosopher. Ordinary men identify happiness
with something obvious and visible, such as pleasure or wealth or
honor—everybody gives a different definition, and sometimes the same
person's own definition alters: when a man has fallen ill he thinks that
happiness is health, if he is poor he thinks it is wealth. And when people
realize their own ignorance they regard with admiration those who pro-
pound some grand theory that is above their heads. The view has been
held by some thinkers[1] that besides the many good things alluded to
above there also exists something that is good in itself, which is the
fundamental cause of the goodness of all the others.

Now to review the whole of these opinions would perhaps be a
rather thankless task. It may be enough to examine those that are most
widely held, or that appear to have some considerable argument in their
favor. . . .

From Aristotle, *Ethics for English Readers,* trans. H. Rackham. Oxford: Basil Black-
well, Publisher, 1952. Reprinted by permission.

Reasons for doubting whether enjoyment, fame, virtue, or wealth is the whole good

To judge by men's mode of living, the mass of mankind think that good and happiness consist in pleasure, and consequently are content with a life of mere enjoyment. There are in fact three principal modes of life—the one just mentioned, the life of active citizenship, and the life of contemplation. The masses, being utterly servile, obviously prefer the life of mere cattle; and indeed they have some reason for this, inasmuch as many men of high station share the tastes of Sardanapalus.[2] The better people, on the other hand, and men of action, give the highest value to honor, since honor may be said to be the object aimed at in a public career. Nevertheless, it would seem that honor is a more superficial thing than the good which we are in search of, because honor seems to depend more on the people who render it than on the person who receives it, whereas we dimly feel that good must be something inherent in oneself and inalienable. Moreover, men's object in pursuing honor appears to be to convince themselves of their own worth; at all events they seek to be honored by persons of insight and by people who are well acquainted with them, and to be honored for their merit. It therefore seems that at all events in the opinions of these men goodness is more valuable than honor, and probably one may suppose that it has a better claim than honor to be deemed the end at which the life of politics aims. But even virtue appears to lack completeness as an end, inasmuch as it seems to be possible to possess it while one is asleep or living a life of perpetual inactivity, and moreover one can be virtuous and yet suffer extreme sorrow and misfortune; but nobody except for the sake of maintaining a paradox would call a man happy in those circumstances.

However, enough has been said on this topic, which has indeed been sufficiently discussed in popular treatises.

The third life is the life of contemplation, which we shall consider later.

The life of money-making is a cramped way of living, and clearly wealth is not the good we are in search of, as it is only valuable as a means to something else. Consequently a stronger case might be made for the objects previously specified, because they are valued for their own sake; but even they appear to be inadequate, although a great deal of discussion has been devoted to them. . . .

Reaffirmation that the good is the ultimate and self-sufficient object of desire and that "happiness" is the good

What then is the precise nature of the practicable good which we are investigating? It appears to be one thing in one occupation or profession and another in another: the object pursued in medicine is different from that of military science, and similarly in regard to the other activities.

What definition of the term "good" then is applicable to all of them? Perhaps "the object for the sake of attaining which all the subsidiary activities are undertaken." The object pursued in the practice of medicine is health, in a military career victory, in architecture a building—one thing in one pursuit and another in another, but in every occupation and every pursuit it is the end aimed at, since it is for the sake of this that the subsidiary activities in all these pursuits are undertaken. Consequently if there is some one thing which is the end and aim of all practical activities whatsoever, that thing, or if there are several, those things, will constitute the practicable good.

Our argument has therefore come round again by a different route to the point reached before. We must endeavor to render it yet clearer.

Now the objects at which our actions aim are manifestly several, and some of these objects, for instance money, and instruments in general, we adopt as means to the attainment of something else. This shows that not all the objects we pursue are final ends. But the greatest good manifestly is a final end. Consequently if there is only one thing which is final, that will be the object for which we are now seeking, or if there are several, it will be that one among them which possesses the most complete finality.

Now a thing that is pursued for its own sake we pronounce to be more final than one pursued as a means to some other thing, and a thing that is never desired for the sake of something else we call more final than those which are desired for the sake of something else as well as for their own sake. In fact the absolutely final is something that is always desired on its own account and never as a means for obtaining something else. Now this description appears to apply in the highest degree to happiness, since we always desire happiness for its own sake and never on account of something else; whereas honor and pleasure and intelligence and each of the virtues, though we do indeed desire them on their own account as well, for we should desire each of them even if it produced no external result, we also desire for the sake of happiness, because we believe that they will bring it to us, whereas nobody desires happiness for the sake of those things, not for anything else but itself.

The same result seems to follow from a consideration of the subject of self-sufficiency, which is felt to be a necessary attribute of the final good. The term self-sufficient denotes not merely being sufficient for oneself alone, as if one lived the life of a hermit, but also being sufficient for the needs of one's parents and children and wife, and one's friends and fellow-countrymen in general, inasmuch as man is by nature a social being.

Yet we are bound to assume some limit in these relationships, since if one extends the connection to include one's children's children and friends' friends, it will go on *ad infinitum*. But that is a matter which must be deferred for later consideration. Let us define self-sufficiency as the quality which makes life to be desirable and lacking in nothing even when considered by itself; and this quality we assume to belong to happiness. Moreover, when we pronounce happiness to be the most desirable of all things, we do not mean that it stands as one in a list of good things—were it so, it would obviously be more desirable in combination with even the smallest of the other goods, inasmuch as that addition would increase the total of good, and of two good things the larger must always be the more desirable.

Thus it appears that happiness is something final and complete in itself, as being the aim and end of all practical activities whatever.

> *For a more specific conception of the kind of "happiness" which is the good, we do well to examine whether nature intended man for anything, as it intended the eye for sight. What is distinctive of man is reason, so the happiness which is the good must be the exercise of reason in living.*

Possibly, however, the student may feel that the statement that happiness is the greatest good is a mere truism, and he may want a clearer explanation of what the precise nature of happiness is. This may perhaps be achieved by ascertaining what is the proper function of man. In the case of flute players or sculptors or other artists, and generally of all persons who have a particular work to perform, it is felt that their good and their well-being are found in that work. It may be supposed that this similarly holds good in the case of a human being, if we may assume that there is some work which constitutes the proper function of a human being as such. Can it then be the case that whereas a carpenter and a shoemaker have definite functions or businesses to perform, a man as such has none, and is not designed by nature to perform any function? Should we not rather assume that, just as the eye and hand and foot and every part of the body manifestly have functions assigned to them, so also there is a function that belongs to a man, over and above all the special functions that belong to his members? If so, what precisely will that function be? It is clear that the mere activity of living is shared by man even with the vegetable kingdom, whereas we are looking for some function that belongs specially to man. We must therefore set aside the vital activity of nutrition and growth. Next perhaps comes the life of the senses; but this also is manifestly shared by the horse and the ox and all the animals. There remains therefore what may be designated the practical life of the rational faculty.

But the term "rational" life has two meanings: it denotes both the mere possession of reason, and its active exercise. Let us take it that we here mean the latter, as that appears to be the more proper signification of the term. Granted then that the special function of man is the active exercise of the mind's faculties in accordance with rational principle, or at all events not in detachment from rational principle, and that the function of anything, for example, a harper, is generally the same as the function of a good specimen of that thing, for example a good harper (the specification of the function merely being augmented in the latter case with the statement of excellence—a harper is a man who plays the harp, a good harper one who plays the harp well)—granted, I say, the truth of these assumptions, it follows that the good of man consists in the active exercise of the faculties in conformity with excellence or virtue, or if there are several virtues, in conformity with the best and most perfect among them.

Notes

1. Plato and the Academy.
2. A mythical Assyrian king; two versions of his epitaph are recorded, one containing the words "Eat, drink, play, since all else is not worth that snap of the fingers," the other ending "I have what I ate, and the delightful deeds of wantonness and love in which I shared; but all my wealth is vanished."

8

The Morals of the Catholic Church

Saint Augustine

Chapter V. Man's chief good is not the chief good of the body only, but the chief good of the soul

Now if we ask what is the chief good of the body, reason obliges us to admit that it is that by means of which the body comes to be in its best state. But of all the things which invigorate the body, there is nothing

From "The Morals of the Catholic Church," *Basic Writings of Saint Augustine*, ed. by Whitney J. Oates. Copyright 1948 by Random House, Inc. Reprinted by permission.

better or greater than the soul. The chief good of the body, then, is not bodily pleasure, not absence of pain, not strength, not beauty, not swiftness, or whatever else is usually reckoned among the goods of the body, but simply the soul. For all the things mentioned the soul supplies to the body by its presence, and, what is above them all, life. Hence I conclude that the soul is not the chief good of man, whether we give the name of man to soul and body together, or to the soul alone. For as, according to reason, the chief good of the body is that which is better than the body, and from which the body receives vigor and life, so whether the soul itself is man, or soul and body both, we must discover whether there is anything which goes before the soul itself, in following which the soul comes to the perfection of good of which it is capable in its own kind. If such a thing can be found, all uncertainty must be at an end, and we must pronounce this to be really and truly the chief good of man.

If, again, the body is man, it must be admitted that the soul is the chief good of man. But clearly, when we treat of morals—when we inquire what manner of life must be held in order to obtain happiness—it is not the body to which the precepts are addressed, it is not bodily discipline which we discuss. In short, the observance of good *customs* belongs to that part of us which inquires and learns, which are the prerogatives of the soul; so, when we speak of attaining to virtue, the question does not regard the body. But if it follows, as it does, that the body which is ruled over by a soul possessed of virtue is ruled both better and more honorably, and is in its greatest perfection in consequence of the perfection of the soul which rightfully governs it, that which gives perfection to the soul will be man's chief good, though we call the body man. For if my coachman, in obedience to me, feeds and drives the horses he has charge of in the most satisfactory manner, himself enjoying the more of my bounty in proportion to his good conduct, can any one deny that the good condition of the horses, as well as that of the coachman, is due to me? So the question seems to me to be not, whether soul and body is man, or the soul only, or the body only, but what gives perfection to the soul; for when this is obtained, a man cannot be either perfect, or at least much better than in the absence of this one thing.

Chapter VI. Virtue gives perfection to the soul; the soul obtains virtue by following God; following God is the happy life

No one will question that virtue gives perfection to the soul. But it is a very proper subject of inquiry whether this virtue can exist by itself or only in the soul. Here again arises a profound discussion, needing lengthy treatment; but perhaps my summary will serve the purpose. God will, I trust, assist me, so that, notwithstanding our feebleness, we may give instruction on these great matters briefly as well as intelligibly. In either case, whether virtue can exist by itself without the soul, or can exist only in the soul, undoubtedly in the pursuit of virtue the soul follows after something, and this must be either the soul itself, or virtue, or something else. But if the soul follows after itself in the pursuit of virtue, it follows after a foolish thing; for before obtaining virtue it is foolish. Now the height of a follower's desire is to reach that which he follows after. So the soul must either not wish to reach what it follows after, which is utterly absurd and unreasonable, or, in following after itself while foolish, it reaches the folly which it flees from. But if it follows after virtue in the desire to reach it, how can it follow what does not exist? or how can it desire to reach what it already possesses? Either, therefore, virtue exists beyond the soul, or if we are not allowed to give the name of virtue except to the habit and disposition of the wise soul, which can exist only in the soul, we must allow that the soul follows after something else in order that virtue may be produced in itself; for neither by following after nothing, nor by following after folly, can the soul, according to my reasoning, attain to wisdom.

This something else then, by following after which the soul becomes possessed of virtue and wisdom, is either a wise man or God. But we have said already that it must be something that we cannot lose against our will. No one can think it necessary to ask whether a wise man, supposing we are content to follow after him, can be taken from us in spite of our unwillingness or our persistence. God then remains, in following after whom we live well, and in reaching whom we live both well and happily.

9

Satisfaction, Fulfillment, and the Good

Brand Blanshard

The kind of consideration on which we are embarking is so different from our earlier analyses that it may be well to set down briefly and at once the view of goodness that will gradually emerge from it. We shall hold that

From Brand Blanshard, *Reason and Goodness*. London: Unwin Hyman Limited, 1961. Reprinted by permission.

only experiences are directly or immediately good. When they are good intrinsically, they perform a double function: they fulfill an impulse, drive, or need, and in so doing they give satisfaction or pleasure. Both components, fulfillment and satisfaction, are necessary, and they vary independently of each other. But both are always partial in the sense that they apply to a limited set of needs; and they are always provisional or incomplete, so that goodness is a matter of degree. It is to be measured against an ideal good, which is the kind of life which would fulfill and satisfy wholly.

.　.　.　.　.

We have said that our desires grow out of our satisfactions, and that our inventory of ideal goods is drawn up by our desires. We must qualify this later to make clear the sense in which the desirable exceeds the desired.

.　.　.　.　.

We desire to eat food, and not merely to have the pleasure of eating; we desire to hear music, and play games, and understand, not merely to gain satisfaction out of objects and activities themselves indifferent. And if our goods reflect the content of our desires, we see again that those goods are not exhausted in pleasure or satisfaction. They consist of satisfying experiences as wholes.

Our major goods answer to the main types of impulse-desire. One major drive has been much studied by philosophers, namely the impulse to know, and its career may serve as a suggestion of what we should probably find if we studied other drives in similar detail. Instinctive curiosity, which passes on at a higher level into the desire for understanding, is a distinguishable impulse with an end of its own.

.　.　.　.　.

The desire to know is the desire to see things as they are. Now such seeing is and must be more than a mere registry of fact. What we want is not only to know, but to understand, that is, lay hold of the connections among facts that make them intelligible. Thought at its simplest is judgment, and in the most rudimentary judgment we are grasping a linkage between terms. That linkage as it stands may be unintelligible to us. "That cat purrs"; with this we have entered the field of judgment. But from the first there is awakening within such judgment the spirit of inquiry; we have connected the purring with the cat; can we connect it more specifically with anything else about the cat? In short, why does the cat purr? With that question we have begun our career in science and philosophy. The intellectual quest from the lowest judgment to the most complicated theories of modern physics may be regarded as one persist-

ing effort to answer the question, Why? To answer it calls for a continually widening knowledge and a continual reordering of that knowledge in
the direction of consistency and interdependence of parts, for the end
sought by the theoretic impulse is an intelligible whole. Every advance
toward such understanding brings satisfaction. We repeat, however, that
it is not satisfaction merely that the impulse is seeking, but that which will
satisfy it, namely this particular kind of light. But neither is it light
merely; for what is the point of a knowledge in which one takes no
satisfaction? What is sought is something which, because of its special
character, fulfills the aim of the cognitive impulse, and which because of
this fulfillment, satisfies. This double service of fulfilling and of satisfying
is what makes knowledge good. It is also what makes anything good.

. . . . .

*The good, in the sense of the ethical end, is the most comprehensive
possible fulfillment and satisfaction of impulse-desire.* By a comprehensive
fulfillment I mean one that takes account not only of this or that desire, but
of our desires generally, and not only of this or that man's desires, but of all
men's. That there is and must be such a good, supreme over all others, we
can see by considering how conflicts are resolved. Two distinct and important trends of impulse-desire have been repeatedly mentioned, the cognitive and the aesthetic, each with its special type of good, measured by its
own immanent standard. Suppose these two conflict. Suppose that a youth
has interests and talents for philosophy on the one hand, and music on the
other, and feels drawn in both directions, but that one or the other must be
sacrificed, if proficiency is to be reached in either. In any concrete case he
would no doubt need to take into account the importance of these two
activities to his community, but by way of taking one difficulty at a time, let
us exclude these considerations. The youth must then decide the conflict
by considering his good as a whole. He will have to ask himself the question, How may I, as a person who can gain some measure of fulfillment
through each of these channels, gain most of this on the whole—by making
myself a philosopher, by making myself a musician, or by making myself
some sort of hybrid between them? And if he is really free in the matter,
the obvious way to settle it is to take stock of his interests and powers. If he
has a very strong interest and bent for speculative analysis, and a feeble
enjoyment and skill in music, then, as between these two, the choice will
be easy; his good will be in philosophy, since he would find a completer
fulfillment there. If the reverse holds, the choice will similarly go to music.
But the case in fact may well be harder; he may discover an approximately
equal bent for each. Then, if he must drop one or the other, he will
consider which of the two would carry with it the larger range of subsidiary
satisfactions. The two walks of life will involve different incomes, associ-

ates, surroundings, holidays, hours of work and freedom; the one activity may stimulate and support more fruitfully than the other all sorts of minor interests, scientific, political, and literary. In practice he will probably select one or the other as his principal business and try to keep the other alive in a secondary role.

Though this account of the good lays stress on desire, it is not so much on *de facto* desire, the want or wish of the moment, as on reflective desire, the desire that emerges after correction by thought and experience. The good is what brings fulfillment and its attendant pleasure to desire of this self-amending kind. These desires arise because we are the sort of beings we are, and wisdom lies in making them more accurately and fully expressive of human nature; "our aims," said Emerson, "should be mathematically adjusted to our powers." There is nothing novel in such an account of the good; indeed it is as old as the Greeks. We shall not try to show this by chapter and verse. It will perhaps be enough to quote a passage in which A. E. Taylor sets out the drift of the ethical thought of Plato and Aristotle. In philosophy there are no authorities, but at least there is some comfort in having such august names on one's side. Taylor notes that

> εὐδαιμονία [eudaimonia], for both of them, is not primarily getting something which I desire; it is living the kind of life which I have been constructed to live, doing the "work of man," and if we want to know what life rather than any other should be pronounced εὐδαιμών, we have to begin by asking what is the "work" which man, and only man, in virtue of his very constitution, can do. It is true, no doubt, that Plato holds that all of us also do desire εὐδαιμονία, if only most of us were not as unaware as we are of the real nature of our most deep-seated desires. But the very reason why we all have this insuperable *desiderium naturale* for a certain kind of life is that it is the life we have been constructed by God or by Nature to lead. We are unhappy, without clearly knowing why, so long as we are living any other kind of life, for the same reasons that a fish is unhappy out of water. The true way to discover what it is that we really want out of life is to know what kind of life we have been sent into the world to lead. We do not lead that life as a "means" to the "enjoyable results" of doing so, any more than the fish lives in the water, or the bird in the air as a means to the pleasure of such a life; we enjoy the pleasure (as the tenth book of the *Nicomachean Ethics* explains) because we are living the kind of life for which we were made.[1]

Note

1. *Mind* 48 (1939):280.

II

Computers and
the Ideal Life

Privacy

David Burnham (reading 10) calls attention to the ever-growing network of data bases with information about the private lives of all citizens, and he explores the potential dangers inherent in the unjustified use of this material. For instance, investigators have already used computer records of telephone calls and other transactional information to assemble electronic profiles and to document the movements and contacts of their opponents.

Philip Adler and his colleagues (reading 11) address the issue of employees' privacy being violated because of the ways their companies maintain and use personal information. The essay explores in detail the legal and psychological aspects of this problem, and concludes by addressing some policy guidelines to protect employee privacy.

Thomas Mylott (reading 12) explains the growth of legislation to guard privacy by regulating government and corporate use of personal information. His article, which centers on the protection of privacy, begins with a brief discussion of legislation requiring the government to divulge certain information to the public. In his conclusion, Mylott touches on the need for computer professionals to design data systems compatible with current—and future—legislation.

10

Data Protection

David Burnham

It is a truism that we live in a world radically different from that of our grandparents. One way to measure the great distance we have traveled during this microsecond of human history is to compare the records that documented the life of an American before the turn of the century with the records that document our individual lives today. One hundred years ago, the few records that existed could tell us when a child was born, when a couple were married, when a man or woman died, and what the boundaries of the land purchased by a family were. In those days, of course, only a handful of the American people went to college. Social security, income taxes, and life insurance did not exist. Three-quarters of the population was self-employed.

Today, fewer than 5 percent of the American people work for themselves. And of the remaining 95 percent, almost half are employed by large corporations that collect detailed information about the education, health, family, and work habits of their employees. Today, two out of three Americans have life insurance and nine out of ten are covered by health insurance plans. Insurance companies usually collect large amounts of information about their customers—revealing information such as whether they are seeing a psychiatrist, what drugs they use, and whether they have a drinking problem. Today, 60 million students are enrolled in schools and colleges that generally collect detailed personal and financial histories about both the student and his or her parents.

The vast scale of information collected by government agencies, private corporations, and institutions such as hospitals and universities would not be possible without large centralized computers or, alternatively, linked series of smaller computers. It is also true, however, that some of these organizations did in fact collect some of this information before the computer. With armies of meticulous clerks, there were a few industries and a few countries like Germany and Chile that did compile massive handwritten records about the lives of an amazing number of people. The computer, however, has powers well beyond that possessed

From David Burnham, *The Rise of the Computer State,* chap. 3. Random House, 1983. Reprinted by permission.

by human scribes, no matter how numerous, and thus has fundamentally altered the nature of society's records.

The first important change is that the computer mass-produces what has come to be called "transactional information," a new category of information that automatically documents the daily lives of almost every person in the U.S. Exactly when did you leave your home? Exactly when did she turn on the television? Exactly when did he deposit the check? Exactly when were the calls made from their telephone? How many times have you driven your automobile? In the centuries before the computer, transactional information answering these kinds of questions was almost never collected. And in those very few instances where it was collected, it was not easily available for later inspection.

With the computerized filing systems now available, the larger organizations of our society can easily collect and store this new kind of information. Equally important, they can combine it with automated dossiers containing the traditional kinds of information such as a person's age, place of birth, and the material contained in school and work records.

There is one more important development made possible by the computer: the incredible maze of electronic highways that can move the new and old information about the country in a matter of seconds at an astoundingly low cost. The automatic exchange of information between different data bases was not seriously considered in the first years of the computer age. But as the technology has become more subtle and sophisticated, it gradually is reducing the barriers between these giant repositories, increasing their ability to "talk" with each other.

The contributions of these linked data bases to our daily lives are enormous. The swift granting of lines of credit to a substantial number of American people would not be possible without the computerized data bases maintained by credit reporting companies like TRW and Equifax. The hundreds of millions of checks written each year by tens of millions of Americans could not be processed and cleared without the computerized data bases maintained by the separate banks and the Federal Reserve System. Easy movement about the U.S. would be far harder without the computerized reservation systems of the airlines and car rental agencies. The collection each year of nearly $500 billion in federal taxes from almost 100 million individuals and corporations would be extraordinarily difficult without the computerized data bases of the Internal Revenue Service and the large corporations [that] each year employ more and more Americans.

There are a variety of reasons why understanding the true significance of all of these changes is very hard. First, there is the fundamental difficulty of putting anecdotal flesh on the bones of the abstract truth that information is power and that organizations increase their power by learning how to swiftly collect and comprehend bits and pieces of information.

This difficulty is greatly multiplied by the sheer force of the tools of the new information age: the machines that can locate a single item in a file of millions in the blink of an eye or that can swiftly develop statistical trends by massing these single items, and the communication links that can shuttle the collected information about the world at almost the speed of light.

When thinking about the impact of these technical achievements, allow your mind to wander beyond the traditionally narrow boundaries of the computer debate. Consider how the technology is altering the power of large organizations. Consider how the technology is affecting our social values such as the notion of checks and balances, the role of work, and the importance of spontaneity. Consider the far-reaching changes it is bringing to the nation's economy.

What does it mean, for example, that the officials and clerks of the U.S. government, each year armed with more and more computers, have collected 4 billion separate records about the people of the U.S., 17 items for each man, woman, and child in the country? What does it mean that an internal communications network serving just one multinational corporation now links more than 500 computers in over 100 cities in 18 countries and has been growing at a rate of about one additional computer a week in recent years? What does it mean that 10,000 merchants all over the country are able to obtain a summary fact sheet about any one of 86 million individual Americans in a matter of three or four seconds from a single data base in southern California? What does it mean that a handful of federal agencies, not counting the Pentagon, have at least 3 separate telecommunication networks stretching all over the U.S.?

Two of the world's largest and most complicated systems of linked data bases are controlled by the American Telephone and Telegraph Company and the Federal Bureau of Investigation. AT&T's gigantic network was deliberately developed by the company's scientists, engineers, and businessmen. The FBI system, which seeks to link the computerized data bases operated by a majority of the fifty states, has developed in a more haphazard fashion.

Because the information-collecting and distributing systems of both the telephone company and the FBI began to function long before the birth of the computer, they illustrate the important point that the new tools of information processing usually are extensions of old bureaucracies, not shiny stand-alone machines that can be considered on their independent merit.

A second trait shared by these two very different systems is that neither has been subject to much outside scrutiny during much of their development. AT&T is a private company regulated by 50 state utility bodies of widely varying quality and a federal commission that never had the staff adequately to monitor the company's interstate operations. And very few congressional committees or state legislatures have taken the time or possess the perspective to consider the impact of the gradually growing network linking the FBI to the states.

Computers generate transactional information for many purposes and many organizations. They allow the construction of huge, speedy, and low-cost communication networks to transmit many different kinds of information to thousands of different customers. But the very different information-processing systems developed by AT&T and the FBI provide clear examples of how transactional information and mass networks have enhanced the impact of these two precomputer bureaucracies on all of our lives. First, consider AT&T. Through its millions of miles of cables, microwave highways, and satellite hookups, the American people make 500 million calls a day—four calls, on the average, for each of the nation's 130 million telephones. Thanks to the computerized data bases that are tied into this massive electronic network at a steadily increasing number of junctures, AT&T has become the largest single holder of transactional information in the world. Buried in the computers of the system are records that can be helpful in drawing an amazingly detailed portrait of any single person, group, or corporation that uses the telephone.

The astounding power of these records is not appreciated by the public, the courts, or Congress. But for government and industry investigators, they have become an important tool. A few years ago, for example, the Senate created a special committee to investigate a very sensitive and delicate subject, the relationship between President Carter's brother Billy and the government of Libya. After many months of embarrassed maneuvering, the Democratic Senate committee issued a report on the antics of the brother of the Democratic president. Almost every other page of the committee's 109-page final report contains a footnote to the precise time and day of calls made by Billy Carter and his associates from at least ten different telephones operating in three different states.

The report said that on November 26, 1979, Billy Carter and an associate began driving to Washington from Georgia. Shortly after beginning the trip, the report said, the two men stopped to telephone the Libyan embassy and request a meeting with a high-ranking official. The assertion was supported by a footnote to telephone company records showing that "a five-minute call at 3:43 was charged to Billy Carter's telephone from [a pay telephone in] Jonesboro, an Atlanta suburb." Another footnote said that calls were made from Carter's office telephone in Georgia to the Libyan embassy in Washington "on March 7, March 10, four times on March 11, twice on March 12, three times on March 13, three times on March 14, March 15, and March 17."

The investigations of the special Senate committee were publicly announced and the telephone records that document the report were obtained by a formal legal process. But this sometimes is not the case. One of the top officials in the Nixon White House, for example, claims that, shortly after the automobile accident that claimed the life of Mary Jo Kopechne in Martha's Vineyard, the White House political operatives ordered the FBI to obtain the telephone credit card records of Senator Edward Kennedy. These records, which almost certainly would have revealed who and when Senator Kennedy called immediately after the accident, obviously would have been considered useful to those Nixon advisers who thought Kennedy was a likely opponent in the coming elections. Though reporters for the *New York Times* determined that the records in question disappeared from the files of New England Bell shortly after the accident, they never found documentary evidence confirming the account of a top Nixon lieutenant that they were obtained by the White House.

Both the Billy Carter and Teddy Kennedy cases illustrate why investigators are so interested in transactional information. First, the information can be extraordinarily revealing. Only considering the data that can be collected from a telephone computer, investigators can learn what numbers an individual has called, what time of day and day of the week the calls were made, the length of each conversation, and the number of times an incorrect number was dialed. Considered as a whole, such information can pinpoint the location of an individual at a particular moment, indicate his daily patterns of work and sleep, and even suggest his state of mind. The information can also indicate the friends, associates, business connections, and political activities of the targeted individual.

But there is an even more fundamental consideration at stake. Almost by definition, transactional information is automatically collected and stored in the data bases of the telephone company, the electronic equipment of banks, and the computers of two-way interactive television systems. This means that transactional information can be obtained months after the instance when the particular event that is documented by the records actually occurred.

This ability greatly enlarges the scope of any investigator. Before the computer age, it was extremely hard to develop concrete evidence about the activities and whereabouts of an individual unless someone had been assigned to follow him. In most cases, investigators were limited to pursuing the handful of individuals they believed might undertake a forbidden act in the future. Now they can move back in time, easily gathering concrete evidence about any person of interest long after the forbidden act occurred.

The broad broom of transactional information, however, can sweep up much more than the highly revealing computer tracks of an individual citizen. In at least two instances, for example, evidence has recently come to light where this same kind of computerized information was used by AT&T to track the activities of several large corporations and even the ethnic and economic groups living in a single state.

A significant characteristic of the transactional information collected by AT&T and other major computer systems is that its reach is universal. Transactional information is collected and stored about the telephone use and banking habits of everyone who lives in America, rich and poor, ethical and unethical, white and black, Republican and Democrat. Though the information ultimately may cast a revealing light on the activities of a single individual, it is collected about the activities of all.

The system of data bases that gradually are being linked by the computerized network of the FBI does not share this universal quality. Instead, it collects and distributes information about one segment of the population, the millions of Americans who are arrested each year in the U.S.

. . . . .

The computer panders to the natural human instinct to desire more information about everything. But there are some law enforcement officials who question whether more arrest records are actually going to help them do their work. They contend that the value of the arrest record has been greatly diluted, partly because so many Americans have been arrested for so many significant causes that it is hard to separate the wheat from the chaff. They further believe the records are of questionable value because they frequently do not disclose whether the case was immediately dismissed, whether the defendant was found guilty or innocent, and what sentence, if any, was imposed. Beyond the pragmatic judgment about the utility of providing the patrolman instant access to arrest records, they see a larger ethical and constitutional question: do the American people want their police making arrests for *current* activities partly on basis of *past* behavior?

The idea that a national rap sheet system would make an important contribution to our work here is just a bunch of baloney. Our problem is not to find out who the guy is. Our biggest problem is, once we catch him coming out of a house with the goods, how do we keep him in jail and how do we make sure he stays in jail? If anything, we have over-information-oriented and over-computerized this department.

These critical comments from the police are drawn from a 1981 report to the Office of Technology Assessment [OTA] as part of a congressional effort to determine the impact—positive and negative—of the FBI's proposed system. This particular report was prepared by Dr. Kenneth C. Laudon, a professor at New York's John Jay School of Criminal Justice. During its preparation, Dr. Laudon interviewed over 140 experienced criminal justice officials working in four states and six cities. Dr. Laudon found that the prosecutors and judges he talked with were even more doubtful about the utility of the FBI plan than the police.

The complaints of the policemen, prosecutors, and judges interviewed by Dr. Laudon went well beyond simply questioning the utility of a summary record that has been lifted out of its original legal and social context. Over and over again, the users complained that the records being moved along the experimental criminal history segment of the FBI's communication network were incomplete and inaccurate. The subjective judgments of the police officials, prosecutors, and judges about the poor quality of the information was supported by a second investigation undertaken by Congress's Office of Technology Assessment.

. . . . . .

At the request of OTA, Dr. Laudon also checked a sample of 400 arrest warrants from the 127,000 contained in the FBI's "hot file" on a single day in August of 1979. Upon comparing the information on the FBI's warrant notices with the information in the local court records, he found that 10.9 percent of the sample already had been cleared or vacated, 4.1 percent showed no record of the warrant at the local agency, and a small additional number of warrants had other problems. Again, assuming the validity of the sample, it appears that on that single day in 1979, 17,340 Americans were subject to false arrest because the FBI computer incorrectly showed they were wanted when the warrants in question had been cleared or vacated.

Official records, of course, were subject to improper use long before the computer became an important part of every major public and private bureaucracy in the U.S. But an executive in a New York state agency explained how the computer has enlarged the opportunity to abuse.

"Technology may not be the only villain, but it is one villain," the official said. "Before high technology, you could actually control information better; at least, it could not spread very far because it was impractical to transmit it. With the computer system, you can't control it any more, largely because of the automatic interfacing of the system, which makes it difficult for even us to know who's getting our information."

The Office of Technology Assessment asked a team at the Bureau of Governmental Research and Service at the University of South Carolina to try to determine exactly who was now receiving criminal history information and to assess the social impacts of computerizing these histories. According to the South Carolina researchers—Lynne Eickholt Cooper, Mark E. Tompkins, and Donald A. Marchand—the use of criminal records outside the traditional confines of the criminal justice system is enormous and growing.

All applicants for federal positions, all military recruits, many of the hundreds of thousands of citizens working for private contractors who are doing jobs for the federal government, and all new employees of federally chartered and insured banks are among the millions of persons who have long been subject to criminal history record checks for many years, the researcher reported.

In addition, with the recent growth in the number of states and cities requiring licenses and permits for almost any kind of job, the population subject to criminal record checks has further exploded. At the last count, more than 7 million Americans must obtain licenses to earn their living. In California, for example, 47 separate licensing boards, 50 state agencies, and 32 out-of-state agencies have access to the criminal history records stored in the state's computerized record system. In New York the use of criminal records by law enforcement agencies has declined in recent years, while its use by private employers has gone up.

Through a variety of federal and state laws, society has made the collective decision that individuals *convicted* of certain crimes may be properly denied certain privileges. Sometimes such individuals lose the right to vote. Sometimes they may not run for public office. Sometimes they are prohibited from bidding on government contracts.

The U.S. also pays a good deal of lip service to the principle that an arrest or investigation, *without a conviction,* should not be sufficient grounds to deny an individual an honorable place in the job market. In late 1981, President Reagan provided an eloquent testimonial to this principle when asked why he had not asked three officials in his administration to remove themselves from office after they were accused of illegal activities. The three men were Labor secretary Raymond Donovan, CIA director William Casey, and National Security Council Director Richard Allen.

"I believe in the fairness of the American people, and I believe that in recent years there has been a very dangerous tendency in this country for some to jump to the conclusion that accusation means guilt and conviction," Mr. Reagan told reporters. "And I think it is high time we recognize that any individual is innocent until proven guilty of a wrongdoing, and that's what we are going to do."

Mr. Reagan's eloquent statement has considerable backing in the laws and regulations of the U.S. The Civil Service Reform Act of 1978 includes a provision limiting federal suitability checks to conviction records, and only for those crimes that are reasonably related to job performance. Three states have laws explicitly barring employer discrimination against ex-offenders. Many state and local human relations offices have issued rules barring private employers from requesting arrest information and limiting the use of conviction information.

Despite the various restrictions, however, there is considerable evidence that the president's policy for the high officials of his administration frequently does not obtain for the average job hunter. According to one congressional survey, for example, the 50 states handed out a total of 10.1 million criminal history records in 1978. Two million of the records— one out of five—went to private corporations and government agencies that were not part of the criminal history system.

The Labor Department study that concluded that 40 million Americans have an arrest record also estimated that just under 26 million of them were in the job market. With so many people applying for so many different jobs, it is nearly impossible to generalize about the impact of criminal records on their lives. Certainly, the bank is on solid ground when it decides not to hire a convicted bank robber. Certainly, no one would argue about a hospital personnel director who decides not to hire a convicted narcotics dealer.

But there is good evidence that some employers assume that any kind of record, no matter what the offense, no matter what the outcome, is a powerful mark against the individual. More than a decade ago, two sociologists named Richard Schwartz and Jerome Skolnick attempted to measure this kind of bias in an experiment in which a number of employers were shown the employment folders submitted by 100 men who were looking for a menial job in the Catskill area of New York. The applicants were broken down into four groups. The first group had no criminal records. The second group had been arrested for assault and acquitted of the charge with a letter from the judge explaining the presumption of innocence. The third had been arrested for assault and acquitted, but there was no letter from the judge. The fourth group had been convicted.

The employers were asked whether they would be willing to offer the individuals in each group a job. Thirty-six percent said they would hire the men with no record, 24 percent said they would hire the men who had been acquitted and had a letter from the judge, 12 percent said they would hire the men who had been acquitted but had no letter, and 4 percent said they would hire the men who had been convicted.

The findings of this simple study, replicated in a number of subsequent research projects, are a dramatic illustration of the powerful impact of a criminal record, even if the individual under consideration has been acquitted. The implications of these findings—when considered in terms of the number of Americans with records and the accuracy and completeness of these records—are staggering.

Computerized records being rocketed around the country would be a greater help to the police and less damaging to the public if they were accurate and complete. In the pursuit of this worthy goal, after failing to reach a compromise on comprehensive legislation to govern the NCIC [National Crime Information Center], Congress adopted a brief amendment to the Crime Control Act of 1973. The 139-word amendment, which all parties regarded as an interim measure, contained three very simple principles. First, to the maximum extent feasible, criminal history information disseminated with the help of federal funds should contain disposition, as well as arrest data. Second, procedures would be adopted to require that the information is kept current. Third, the information would be used for law enforcement, criminal justice, and "other lawful purposes." As already noted, the exception for other lawful purposes has become something of a gusher.

The 1973 amendment also offered the citizen an apparent remedy. "An individual who believes that criminal information concerning him in an automated system is inaccurate, incomplete, or maintained in violation of this title, shall upon satisfactory verification of his identity, be entitled to review such information and to obtain a copy of it for the purpose of challenge or correction." Almost three years later, the government followed up on the 1973 amendment by issuing a vaguely worded regulation that nominally required the state to develop policies and procedures to ensure the privacy and security of criminal history records. Individual access and review was one of five specific areas the federal government said the states must cover in their regulations.

Ever since the computer became a major force in the administration of large government and business organizations in the mid-1960s, the individual's right to see and correct his own computerized record has been held up as a miracle cure for many of the potential abuses of the computer age. Alan F. Westin, in his pioneering 1967 book *Privacy and Freedom,* was among the first to find wonderful powers in the cure of public access. The principle was subsequently embraced by both the Privacy Act of 1973 and the 1977 report of the Privacy Protection Study Commission.

But the OTA report indicating that more than half of the millions of criminal history records are now circulating in the U.S. are incomplete, inaccurate, or ambiguous is compelling evidence that the remedies prescribed by law and regulation are not very effective. And when the congressional researchers looked at the actual record-keeping practices of four states and six urban areas, they found strong evidence that the promises made in the 1973 Crime Control Act amendment are largely an illusion.

. . . . .

The astounding finding that only one out of five states has ever sought to audit and purge the information in its criminal history files may explain why so many of the records are inaccurate or incomplete. It also demonstrates how difficult it is for the federal government to force state and local agencies to meet a standard established by Congress.

Computer scientists and manufacturers purport to believe that their machines are neutral. This is true, of course, as long as the technology remains in the showroom. The neutrality evaporates, however, when powerful officials running powerful bureaucracies harness the computers to achieve their collective goals. Often both the goals and methods of achieving them are in the public interest. History tells us, however, that the organizations of fallible men sometimes lose their way.

During the 1960s and 1970s the leadership of the FBI and a number of police departments throughout the country came to believe that their responsibilities went beyond arresting those who had committed criminal acts. Their job, some policemen thought, extended to trying to channel the political thoughts and life-styles of the American people along certain narrow paths. Police departments in such cities as Chicago and Los Angeles enormously expanded their surveillance of individuals who did not share conventional police values. The U.S. Army ordered its intelligence agents to attend all sorts of political rallies. The FBI, with the explicit approval of J. Edgar Hoover, undertook a secret program in which it deliberately sought to have critics of the federal government fired from their jobs or otherwise discredited by mailing false and anonymous reports to their neighbors and colleagues.

The FBI's effort to expand its activities beyond the precise boundaries of criminal law did not stop with the anonymous smearing of thousands of Americans who did not happen to share Mr. Hoover's view of the universe. From 1971 to 1974 the FBI harnessed the computerized data bases of the National Crime Information Center to pick up the movements of persons who were not criminal suspects.

Under the public relations establishing the NCIC and how it would operate, the FBI declared that only the names of persons who had been formally charged with a crime would be listed in the NCIC computer. In July of 1975, however, John Tunney, then chairman of the Senate Subcommittee on Constitutional Rights, announced the discovery that the bureau had violated its own regulations by using the NCIC "to keep track of individuals who might be of interest to the FBI for whatever purpose, including possible political reasons."

Three months later, the Justice Department confirmed that from 1971 to 1974 the FBI had instructed its NCIC's computers to sound an alarm any time a local law enforcement agency sent a message to Washington indicating that any one of 4,700 individuals was arrested. When it is remembered what a large part of the population have arrest records, and how many persons were being arrested during that particular period for taking part in civil rights and antiwar demonstrations, the automated flagging mechanism can be viewed as a potentially powerful surveillance device.

One indication of how embarrassed the Justice Department was about the FBI's secret operation of a surveillance system outside the limits set down in its own regulation was the way the project was abruptly halted on the precise day that the Senate subcommittee staff made its first inquiry. But [by 1981] there [were] indications that the FBI and other law enforcement agencies of the Reagan administration [wanted] to resurrect the use of the NCIC computers to track persons who [were] not the subject of a formal arrest warrant.

The first hint of this possible shift in NCIC policy came at a meeting of the NCIC advisory board in June of 1980 in a talk by Kier T. Boyd, an FBI inspector and the deputy assistant director of that part of the bureau that controls the FBI's very large computerized telecommunications system. According to the minutes of the advisory board meeting, Mr. Boyd openly advocated the start-up of the same kind of surveillance that had been so suddenly halted in 1974.

The official began by acknowledging that the tracking of the early 1970s had run into a wave of adverse reaction because "of the political climate of that time" and that the policy under consideration in the spring of 1981 "smacks of the tracking system which in the past has not been very well received by certain quarters of Congress." In an apparent reference to the growing power of conservatives in the Congress, however, Mr. Boyd said he now felt "that the climate is very definitely changing and now we have an opportunity to raise the system to a new level."

In the summer of 1982, the Justice Department approved the request of the Secret Service to use the FBI computer to keep track of persons it suspects may be a danger to officials. But Representative Edwards, [a] one-time FBI agent, asked the Justice Department not to implement the plan until a thorough public examination of the issues had been completed and possible legislation considered.

Without a law, Mr. Edwards said in a letter to Attorney General William French Smith, what assurances do we have that this system will not evolve into the sort of system maintained by the Secret Service in the 1970s when "dangerousness" and "threat" were interpreted to include political dissent? In 1972, the Secret Service had nearly 50,000 individuals on its lists, including such "threats" as Jane Fonda, Tom Hayden, Ralph Abernathy, Cesar Chavez, Benjamin Spock, and Walter Fauntroy. Among the organizations listed were the NAACP, the Southern Christian Leadership Conference, and the John Birch Society.

It is very hard to challenge a Secret Service request for a tool that it declares will provide improved protection for the president. But it is easy to understand that the principle of restricting the FBI's computerized telecommunications network to tracking criminal suspects is an important one. It also is easy to understand that changes in a powerful instrument of federal surveillance should not be made without public discussion, congressional hearings, and legislative authorization. Once an exception has been made for one category of persons, each succeeding category will be that much easier.

— 11 —

Employee Privacy

Philip A. Adler, Jr., Charles K. Parsons,

and Scott B. Zolke

Picture the following scenario. A supervisor's relationship with a subordinate has deteriorated to an antagonistic stage. The supervisor learns that the employee has been diagnosed by a physician as being paranoid. Seeking to strengthen his position and credibility in the matter, the supervisor circulates an interoffice memorandum indicating the physician's

Reprinted by permission of the publisher from "Employee Privacy: Legal and Research Developments and Implications for Personnel Administration," by P. A. Adler, Jr., C. K. Parsons, and S. B. Zolke, *Sloan Management Review,* Winter 1985, pp. 13–21. Copyright © 1985 by the Sloan Management Review Association. All rights reserved.

assessment of the employee's mental state. The employee challenges this action in court as libelous and an unnecessary invasion of privacy.

Can the previous situation occur? In fact, it has occurred and a state supreme court ruled in the employee's favor by stating that the employee's privacy had been violated by the interoffice memorandum.[1] This recent case, though another appeal is still possible, indicates the difficulty in maintaining the delicate balance between employer managerial interest and employee personal privacy. It also establishes another benchmark in the growing relationship between law and employee privacy which is now an integral area of concern within the field of personnel administration.

The issue of employee privacy has experienced a dramatic increase in attention from human resource administrators during the past few years. After a relatively slow start, perhaps triggered by an Orwellian philosophy revolving around fear of "Big Brother," concern for personal privacy has rapidly moved from the realm of science fiction to the reality of governmental law and corporate ethics, becoming, in the process, a central issue of employee relations.

A right to privacy is routinely associated with citizenship in the United States. However, there is a growing debate concerning the extent to which the government should become involved as an explicit protector of this right. This debate is carried on in books, articles, technical reports, and other documents. Stone, Gardner, Gueutal, and McClure have reported that over 2,100 written documents have been recently published on the privacy issue.[2] In addition, both state and federal governments have become involved to some extent, continuing to gather information and to consider advocacy positions in the privacy area.

The fact that personal information is required in order to conduct business effectively in an increasingly complex society must be considered a principal contributor to the privacy issue. The questions of what personal data is required to effectively conduct business affairs can be considered from a triangular perspective:

1. What personal data is actually required?
2. How will such data be obtained?
3. Under what circumstances will this data be released?

The utilization of this data usually falls into two categories. The first category involves the use of personal information about the employees as an input to managerial decisions regarding hiring, employee work assignment, compensation and benefits, promotions, and terminations. The second category concerns the release of employee personal data to external individuals and organizations for such purposes as job references, credit checks, and the like.

There appear to be two major sources of ambiguity when formulating and interpreting management policy in the area of employee rights to privacy. The first is "What are the specific statutes and legal framework under which employees have a formal right to privacy?" The second is "What information handling practices and managerial actions are likely to be perceived as privacy violations by employees?" We will address these two topics and discuss their implications for management policy-making.

Legal Background of the General Privacy Issue

"There can be no public without full publicity in respect to all consequences which concern it. Whatever obstructs and restricts publicity, limits and distorts public opinion and checks and distorts thinking on social affairs."[3] Nevertheless, many societal functions would be ineffective were they not exercised in confidence.

The First Amendment serves to ensure the free flow of information and ultimately protects the idea of an informed public. This concept finds it origin as far back as the Magna Carta. However, the First Amendment freedoms are not absolute, and are tempered by three important restrictions: (1) the privilege to withhold records in safeguarding the national security; (2) compliance with statutory law; and (3) the prevention of unwarranted invasions of privacy.

Though commonly assumed to be a right of citizenship, the U.S. Constitution does not make specific mention of the right to privacy. However, as Duffy notes, amendments 1, 2, 3, 4, 5, 9, and 14 have been interpreted as collectively providing some privacy protection. In addition, several states (Alaska, Arizona, California, and Washington) have enacted some degree of constitutional protection.[4] Rather than review these protections state by state, we suggest that interested readers further consult their particular state codes.

Although there is no general explicit constitutional right to privacy, many state and federal courts have recognized certain protected "zones" of privacy. The most widely recognized "zones" have centered on sex and marriage.

This growing concern for privacy has been the subject of many disputes, especially given the limited applicability of the Privacy Act of 1974. In resolving the tensions between the opposing needs of disclosure and confidentiality, the courts have been without any uniform rules.

The Privacy Act requires that: (1) federal agencies inform individuals (including federal employees) that there are personal data record-keeping systems containing information about them; (2) agencies permit individuals to copy, correct, and amend the recorded personal information; and (3) the type of data agencies may collect about an individual be limited. With certain exceptions, the agency must obtain written consent from the individual before making disclosures of personal information to outsiders.

The 1974 Privacy Act created the Privacy Protection Study Commission, which was to recommend to Congress principles and practices that should be required of private businesses. These recommendations, reported in 1977, include that the 1974 Privacy Act *not* be extended to the private sector, that employees be given some new privacy protection, and that private sector organizations adopt some privacy policy protection on a voluntary basis.

As of 1981, nine states (Arkansas, California, Connecticut, Indiana, Massachusetts, Minnesota, Ohio, Utah, and Virginia) had passed privacy acts similar to the federal Privacy Act of 1974. Two other states (New York and Colorado) have passed less comprehensive laws. These laws, like their federal counterpart, apply to government agencies only. Eight states (California, Connecticut, Maine, Michigan, North Carolina, Oregon, Pennsylvania, and Wisconsin) now have legislation giving employees the right to inspect their own personnel file. Michigan's statute is the most detailed and provides safeguards as summarized by Duffy:

> [E]mployees may, upon written request, examine and copy personnel records at a reasonable time and place. (The term personnel records appears to be broadly defined although the Act specifically excludes certain reference letters, comparative evaluations, medical reports, and investigative or grievance files maintained separately.) Employees may request that information in their personnel files may be amended or corrected. If the request is denied, they may file dissenting statements that are to be kept in a file with the disputed information. If the employer maintains investigative files, employees must be notified of their existence at the end of the investigation or after two years. Investigative files must be destroyed if no action is taken.

Duffy goes on to note that other states have not been as comprehensive, but specific information items have drawn attention. Polygraphs and other so-called truth verification devices have at least some limitations in twenty-one states with Massachusetts, New York, and Rhode Island absolutely prohibiting their use in the employment context. Other statutes limit the use of arrest records and restrict access to medical records and financial data by outside parties.[5]

In addition to the federal and state statutes, many courts now recognize an independent tort for invasion of privacy. The legal concept is founded upon that body of unwritten principles originally based on the usages and customs of the community, recognized and enforced by the courts. However, invasion of privacy is a separate and distinct entity from defamation.

Invasion of privacy occurs where information not reasonably related to a legitimate interest is disclosed. Defamation applies to those instances where false or misleading information is disclosed. In the context of employee–employer relationships, most courts will recognize the furtherance of the employer's business or the public good as constituting a legitimate interest.

Cases do exist where the disclosure of defamatory information, if reasonably believed to be true and necessitated by a need to protect a business interest, will not be actionable. On the other hand, disclosure of true, embarrassing, private facts to people who do not possess a legitimate interest in the subject matter, will constitute an invasion of privacy.[6]

One federal court has already recognized that an employer can be liable for negligence in the maintenance of a personnel file.[7] Alone the same line of reasoning, Representatives Barry Goldwater, Jr. and Edward Koch introduced legislation in 1975 which would have set privacy guidelines applicable in the private sector. The "Comprehensive Right to Privacy Act" (H.R. 1984) was never voted upon; however, the findings of the seven-member Privacy Protection Study Commission are worthy of mention. The Commission recommended that legislation:

1. Require businesses to publish notice of their file and information systems;

2. Require businesses to collect, directly from the individual, only personal information necessary to accomplish proper business purposes;

3. Restrict the transfer of information between businesses;

4. Require businesses to ensure that personnel files are kept accurate and timely;

5. Allow individuals to inspect files;

6. Require businesses to provide written notice to any person concerning whom it has information;

7. Provide strong penalties for violations; and

8. Establish a board to oversee compliance.

It is suggested that regardless of which of these recommendations (if any) are codified by law, some type of uniform code is needed to clarify for the courts and personnel administrators those actions which are permissible.

Consistent with the Commission's findings, most states' privacy statutes provide the employee with three procedural safeguards:

1. The right to be informed of the existence of the personnel file;
2. The right to inspect the personnel file; and
3. The right to correct inaccuracies in the personnel file.

Position of the Courts

In the decisions regarding privacy, most courts have concerned themselves with the preservation of two competing interests: the interest in avoiding disclosure of personal matters and the interest in providing for the free flow of information in making important business decisions. William Prosser, an authority on tort law in American jurisdictions, wrote that privacy could be categorized into four areas, two of which directly apply to the employment relationship: intrusion into private affairs and public disclosure of embarrassing private facts.[8]

Accordingly, when controversies arise regarding personnel files, employers will first have to be able to establish that they only retain information reasonably related to their business and that they have taken reasonable steps to preserve the privacy of their employees. For example, an employer's obligation to protect the privacy interests of its employees often conflicts with the needs of a labor union, which seeks to review personnel files in order to protect its members from discriminatory hiring and firing practices. These conflicting interests require a careful balancing of the respective harms should such information be disclosed or withheld.

The privacy question has further been clouded by administrative investigations. When an employer receives a request to produce the personnel records of its employees, the employee has no standing from which to attack the request. It becomes incumbent upon the employer either to comply with the request, thereby breaching the employee's privacy, or to assert the employee's privacy rights and run the risk of being held in contempt for refusal to comply. These are the types of problems, absent any uniform legislation, with which today's personnel administrators are faced.

. . . Several noteworthy decisions indicate the direction that the United States Supreme Court is taking with respect to the privacy question. The most significant privacy case of the 1970s was *Whalen* v. *Roe*.[9] There, a state health agency required employers to disclose employee medical records in order to document the outbreak of contagious diseases. Roe contested this practice and argued that such disclosure was an unwarranted invasion of privacy. The United States Supreme Court rejected Roe's argument and held that the public interest in health was more important than the individual interest in keeping medical records private. Further, there was no showing of any immediate, physical, tangible injury, and hence the disclosure was upheld.

Privacy is often measured in terms of the expectations of the individual versus the public's interest in gaining access to the information. A recent decision by the United States Court of Appeals for the Sixth Circuit permitted a similar disclosure as in *Whalen* v. *Roe,* but required the National Institute for Occupational Safety and Health to provide adequate security measures to protect the privacy of those employees whose medical records were reviewed.[10]

It would appear that where a federal agency, which is governed by strict procedures for the safekeeping of records, seeks access to confidential employee records, the courts will permit disclosure as long as there exists a reasonable relation to a job-related health hazard, and adequate safeguards are employed. However, one federal court has held that the interests in protecting employee health are superior to the privacy interests of the individual. As justification for such a position, the federal court held that the fact that the employees were willing to divulge sensitive information to their employer constituted a dilution of their privacy interests.[11]

Many times, before an agency can obtain employee records, the employee is asked to consent to disclosure. This is especially true in cases dealing with psychological testing data. The United States Supreme Court has ruled that where there exists substantial evidence of actual adverse effects to the employee from disclosure, a governmental agency cannot compel the employer to release sensitive information.[12] Hence, although the employee may have diluted his privacy interest by permitting his employer to accumulate sensitive information, any showing of actual harm will prevent forced disclosure.

As seen from these cases, the risk to the affected employee is a key component in determining whether information can be disclosed. Most government agencies have policies and procedures which minimize the risk that confidential information will be disclosed to noninterested parties. Should the disclosure pose such an adverse effect on the employee that to do so would offend reasonable sensibilities, most courts will refuse to compel an employer to comply with a request. On the other hand, as long as the agency has adequate safeguards, and the need for the information is related to health and safety, disclosure will be mandated.

Most courts have held that there is no infringement of privacy when ordinary sensibilities are not offended. So long as there is not a flagrant breach of decency and propriety, the courts will recognize that no individual can expect complete noninterference from the society in which he or she lives. Accordingly, information regarding salary, business connections, age, experience, education, and criminal convictions will not constitute an unwarranted intrusion into an individual's right to privacy.

Major Privacy Ambiguities

There continues to be confusion in the minds of many individuals, including personnel administrators, as to the legal status of employee privacy. The dividing line between legal and ethical controls affecting the privacy of employees is not as vague, however, as is frequently believed. In fact, there are actually few direct legal constraints pertaining to protecting employee privacy. At present, there are no broad federal regulations protecting the privacy of private sector employees. As noted earlier, the privacy of federal employees is protected by the Privacy Act of 1974. This federal law has been held by the courts to cover, in certain instances, state, county, and municipal employees.

Employees in the private sector must still depend primarily upon the traditional legal implications of slander and libel, as well as ethical considerations, for the protection of personal data accumulated by their employers. However, due to the uncertainty of some personnel administrators about the existence and coverage of "privacy laws," private sector employee privacy is often artificially protected by "government regulations" which, in reality, have no direct legal impact upon privacy matters. Some personnel administrators appear unaware that their "legal" concern for protecting employee privacy is actually based on governmental nondiscrimination laws, rather than on laws directly concerned with the employee privacy issue.

Accordingly, private sector employees may gain a certain subtle degree of privacy protection from their employers simply because of confusion about nondiscrimination laws with which managers desire to comply. These personnel administrators believe they are complying with "nonexistent" privacy laws, when they actually are basing their actions on familiarity with provisions of governmental nondiscrimination statutes.

In fact, a recent federal court decision has stated that an employer's invasion of employee privacy may constitute a civil rights violation.[13] However, this issue arose because of the existence of pertinent civil rights legislation, rather than as a result of privacy laws per se. Thus, even if compliance occurs for the wrong legal reasons, personnel administrators "properly" protecting employee privacy will find themselves legally correct.

For another example, consider a recent case before the District Court of Appeals in Ohio. There, a jury verdict of $10,000 was sustained in an invasion of privacy case brought by an employee against his privately held corporate employer. The controversy centered on the employee's answer to a question regarding whether he had ever been convicted of a criminal offense. He answered in the negative and failed to note that he had previously been convicted of armed robbery in a juvenile delinquency proceeding. Several months after he was hired, the personnel director learned of the juvenile proceedings from the local police. His employment was terminated as a result of the alleged false statement.

At the grievance hearing, wherein the nature of the juvenile conviction was disclosed, the personnel director, a union representative, and other employees were present. The employee's lawsuit sought damages for the invasion of his privacy, alleging that his employer should not have disclosed anything more than the existence of a juvenile delinquency conviction.

In affirming the verdict, the appellate court noted that the right of privacy is concerned with a person's peace of mind that his private affairs will not be made public. Although an employer does have a qualified privilege to make limited communication to parties who possess a legitimate interest in the information, the publication of the nature of the juvenile offense, that is, the armed robbery, went beyond that privilege. This decision is especially significant inasmuch as the appellate court held that there exists no requirement that a plaintiff prove malice to sustain an invasion of privacy action.[14]

Currently, because of various court interpretations of the privacy issue, there is considerable question as to what is proper employee privacy protection under civil rights and nondiscrimination laws. The federal courts are clearly increasing their attention to the employee privacy issue, even if under the guise of civil rights laws, and consequently, personnel administrators should be aware of a rapidly increasing number of pending cases and plaintiffs' awards in this regard. Accordingly, what may be "traditionally" accepted as proper protection of employee privacy is subject to sudden legal change.

Although the courts seem to recognize that employers must have a reasonable opportunity, relevant to possessing employee personal data, to conduct their business affairs, personnel administrators should not take this to mean they have a broad license regarding the collection and protection of such data. Even if employee personal data is kept secret upon acquisition by [an] employer, the mere obtaining of some of that information may be a civil rights violation. Thus, employers should be most cautious of asking the "wrong" questions (per court desisions) of employees, regardless of their reasons why such information is necessary to the conduct of "normal" business activity.

Summary of Legal Background

There are multiple indicators that protection of employee privacy will receive even more emphasis in the near future. First, as noted above, the right to privacy is commonly assumed to be a basic right of U.S. citizens. However, there are several trends in our society indicating that the protection of these rights is becoming more problematic. Many people have heard of the Privacy Act of 1974, but it is a common misconception that the act applies to a wide range of employment situations. In fact, the act only covers employees of federal agencies. It does prohibit certain information dissemination practices without employee approval, and requires that employees be permitted to review their personal employment information files upon request; however, these protections have not yet been extended to the private sector.

Second, several states have passed some form of law to protect privacy rights, but these vary in comprehensiveness. Although most of the state laws are relatively new, it is probable that both interpretation and enforcement of the laws will be inconsistent across these states.

Third, there are commercial and governmental interests in information about people, and employers provide a potentially rich source for this information. For instance, the concept of credit rating is based on the idea that credit bureaus should gather individual financial data, including present salary and salary history.

Fourth, there is also a growing capacity within organizations to gather, save, and rapidly transmit employee information through computers and telecommunications devices. The improvement in technological capacity has been accompanied by the creation of centralized data bases within organizations that link together many pieces of information about an employee which, in the past, had been scattered throughout the organization. All of these above conditions are viewed by some as threats to the basic right of privacy, leading to the continuing debate on workplace privacy.

On a national level, the debate will be fueled by incidents of employee privacy violation that gain national attention. For a given organization, the amount of conflict surrounding issues of employee privacy will be determined by the accumulation of privacy violations. Therefore, it becomes very important to understand what factors and information handling practices affect employee perceptions of privacy violations. We will now review the recent research on perceptions of privacy violation.

Research on Perceptions of Privacy Violation

Research on employee perceptions of privacy is only beginning to accumulate. However, the studies that do exist suggest that there are a number of factors that affect perception of privacy violation. The first point of interest is whether or not employees know how their employers handle employee personal information. In a survey of 2,047 employees from five companies, Woodman, Ganster, Adams, McCuddy, Tolchinsky, and Fromkin found that across sixty information items (e.g., demographics such as age and sex, financial situation, medical information), employee beliefs about whether or not the company retained information were accurate an average of 68 percent of the time.[15] Management in each of the five companies had reported actual practices.

Employees tended to overestimate the maintenance of affiliation information (union, religious, political) and underestimate the maintenance of medical information. These inaccuracies are understandable because 85 percent of the employees reported that they had never inspected their personnel file. When asked whether or not they had experienced an upsetting incident in their company's handling of their personal information, 7 percent said yes to a disclosure within the firm and 3 percent said yes to some disclosure outside the firm. As a general reaction to the company's information handling practices, 44 percent were satisfied, 22 percent were dissatisfied, and 34 percent reported [that they were] uncertain.

One of the most disturbing findings from this study is the general lack of knowledge about what is maintained and what is not. A lack of knowledge sets a fertile ground for rumors based on isolated incidents of privacy violations.

As part of the same survey, Tolchinsky, McCuddy, Adams, Ganster, Woodman, and Fromkin reported which information-handling practices employees felt were most disturbing.[16] Each respondent in the survey responded to one of sixteen hypothetical job situations that reflected different practices. The four factors studied were: (1) type of information (personality or performance); (2) permission for disclosure (yes or no); (3) consequences of disclosure (positive or negative outcome for employee); and (4) location of disclosure (released to sources within the organization [internal] or sources outside the organization [external]).

All factors were found to be statistically significant, with greater perceived privacy invasion occurring: (1) for release of personality rather than performance information, (2) for permission not obtained prior to release, (3) for a negative rather than a positive consequence, and (4) for information released to an external rather than an internal party. Probably more interesting was the finding that permission was the most important factor.

In fact, when previous permission is obtained, there is no difference between releasing information to internal or external parties. Also of interest was the finding that the outcome of the information release (positive or negative) was a relatively minor factor compared to whether or not permission was obtained prior to disclosure. Clearly, an individual's perceived control over personal information (permission obtained before release) is more important than the eventual outcome as far as perceptions of privacy are concerned.

This latter conclusion was also supported in a study by Fusiller and Hoyer, who used a hypothetical personnel selection situation. Respondents felt a greater degree of privacy invasion when personal information from their current employer was released without their consent than when permission was obtained.[17] In addition, if the respondent received a job offer, perceived privacy violation was lower than if he or she did not get the offer. In this study, the location of the disclosure (within the university or outside the university) and type of information (personality or performance) did not make a difference.

Though the current article has focused on the employer-employee relationship, there are many other institutions that gather, retain, and disseminate personal information (e.g., insurance companies, the Internal Revenue Serivce, lending and other credit-granting institutions, law enforcement agencies). The question about public perceptions of the relative likelihood of privacy violations in different institutions was the topic of a study. In interviews with 193 subjects, Stone et al. asked about individuals' values, beliefs, attitudes, experiences, and future intentions concerning information-handling practices in one of six different types of institutions, one of which was an employer. The researchers found that individuals felt more confident about their ability to control personal information maintained by their employer than their ability to control that maintained by any other type of institution. However, this difference in confidence did not produce significant differences in willingness to support legislation that would control information-handling practices in these different institutions. The researchers also noted that the average scores for the legislation support scale were relatively high across all institutions.[18]

Managerial Implications and Conclusions

From our review of both the legal and psychological literature on workplace privacy, it is clear that the handling of employee information in organizations has important implications for personnel administrators. Present case law leaves many questions unanswered and the private sector is not covered by legislation in most states. Therefore, the enactment of uniform guidelines incorporating the recommendations of the Privacy Commission would be helpful to personnel administrators.

Regardless of further governmental standards, personnel administrators essentially have two broad internal roles to perform regarding the protection of employee privacy. The first role relates to privacy considerations involved in carrying out their standard staff unit activities such as hiring, training, compensating, and grievance administration. The second role relates to providing staff technical guidance in the privacy arena for managers throughout their organization. Clearly, the performance of the first role establishes the basic pattern for handling employee privacy within that organization, since managers will likely follow the tone of the "examples" on this issue set by their personnel administrators.

Personnel administrators also have an external role to perform for their organizations relevant to protecting privacy of employees. This role involves the furnishing of employee personal information to third parties regarding job references, credit checks, legal investigations, insurance risks, etcetera. Ironically, this external role, with its lesser direct bearing on internal organizational operations, may create the most difficult and potentially dangerous employee privacy challenges for management. We offer some of the more obvious implications and suggest that concerned readers follow up with more detailed suggestions provided by Duffy or Noel.[19]

1. Develop a formal information practices plan. To avoid sensitizing people, do not use the phrase "privacy plan." The plan should be written and circulated to all employees. Supervisors and managers must be made especially aware of their responsibilities concerning their own behavior and that of their subordinates.

2. Get employees involved and have them review their personnel file on a scheduled basis. Have employees formally acknowledge this review.

3. Be careful about personal data not directly related to job performance. Eliminate outdated information. Identify the purpose of each piece of data and who can have access to it.

4. Be careful about outside release. Have employees formally acknowledge what data can and cannot be released to which type of institution. It is the party that releases the information that is at risk, not the party that seeks the information.

Finally, the current dynamics of the employee privacy issue preclude the statement of exacting positions that will eliminate risk in this sensitive area. However, this fluidity does not prevent personnel administrators from understanding general positions and trends in the area.

Notes

1. See "SJC Outlines Rules on Employer Role in Workers' Privacy," *Boston Globe,* July 7, 1984. Decision upheld August 6, 1984 in U.S. Court of Appeals, First Circuit.

2. See E. F. Stone, D. G. Gardner, H. G. Gueutal, and S. McClure, "A Field Experiment Comparing Information Privacy Values, Beliefs, and Attitudes across Several Types of Organizations," *Journal of Applied Psychology* 68 (1983): 459–68.

3. See J. Dewey, "The Public and Its Problems," in L. W. Levy, ed., *Civil Liberties in American History,* (New York: Da Capo Press, 1941).

4. See D. J. Duffy, "Privacy vs Disclosure: Balancing Employee and Employer Rights," *Employee Relations Law Journal* 7 (1982): 594–609.

5. Ibid.

6. See *Quinones* v. *United States, 492 F. 2d 1269 (3rd Cir., 1974)*; W. Prosser and W. P. Keeton, *On Torts,* 5th ed (St. Paul, MN: West Publishing, 1984), pp. 856–57.

7. See *Bulkin* v. *Western Kraft East, Inc. 422. Supp 437 (E.D. Pa. 1976).*

8. See W. Prosser, "Privacy," *California Law Review* 48 (1960): 383.

9. See *Whalen* v. *Roe,* 429 U.S. 589 (1977).

10. See *General Motors Corp.* v. *Director of NIOSH,* 636 F. 2d. 164 (6th Cir., 1980).

11. See *United States* v. *Allis-Chalmers Corp.* 498 F. Supp. 1027 (E.D. Wis. 1980).

12. See *Detroit Edison Co.* v. *National Labor Relations Board,* 440 U.S. 301 (1979).

13. [See] *Phillips* v. *Smalley Maintenance Services,* 711 F. 2d 1524 (11th Clt., 1983), 436 So. 2d 705 (Ala. 1983).

14. See *Chambers* v. *Terex Corp. Cuyahoga County, Ohio,* 8th District Court of Appeals, March 31, 1984.

15. See R. W. Woodman, D. C. Ganster, J. Adams, M. K. McCuddy, P. D. Tolchinsky, and H. L. Fromkin, "A Survey of Employee Perceptions on Information Privacy in Organizations," *Academy of Management Journal* 25 (1982): 647–63.

16. See P. D. Tolchinsky, M. K. McCuddy, J. Adams, D. C. Ganster, R. W. Woodman, and H.L. Fromkin, "Employee Perceptions of Invasion of Privacy: A Field Simulation Experiment," *Journal of Applied Psychology* 66 (1981): 308–13.

17. See M. R. Fusilier and W. D. Hoyer, "Variables Affecting Perceptions of Invasion of Privacy in a Personnel Selection Situation," *Journal of Applied Psychology* 65 (1980): 623–26.

18. See E. F. Stone, D. G. Gardner, H. G. Gueutal, and S. McClure, "A Field Experiment Comparing Information Privacy Values, Beliefs, and Attitudes Across Several Types of Organizations," *Journal of Applied Psychology* 68 (1983): 459–68.

19. See Duffy (1982), pp. 594–609; A. Noel, "Privacy: A Sign of Our Times," *Personnel Administrator* 26 (1981): 59–62.

—— 12 ——

Data Regulation

Thomas R. Mylott III

As obsolescence and foreign competition choke American factories, new industries based on information spring up wherever electricity and telephones are available. Some forecasters see information technology and related industries as the engine to haul the U.S. economy into the twenty-first century. While information industries grow, so too do the problems associated with information. Many different enterprises can profit from information technology. Electronic fund transfer (EFT) networks, credit bureaus, and medical-information bureaus are only a few examples.

While government and private industry need increasing amounts of information and put that data to an increasing variety of uses, anxiety about the use of that information has grown, too. The anxiety has two faces. One is privacy, the concern that private industry and the government know too much about us. The other is openness in government. Many people want to know what the government is up to and what information it possesses. In many ways these concerns, privacy and freedom of information, are at odds with one another. Any legal framework that intends to grapple with the whole beast of information regulation must deal with these opposing forces.

At the present time, there is no comprehensive federal, state, or local approach to the regulation of data. Instead, there are laws that deal with

From *Computer Law for Computer Professionals,* by Thomas R. Mylott III, pp. 102–8. © 1984. Used by permission of the publisher, Prentice-Hall, Inc., Englewood Cliffs, N.J.

narrow concerns and thus are of limited value. But already there are two federal laws that express these conflicting concerns. One is the Freedom of Information Act (FOIA). The other is the Privacy Act of 1974.

FOIA And Data Regulation

The FOIA regulates dissemination of and access to data. The general intention of FOIA is to permit access to data collected by the executive branch of the federal government. That includes the oval office and various agencies. The law's premise is that anyone who wants information that is in the possession of the executive branch should be able to get that information, unless it comes under one of nine exemptions. Generally, all agency opinions, records, and studies that have not been published by the agency in the *Federal Register* are available. FOIA permits access to anyone—individuals, corporations, and foreign governments, for that matter.

The exemptions to the statute are not mandatory, but discretionary. In other words, the agency is not required to withhold information that comes within the exemptions, but rather the agency may withhold such information. Two interesting exemptions are trade-secret and financial information. An agency may also withhold personnel and medical files if disclosure would be an invasion of someone's privacy. An overriding motive for passage of this statute was the belief that the governemnt possessed volumes of information on private citizens that citizens should know about and should have a means of acquiring. However, corporations and not individuals file the majority of requests under the FOIA.

FOIA could serve as a model for legislation at the state and local level. Systems created for any level of government might someday have to accommodate the statute. What FOIA means to system design is unclear. Freedom of information is not directed at the data-processing industry, but rather at all executive branch data collection.

Any statute that requires disclosure of information can easily mandate the form of that disclosure. You have probably assumed a hard-copy disclosure is all that FOIA requires. The day is sure to arrive, though, when the requested data is so voluminous or the purposes of the request are such that someone would demand the data in machine-readable format. You might immediately wonder: EBCDIC or ASCII? (Government data is usually in ASCII.) If the information is in a data base, should you have to provide the data base and the indices or do you have to retrieve relative records and give the requestor a fixed-block, sequential file? Regardless of whether you are requesting or supplying the information, seemingly insignificant issues like the data format have substantial implications in system design from a FOIA standpoint.

So far, we have considered FOIA as a principle applying to government only. But [at] least one other statute provides similar disclosure requirements for private enterprise. As regulation of government and private information grows, those who design and maintain information systems will have to accommodate these regulatory tentacles.

Privacy Act and Data Regulation

On the other side of the fence from the Freedom of Information Act stands the Privacy Act of 1974. This statute also applies to the federal government, specifically, federal agencies in the executive branch. The purpose is to enable an individual to determine what information pertaining to that individual an agency is collecting, using, maintaining, or disseminating. Another purpose is to prevent data maintained for one purpose, such as income-tax administration, from being used or made available for some other governmental purposes. In theory, the Privacy Act should have prevented the Internal Revenue Service from supplying its information to enforce Selective Service registration. Presently, the opponents and proponents are slugging it out in court.

The Privacy Act gives you access to the data that the government has collected about you. It permits you to get a copy of that data. There are also some normative requirements in the statute that require agencies to strive for accuracy and currency of data and that specify that the only data gathered should be data that is used for some necessary and lawful purpose of the agency. Federal agencies must also have adequate security and safeguards to prevent misuse of data. There are criminal penalties for violations.

There have been few lawsuits interpreting the statute. So at the present time, the legal criteria by which to evaluate an agency's standard of care in maintenance and security of information under the Privacy Act is unknown. The statute itself was a giant leap forward in that it gave you the legal right to have some control over information that the government gathers on you. The federal act could serve as a model for states' privacy laws, and it also could be the intellectual basis for statutes that apply to private industry. An extension of the statute's conceptual basis could be a law that specifies that any information maintained on an individual may be stored for certain purposes only and that the subject of the information must have the ability to access the information or have a copy of it. This could pose quite a burden on certain industries, particularly if everyone began demanding access to their files.

The Privacy Act, however, does not require the government to notify you that it is collecting information about you. A prerequisite to your demanding to inspect information is knowing who is gathering the data in the first place. While information systems currently do not have to inform the subject of data collection, such a requirement is a logical consequence of giving individuals meaningful access to that data.

Always compelling information gatherers to give notice that they are collecting information could become a tremendous burden to the government, as well as to private enterprise. To be rational, the benefits of providing notice must equal or exceed the cost. Unfortunately, you can quantify only the cost. Adding up the benefits in dollars is like trying to put a price tag on freedom of speech—it is very hard to do. Instead, society must weigh quantifiable costs with social values. This is not a problem limited to notification. It is a problem that pervades all privacy issues. Nearly all defenses of privacy will cost something. Some may even undermine efficiencies established by computer information processing. The decision of how far you go in protecting privacy is partially an economic one. Since you cannot quantify privacy itself, the final analysis must rely on societal values, which have no list price.

Fair Credit Reporting Act and Data Regulation

The third federal statute of interest is the Fair Credit Reporting Act (FCRA). Unlike the FOIA and the Privacy Act, this statute applies to private industry. In general, the Fair Credit Reporting Act regulates credit bureaus. It regulates credit bureaus in several ways. First, it determines the purposes for which credit reports may be maintained. Within certain limits, the statute allows data to be maintained pursuant to a court order or government agency order. FCRA also allows others to maintain information and to use the information for extending credit, for underwriting insurance, for employment purposes, for determining the eligibility of government licensing where financial responsibility has to be determined, and for certain other legitimate business needs in connection with consumer business transactions.

The statute also specifies that certain types of data may not be maintained. In particular, obsolete data is forbidden. The FCRA defines obsolete data. For instance, a credit-evaluating system may maintain no record of a bankruptcy more than 10 years old. And it may maintain no records of suits or judgments older than 7 years or those for which the statute of limitations have expired, whichever is longer. If there has been a tax lien on the consumer that has been paid and is older than 7 years, that information also may not be present. Credit bureaus must also purge from their files records for consumer accounts that were placed in collection and then written off more than 7 years ago. Arrest information or criminal records cannot be maintained more than 7 years after the disposition, release, or parole of the individual. Other adverse data about an individual cannot be more than 7 years old. All the restrictions on the maintenance of so-called obsolete data do not apply to credit transactions of $50,000 or more, life insurance that is greater than or equal to $50,000, or employment for a job that pays a salary greater than $20,000 a year. Inflation may transform these exceptions such that they gobble up the statute.

FCRA also applies to investigative works. If an investigative report is for employment purposes and if the information is likely to have an adverse effect, then the consumer must be notified of the report and to whom the data is going. Also, whoever maintains information must employ strict procedures designed to ensure that the information reported is complete and current. FCRA further requires that whoever maintains the data use reasonable procedures to avoid violations. There must be a disclosure by the reporting bureau to the consumer of certain requests for information.

When a consumer disputes the data's accuracy, the statute sets up various procedures. If the consumer complains to the reporting agency of an error, the data must be reinvestigated unless there are reasonable grounds to believe that the subject's claim is fruitless or irrelevant. If, after the reinvestigation, data is discovered to be inaccurate or unverifiable, FCRA instructs the bureau to delete the data. If there is a dispute about the validity of data, the consumer has the right to place a statement concerning the dispute in the file. The bureau is required to inform those to whom it disseminates this information of the dispute, again unless the bureau believes that such dispute is fruitless or irrelevant. If information is deleted because of a consumer dispute or if a consumer places a statement in the file, the bureau must, at the consumer's request, notify all third parties who received the report for employment purposes within the previous two years and anyone who has received any report within the previous six months.

FCRA also regulates the recipients of information from credit bureaus and similar types of agencies. If credit is denied or if there is an increased charge to the consumer for receiving credit because of a report from other than a reporting agency, the user who denied credit must notify the consumer within 60 days after such notice. If the consumer requests the reasons for denial, the user must also disclose the nature of the information. If the data is from a reporting agency, the user of the data must supply the consumer with the name and address of the agency. There are criminal penalties for violations of the FCRA, such as obtaining a consumer's file under false pretenses or the unauthorized disclosure of such information by an agency.

Other Federal Regulation of Data

The federal government also regulates information in some ways that are not quite as noticeable as the FOIA, Privacy Act, or FCRA. A good example of this is IRS Revenue Ruling 71-20. This revenue ruling is a determination by the IRS about what constitutes records for the purposes of the Internal Revenue laws. Revenue Ruling 71–20 determines that "punch cards, magnetic tapes, disks, and other machine sensible data media used for recording, consolidating, and summarizing accounting transactions and records within a taxpayer's automatic data-processing system are records within the meaning of (the Internal Revenue laws) and are required to be retained so long as the contents may become material in the administration of any Internal Revenue law."

Present Data Regulation
and the Computer Professional

What does current information regulation mean to the data-processing professional? First, clearly the government has already begun to regulate information. If you are designing or operating a system to which certain statutes already apply, especially in the credit bureau or federal agency area, then there are laws that already apply to you. Possibly you also may be subject to certain state privacy regulation or information regulation. Data-processing professionals must be aware that regulation of information is spreading from regulation of only the government's use of data to the private sector's use of data. The regulation will increase. Equally as likely is that there will be no comprehensive federal, state, or local data-regulation policies. What will happen is that private enterprise will be subject to many overlapping, sometimes conflicting, and certainly confusing laws regulating information. The only way for computer professionals to avoid the goulash of regulation already simmering is to work with legislators to establish a sound foundation on which to enact data-regulation statutes.

Therefore, for the time being, you need to be aware of two facts. First, your current and proposed systems may already be subject to data regulations. Second, regulation of information will continue and expand.

All systems that maintain information about individuals are candidates for scrutiny. Such systems may already be subject to certain federal, state, or local regulations. Any system that maintains individual information of a financial nature that you disclose to third parties is probably covered by the Fair Credit Reporting Act and must meet the FCRA's standards. Perhaps your particular systems are covered by some law this chapter has not discussed. An in-depth study of all data regulation is beyond the scope of this [essay]. More likely, though, there is no specific statute that covers your system. However, systems that maintain data on individuals should be scrutinized not only for what current laws apply to them but also for the areas in which future legislation may affect them. Regulatory possibilities should become considerations in the operation of current systems and the design of prospective systems.

For instance, if your system maintains information primarily for internal purposes, that is, you are not in the business of selling this data, then your systems most likely to be regulated first are payroll and personnel. As innocuous or mundane as these systems seem, they have tremendous impact on employees. Keep in mind the possibility that employees may gain the right to examine the validity of data, to receive notice of any transactions affecting their data, and to dispute what remains in the files, as well as having control over what information may be stored in the first place.

As information matures into an asset of growing value to our society, the threats to personal privacy will probably catch you unaware. Perhaps you will be denied credit erroneously, or maybe someone you know will be hurt or arrested because an invalid license plate number was in a police information system.

Foul-ups such as these have happened and may happen again. Those who are not computer professionals recoil at these mishaps and clamor for laws to prevent recurrences. Computer professionals are in the best position to reduce the mistakes and to design systems that accommodate the concerns of freedom of information, privacy, technical feasibility, and economic sense. Yet none of this will happen unless computer professionals are willing to join with the rest of society in a joint effort to address these concerns.

Quality of Personal Life

Sherry Turkle (reading 13) takes on the difficult task of exploring the relationship between computer science and psychoanalysis. Because computers offer a model of what it means to have memory, she argues, the existence of computers with memories—and artificial intelligence (AI)—undermined behaviorist psychology. She believes that the recognition that AI includes active and interactive inner agents can be a starting place for useful exchanges between computer science and psychoanalysis.

Marvin Minsky (reading 14), who also compares people with computers, presents his argument as a series of questions. For instance, he asks whether computers can do only what they are told, be creative, know what something means, or know about the real world. Using cogent analogies and his own model of how people learn, he challenges his readers' assumptions that computers could never have creativity, common sense, and other traits we consider peculiarly human.

Herbert and Stuart Dreyfus (reading 15) forcefully sound the alarm against the "computer-type rationality" that is pervading our society, replacing experts and expert intuition and making us servants of machines. Human thought, they argue, is holistic in nature and inseparable from emotions, skills, and our situation in society and the world. The Dreyfuses therefore condemn the "false optimism and unrealistic expectations" of computer advocates, regarding this view of computer intelligence as an outgrowth of historical rationalism that can be traced back to Socrates.

Artifical Intelligence and Psychoanalysis: A New Alliance

Sherry Turkle

Artificial intelligence and psychoanalysis appear to be worlds apart. Psychoanalysis looks for what is most human: the body, sexuality, what follows from being born of a woman and raised in a family. Artificial intelligence looks deliberately for what is least specifically human: the foundation of its theoretical vision is the thesis that the essence of mental life is a set of principles that could be shared by people and machines.[1]

There is another way in which they appear worlds apart. Artificial intelligence seems scientifically ascendant and has increasingly determined the agenda for academic psychology through its influence on cognitive science. In contrast, psychoanalysis is rejected by academic psychology and in conflict with dominant biological trends in psychiatry. Although there have been recent flurries of interest in Freudian theory, they have come from the worlds of literary analysis and philosophy. To scientific circles, psychoanalysis appears a frozen discipline—frozen in the scientific language of another time, frozen in the psychological assumptions of another culture.

In this essay I propose that if psychoanalysis is in trouble, artificial intelligence may be able to help. And I suggest the nature of this help by arguing that one of the ways computers influence psychological thinking is through a route that is not essentially technical. Rather, computers provide sciences of mind with a kind of theoretical legitimation that I call sustaining myths. Indeed, the early impact of the computer on psychology was clearly of this nature.

Sustaining Myths

As recently as the 1950s, behaviorism dominated American academic psychology, its spirit captured by saying that it was permissible to study remembering but considered a violation of scientific rigor to talk about

From Sherry Turkle, "Artifical Intelligence and Psychoanalysis: A New Alliance," pp. 241–67. *Proceedings of the American Academy of Arts and Sciences,* Vol. 117, No. 1, Winter 1988. Reprinted by permission.

"the memory." One could study behavior but not inner states. The study of mind had to be expressed in terms of stimulus and response. In today's jargon, what lay between was a black box that must not be opened even speculatively.

By the end of the 1960s, the behaviorist hegemony was broken, as were inhibitions about the study of memory and the inner processes of mind. Indeed, within academic psychology scarcely a trace remained of behaviorist methodology. Behaviorism had not been refuted by a critical experiment. There had been many factors influencing this scientific revolution, including the political and cultural climate of the 1960s. And one of the most central was the computer.

The computer's role in the demise of behaviorism was not technical. It was the very *existence* of the computer that provided legitimation for a radically different way of seeing mind. Computer scientists had of necessity developed a vocabulary for talking about what was happening inside their machines, the "internal states" of general systems. If the new machine "minds" had inner states, surely people had them too. The psychologist George Miller, who was at Harvard during the heyday of behaviorism, has described how psychologists began to feel embarrassed about not being allowed to discuss memory now that computers had one:

> The engineers showed us how to build a machine that has memory, a machine that has purpose, a machine that plays chess, a machine that can detect signals in the presence of noise, and so on. If they can do that, then the kind of things they say about the machines, a psychologist should be permitted to say about a human being.[2]

. . . . . .

. . . Computer programs provided a way to discuss beliefs and rules as causing behavior. Why did pawn take pawn? Earlier psychologies would have rejected "because the pawn blocked the bishop" as a casual explanation. It would merely be giving the chess player's "reasons." But if mind is program, reasons become explanations. A large part of the computer's appeal for psychologists is that it allows them to open the black box that is the mind. Once that box is open, the computer suggests ways to fill it with concepts that are close to commonsense understandings.

Indeed, the crux of Miller's story about memory is that computers gave psychologists permission to investigate something that "everyone knows" but that had been banished from science—the idea that people have memories. In the past two decades, cognitive science has been dominated by the computer legitimating the study of something else that "everyone knows," this time the idea that people have information and use rules, and that much of this information can be formulated in words. In the late

1950s Allen Newell and Herbert Simon built a computer program called the General Problem Solver (GPS), which was guided by something very close to verbal reasons recoded as computational rules. Questions such as "Why did GPS do such and such?" could be answered by reference to whatever rules it had been given. Why should references to rules not be used to answer questions about what people do when faced with similar problems? The existence of GPS gave credibility to the question.

There is a widespread view that the computer presence tends to move psychology toward more rigorous and quantifiable theories, arguing that the computer, by its nature, requires rules, rigor, and formalism. But the story of the computer's influence on psychology is not so simple. For example, its "first act"—the attack on behaviorism—went in the direction of creating a less rather than a more constrained, a "softer" rather than a "harder" science of mind.

Artificial intelligence is the most explicit channel for the computer's influence on psychology. It asserts a global materialism and also offers particular theories about how mind works. Its dual agenda is to build "machines that think" and to use machines to think about thinking. Its methodological premise is that if one builds a machine that can do something intelligent, the way one gets the machine to do it is relevant to thinking about how people do it as well.

.   .   .   .   .

Traditional Affinities

The very idea of AI—to create mind in machines—subverts traditional notions of the autonomous self in a way that parallels the psychoanalytic enterprise. Most people see the autonomous self as an unproblematic idea because they have a day-to-day experience of having one. Our everyday language captures that experience and expresses the idea of free will; we say, "I act," "I do," "I desire." And even when people have learned through theology or philosophy to question the idea of free will, what they tend to do is make small modifications in their notion of the autonomous self; it becomes a self whose decisions are constrained. Inherent in psychoanalysis is a more radical doubt. The unconscious does not constrain; it constitutes a decentered self. Inherent in AI is an even more threatening challenge: If mind is program, where is the self? It puts into question not only whether the self is free, but whether there is one at all.

Traditional humanism is committed to the notion of an acting, intentional subject. In its challenge to the humanistic subject, AI is subversive in a way that takes it out of the company of rationalism and puts it into the company of psychoanalysis and radical philosophical schools such as deconstructionism. The psychoanalytic subject is decentered in the web of the unconscious; the deconstructionist subject is decentered in language; the computational subject is decentered—indeed, perhaps dissolved—in the idea of program.

These affinities will not reassure the traditional humanist who has gotten used to seeing AI as an enemy. They do not make AI any less of an assault on the idea of the self. But the attack comes from the left, so to speak, rather than from the right. Artificial intelligence is to be feared as are Freud and Derrida, not as are Skinner and Carnap.

The computational "explanation" of the chess move points to another way in which AI is more like Freud than Skinner. Within traditional science, and certainly for behaviorism, the line between subject and object is taken as sacred. But for Freud, his self-analysis, his technique of self-understanding, was indissociable from the development of his general theory. Like psychoanalysts, AI theorists have made a profession of dissolving the line between subjective and objective reflection. The intelligence embodied in the chess move is intelligence derived from personal knowledge of chess. "There's only one place to get ideas about intelligence, and that's from thinking about myself," says AI scientist Roger Schank. "In the end, I have just myself, and if it feels right, that's what I have to trust," says Donald Norman, an AI-influenced cognitive psychologist.[3]

Marvin Minsky, one of the founders and theoretical leaders of AI for the past quarter of a century, has always made it clear that as far as he is concerned, you can make a machine do only what you yourself know how to do. In order to build a program, you have to engage in self-analytic activity. In the early 1960s Minsky worked with a student, Thomas Evans, on an AI program that could pass the familiar visual-analogy tests: *A* is to *B* as *C* is to *D, E,* or *F,* where each letter stands for a geometrical drawing. His method was psychological: think about yourself. And its reference point was psychoanalysis: "What you had to do was something like what Freud did. Tom Evans and I asked ourselves, in depth, what we did to solve problems like this, and that seemed to work out pretty well."[4]

Behaviorism rigorously forbids any reference to personal experience, and most other psychological schools try to ignore the issue. But AI and psychoanalysis have each articulated the need to integrate personal reference into theoretical construction. Each, in its own way, is a science of self-reflection.

But are such affinities superficial? After all, psychoanalysis explores the mind to discover the irrational; artificial intelligence invents machines through the exploitation of the rational. In fact, what stands between psychoanalysis and AI is not AI's "materialism." In the past quarter of a century, psychoanalysts have learned the necessity and the productivity of an intensified dialogue with psychopharmacology and neuroscience. And Freud himself hoped that someday his science of mind would be tied to its physical substrate, even if his own first efforts to make the connection had led to an impasse.[5] What stands between psychoanalysis and AI is the view that AI is synonymous with rationalism, or rather with the kind of rationalism embodied in the idea of information processing.

If AI has seemed somewhat unitary in its implications for thinking about people, it is because what many observers know as AI is really information processing, a rule-driven, hierachical approach to creating intelligence. But information processing is only one part of a larger picture.

The Two AIs

In the mid-nineteenth century George Boole formalized rules of logical inference in an algebraic form systematic enough that he felt entitled to call his work "The Laws of Thought."[6] Of course, Boole's title reached beyond his achievement, which is far from an all-inclusive model of mind. For one thing, Boole's laws need an external agent to operate them.

Boole's laws are something a person could use, but a computational version of Boole breathes life into his equations. An operator in the form of a computer program is placed within the system. Once there, the operator and the laws can be seen as a functioning model, if not of the mind, at least of a part of the mind.

One major branch of AI research can be described as doing just this— pursuing Boole's project in computational form. Information-processing AI gives active shape to formal propositions to create an embodiment of intelligence as rules and reason. Boole formulated algebraic rules for the transformation of logical propositions. Modern computer science has enlarged the logical and propositional to a more general notion of what it calls information, and it has enlarged algebraic transformation to a more general notion of computational processing. Boole would recognize a kinship between his project and Newell and Simon's way of putting these two advances together in GPS and other programs that laid the foundation for information-processing AI.

But artificial intelligence is not a unitary enterprise. Computation is a stuff out of which many theories can be fashioned. It is true to say that there is not one AI but many. And it is helpful to say that there are essentially two. The first is information processing, its roots in logic, the manipulation of propositions to obtain new propositions, the combination of concepts to obtain new concepts. The second comes from a very different style of work, present from the earliest days of the field but now having increasing influence, to the point of being the focus of attention wherever AI is discussed, from research seminars to popular articles. This second is "emergent AI".

Emergent AI has not been inspired by the orderly terrain of logic. The ideas about machine intelligence that it puts forward are not so much about teaching the computer as about allowing the machine to learn. This AI does not suggest that the computer be given rules to follow but tries to set up a system of independent elements within a computer from whose interactions intelligence is expected to emerge. From this perspective, a rule is not something you give to a computer but a pattern you infer when you observe the machine's behavior, much as you would observe a person's. Its sustaining images are not drawn from the logical but from the biological.

Information processing breathes life into Boole by putting an operator into his sytem, but what it operates on shares the static nature of Boole's propositions. In traditional computers, millions of units of information sit in memory doing nothing as they wait for the central processor to act on them, one at a time. Impatient with this limitation, the goal of emergent AI is "pure" computation. Here, the whole system is dynamic, with no distinction between processors and the information they process. Families of neuronlike entities, societies of anthropomorphic subminds, and sub-subminds are in simultaenous interaction. The goal, no less mythic than the creation of a strand of DNA, is the generation of a fragment of mind.

The two AIs, rule-driven and emergent, logical and biological in their aesthetic, fuel very different fantasies of how to build mind from machine. If information-processing AI is captured by the image of the knowledge engineer, hungry for rules, debriefing a human expert in order to embody that expert's method in algorithms and hardware, emergent AI is captured in the image of the computer scientist, up all night watching the twinkling lights of a computer in the hope that the interaction of "agents" within the machine will create intelligence.

Widely associated with the spirit and substance of the field as a whole (here I have called it the literary stereotype), information processing put AI in a distant relationship to psychoanalysis, whose ideas do not easily translate into rules or algorithms.[7] Indeed, I now turn to how popular notions about AI drawn from information processing suggest that AI is all the things that psychoanalysis is not. My thesis follows directly: when the stuff of AI is expanded to include not only information but also active and interactive inner agents, there is a starting place for a new dialogue between the psychoanalytic and the computer cultures.

. . . . .

Emergent AI and Broad Determination

This biological resonance is illustrated by the perceptron, a pattern-recognition machine designed in the late 1950s and a good first example of emergent AI. Information-processing AI is made out of data and rules. Emergent AI is made out of very different stuff, a stuff most easily captured in anthropomorphic language.

Imagine that you have access to the opinions of a thousand simple-minded meteorologists, each of whom has a different unreliable method of weather forecasting. Each bases a judgment on a fragment of evidence that may or may not be related to predicting rain. How do you form a judgment? A narrowly determined method, in an information-processing system, for example, might be autocratic—identifying the meteorologist with the best track record and going with that vote. Another strategy, both more democratic and more broadly determined, would be to let the majority decide. The perceptron refines the democratic strategy by weighing each vote with a number related to the individual meteorologist's past record.

So, for example, to get a perceptron to recognize a triangle, you show it samples of triangles and nontriangles and make the system "guess." Its first guesses are random. But the perceptron is able to take advantage of signals saying whether its guess is right or wrong to create a voting system in which agents who have guessed right get more weight. Perceptrons are not programmed, but learn from the consequences of their actions.

In the narrowly determined method, you would have complete breakdown if the chosen meteorologist went insane. But in the brain, damage seldom leads to complete breakdown. More often it produces a degradation of performance proportional to its extent. In other words, when things go wrong, the system still works, but not as well as before. Information-processing systems lose credibility as models of mind because they lack this feature; the perceptron shows the graceful degradation of performance typical of the brain. Even with some disabled meteorologists on board, the perceptron still produces the best possible decision based on the subset of functioning actors.

In an information-processing model, intelligent behavior follows from fixed rules. In the perceptron there are none. There is no flow chart, no rule-driver path through the system. Nor are there one-to-one correspondences between information and output. What is important is not what an agent knows but its place in a network, its interactions and connections. The perceptron presents a model of mind as a society in which intelligence grows from the cacophony of competing voices.

In an information-processing model, the concept "rain" would be explicitly represented in the system. In the perceptron the decision "it will rain" is born from interactions among agents, none of whom has a formal concept of rain. Perceptrons shows the emergence of what information processing takes as its raw material. Information processing begins with formal symbols. Perceptrons, like Freud's unconscious, operate on a subsymbolic and a subformal level. And most important for the current discussion, perceptrons rely on the interactions of inner agents, objects within the system.

Object theory is a central aspect of emergent AI and forms the link between AI and new directions of psychoanalytic thought. The inner agents in perceptrons make a bridge to the broad determinism of psychoanalysis. But it is only an opening. After perceptrons and the perceptronlike systems of the 1960s, it took another round in the development of computational ideas before inner objects came to occupy center stage. This is the story to which I now turn—the story of a second generation of emergent AI with an emphasis on inner objects and a new pathway for influence on psychoanalysis.

Emergent AI and Computational Objects

The atmosphere in the AI laboratories of the early 1960s was heady. The work of Norbert Weiner, John von Neumann, and Alan Turing had set off shock waves that were still fresh. The first information-processing programs that emulated fragments of human thought had only recently produced their surprise. Perceptronlike models (and there were many of them, including Oliver Selfridge's "Pandemonium" and Warren McCulloch's "neural nets") led researchers to biologically resonant descriptions of artificial mind. Thoughts were on the ultimate nature of intelligence.

Artificial intelligence researchers saw little reason for a more humble style. On the contrary, AI defined itself as an enterprise of mythic proportions: mind creating mind. In doing so, the field drew a certain kind of person into its culture, not unlike the kind of person drawn into the early circle around Freud. There, too, the enterprise was mythic: the rational understanding the irrational. There, too, it was without precedent or academic security. The first generation of AI researchers, with backgrounds as diverse as mathematics, psychology, economics, and physics, like the first generation of psychoanalysts, had not been trained in "the field" because it did not exist. There was no academic discipline. There were only new worlds to conquer.

In the early 1960s emergent models were as much a part of what seemed exciting in AI as information-processing programs. But for almost a quarter of a century, emergent AI seemed swept aside. In its influence on psychology, AI became almost synonymous with information processing. Newell and Simon developed rule-based systems in purest form—systems that simulated the behavior of people working on a variety of logical problems. Such simulations offered the promise of more—the promise of making artificial mind out of rules. And if you can build mind from rules, then mind can be presumed to have had rules all along. Following this logic, researchers made information-processing models the backbone of cognitive science.

The language of information processing—descriptions of "search," "subroutine," "scripts," and "grammars"—became common currency among psychologists who accepted the idea that "toy programs," little pieces of machine-embodied intelligence, were representative of bigger things to come. Computer programs that could play chess, manipulate blocks, or "converse" with imaginary waiters in imaginary restaurants did more than model small pieces of mental functioning. They supported the idea that the means used to build them, all drawn from the information-processing paradigm, might someday capture the essence of mind. This idea was bolstered by the worldly success of a particular kind of information-processing program—the expert system. In it the AI scientist extracts decision rules from a virtuoso in a field (medical diagnosis, for example) and embeds them in a machine that will then do the diagnosis "for itself."

By the mid-1970s AI was no longer marginal. It had its own academic programs, its own journals, its own conferences. It was well funded because of its value in the marketplace and to the military. Expert systems were used to analyze stock prices, data from oil well drillings, materials from chemical samples. Companies competed to hire AI graduates to start in-house departments. The future of the field became part of a heated discussion about Japanese-American industrial rivalry.

Now AI could promise a more traditional kind of career, much as the medicalization of psychoanalysis paved the way for it to become a professionalized psychiatric specialty. In both psychoanalysis and AI, traditional careers meant new pressure to engage in the kind of work that promised visible results. In psychoanalysis the pressure was to "cure," to work on educational problems, to do "applied psychoanalysis." In AI research the pendulum swung from what had been most mythic about the dreams of the 1950s and early 1960s to what people "knew how to do"— gather rules and code them in computer programs.

But even as the information-processing model reached near-hegemony in the late 1970s, the conditions for something very different were developing. First, there was important technical progress. Computer scientists had long strained against the limitations of the von Neumann computer, in which one processor might manipulate the passive data in a million cells of memory. It had always been obvious that, in principle, the distinction between processor and memory could be abolished by making every cell in the computer an active processor. Doing so, however, had always been prohibitively expensive. But now, projects such as the Connection Machine were realistic enough to be funded. There, the plan is to have a million microprocessors put together to make one computer whose memory and computational power are fully distributed. No longer would there be an operator and the passive material it operated on. Computation was "waking up from the Boolean dream."[8]

Along with hardware that presented fresh possibilities were new ideas about how to program it. The development of programming methodologies with suggestive names like "message passing" and "actor models" created the context for thinking about computational agents in communication. Standard computer programs are lists of instructions in the form of imperatives: "add these numbers," "put the result in memory," "get the content of that memory location." Artificial intelligence programs in Lisp or Prolog operate on more abstract data but still consist of instructions for manipulating information. The first quarter of a century of the development of programming was based on a process language for describing how to pass information from one place to another. But researchers now felt the need to deal with a different kind of event: not the *passing* of something but the *making* of something. By a coincidence that turns out to be highly suggestive for the present discussion, computer scientists called their so-far most prominent response "object-oriented programming."

If you want to simulate a line of customers at a post office counter (in order to know, for example, how much longer the average wait would be if the number of clerks were to be reduced by one), you write a program that creates an internal object that "behaves like" a person in a line at the post office. It advances when the person ahead in the line advances; it knows when it has reached the counter and then proceeds to carry out its transaction. The contrast between this object-oriented approach and traditional programming strategies is dramatic. A traditional Fortran programmer would assign x's and y's to properties of the customers and write computer code to manipulate the variables. Object-oriented programming refers directly to the inner objects that represent the customers in line: x's and y's do not appear.

In object-oriented programming, the programmer makes new objects that, once created, can be "set free" to interact according to the natures with which they have been endowed. The programmer does not specify what the objects will actually do, but rather "who they are."

If something of the "feel" of an information-processing program is captured by the image of the flow chart, something of the "feel" of object-oriented programming is captured by the pictures of file folders, scissors, and wastebasket that appear on the screens of computers with an iconic interface. The icons are a surface reflection of a programming philosophy in which computers are thought of as "electronic puppet shows" and "there are no important limitations to the kind of plays that can be enacted on their screens, nor to the range of costumes or roles that the actors can assume."[9] For mathematicians, the algebraic manipulations in traditional programming have a compelling reality. But for most nonmathematicians, the object-oriented approach has a more direct appeal, the appeal of actors on a stage.

By the early 1980s the coexistence of new parallel hardware and new ideas about objects in programming set the stage for the pendulum to swing away from information processing. The beginning of the decade saw the first of a growing series of papers from very different origins— engineers eager to build new parallel machines, computer scientists eager to try mathematical ideas that could guide new efforts at parallel programming, psychologists looking for new models that had a biological (indeed, a neurological) resonance. Emergent AI had not so much died as gone underground. It reemerged with a vengeance and with a new label: "connectionism." Once again, there would be no distinction between the processor and what it processed. There would be no specified set of operations. There would only be communities of agents in direct interaction with each other.

But proponents of the new theory of connectionism go beyond earlier stages of emergent AI in the steps they want to take away from Boole. For example, the perceptron could not itself generate new objects or elucidate how new objects could emerge. Its agents were programmed by a human acting from the outside. Today's connectionists hope to go further by bringing together parallel machines, the maturation of ideas about how to program them, and most important, a new sense of the central problem facing the field, something that had scarcely been formulated during the 1960s: How are objects created?

Psychoanalytic Objects and Society Models

In the focus on inner objects and their emergence and interaction, AI shares preoccupations that are central to contemporary psychoanalytic theory. As was the case in AI, the development of a psychoanalytic object theory is a later development of the field. It was not where the theory began.

Early psychoanalytic theory was built around the concept of drive, demand that is generated by the body and that provides the energy and goals for all mental activity. But later, when Freud turned his attention to the ego's relations to the external world, the significance and structure of these relations could not easily be framed in drive theory.

By 1917 Freud began to formulate a language to handle these matters. He described a process by which people form inner "objects." In *Mourning and Melancholia,* Freud argued that the sufferings of a melancholic arise from mutual reproaches between the self and an internalized father with whom the self identifies. In this paper Freud described the "taking in" of people (in psychoanalytic parlance, objects, and in this case the father) as part of a pathology, but he later came to the conclusion that this process is part of normal development. Indeed, this is the mechanism for the development of the superego—the taking in, or introjection, of the ideal parent.

According to Freud, we internalize objects because our instincts impel us to. In his work the concept of inner objects needed to coexist with the scaffolding of drive theory. But many psychoanalytic theorists who followed him were less wedded to the drive model. They widened the scope of what Freud meant by "object relations" to the point where we now think of them as a distinctive school. Classical Freudian theory has many overlapping concepts to describe internal objects: memory traces, mental representations, introjects, identifications, and the idea of inner structures such as the superego. The object relations approach is more specific about what we contain. It describes a society of inner agents, or "microminds"—"unconscious suborganizations of the ego capable of generating meaning and experience, i.e. capable of thought, feeling and preception."[10] Relationships with people, "brought inside" as inner entities, are the fundamental building blocks of mental life.

Whereas Freud focused his attention on a single internalized object, the superego, object relations theorists described a richly populated inner world. Psychoanalyst Melanie Klein went so far as to characterize the people that the child brings inside (as well as the representations of parts of the body) as having psychological features, personalities. They can be seen as loving, hating, greedy, envious. Psychoanalyst W. R. D. Fairbairn even reframed the basic Freudian motor for personality development in object relations terms. For Fairbairn, the human organism is not moved forward by Freud's pleasure principle, the desire to reduce drive tension, but rather by its need to form relationships. This constitutes a profound recasting of the psychoanalytic view of the self: people are not fundamentally pleasure-seeking, but object-seeking.

The language that psychoanalysts need to talk about objects—how they are formed, how they interact—is very different from the language they need to talk about drives. In his "Project for a Scientific Psychology," Freud tried to use informationlike terms derived from the description of the reflex arc—the pain fiber carries information to the brain, the motor fiber carries information to the muscle—to talk about memory, instincts, and the flow of psychic energy. But information metaphors break down completely when you use them to talk about inner objects. As in object-oriented programming in computer science, so it is in psychoanalysis. When one talks about objects, the natural metaphors have to do with making something, not carrying something.

In classical psychoanalytic theory a few powerful inner structures—the superego, for example—act on memories, thoughts, and wishes. Object relations theory posits a dynamic system in which the distinction between processor and processed breaks down. The parallel with computation is clear: in both cases there is movement away from a situation in which a few inner structures act on a more passive stuff. Fairbairn replaced the Freudian dichotomies of ego and id, structure and energy with independent agencies within the mind that think, wish, and generate meaning in interaction with each other, much as emergent AI sets free autonomous agents within a computer system.

The development of object relations theory has led psychoanalysts to ask if allegiance to Freud depends on accepting his drive model. Some have tried to preserve Freud's original drive language but to use it in a way that accommodates a new emphasis on object relations—for example, by assigning objects a role in relation to the discharge of drive: they may inhibit, discharge, facilitate, or serve as drive's target. But this reworking of language is less a solution than an attempt to gloss over the problem. It only works if inner objects do not have elaborate properties or if their creation is seen as an occasional event.

But when objects become central to one's understanding of the psyche there is greater pressure to move away from drive theory. Although drive theory has become increasingly sophisticated and open to the discussion of inner objects, the split between a drive approach and an object relations approach is a central division in psychoanalysis today.[11] The division is parallel to the split between information processing and emergent AI. To use Thomas Kuhn's language, object relations theorists are saying that psychoanalysis can no longer proceed as "normal science,"[12] growing by the assimilation of new data into the old theory. For them, object relations is a paradigm shift within psychoanalysis, much as the hypothesis of emergence—that intelligence grows out of the interaction of multiple agents (it is not what you know but who and where you are)—represents a paradigm shift in AI.

Artificial intelligence theorists Marvin Minsky and Seymour Papert have built a computational model that evokes Fairbairn's object relations theory. Their model takes the mind as a society of interacting agents. These agents are anthropomorphized, discussed in the terms one usually reserves for a whole person, but they do not have the complexity of people. Indeed, their model is based (as was the perceptron concept) on these agents being "dumb." Each knows one thing and one thing only. And, like the "voting agents" in the preceptron, their narrowness of vision leads them to very different opinions. The complex structure of behavior or emotion or thought emerges from the conflict of their opposing views.

The most elaborate presentation of this theory, Minsky's book *The Society of Mind,* describes a vast array of agents: censor agents, recognition agents, and anger agents, to name only a few.[13] Not surprisingly, Minsky recognized Freud, who also wrote extensively about censor agents, as a colleague in "society" modeling. More surprisingly, Minsky sees censors as key actors, not only for modeling human thought but also for making intelligent machines.

Minsky's idea of the censor is a dramatic example of the developing resonance between psychoanalysis and emergent AI. Freud's censor protects people from painful thoughts. The extension of this idea to cognitive functioning and to the "thoughts" in a machine does not depend on the assumption that the agents or the system as a whole feels pain. To function coherently, according to Minsky, an intelligent system must develop a certain inattention to its contradictory agent voices. Minsky's formulation is that there cannot be intelligence, artificial or otherwise, without repression. Allen Newell has talked about the necessity for censors in large and complex information-processing systems. But with clear, unambiguous rules stated in advance, an information-processing computer can also do without them. Censors may turn out to be practical, but they are not theoretical necessities in the information-processing paradigm. In the case of society theory, however, censors are intrinsic. Since there cannot be intelligence without contradiction and conflict, only the presence of censors allows intelligence to emerge.

In this, society and Freudian theory join on an important point. Freud did not "discover" the unconscious. His contribution was the elaboration of a *dynamic* unconscious. What is unconscious is not simply forgotten, old, or irrelevant to current functioning. It is repressed. Powerful forces keep it down, and for good reason. Similarly, for Minsky, what is repressed in the computational machine and in what he has called the human "meat machine" *needs* to be repressed.

Freud wrote about the effects of the repression of frightening, emotion-laden experience. Minsky extends Freud's ideas to the cognitive domain.

"A thinking child's mind . . . [needs no one] to tell it when some paradox engulfs and whirls it into a cyclone." Paradox, argues Minsky, is as dangerous as the primal scene. The child knows it is in the presence of a threat when it is asked to sketch the nonexistent boundaries between the oceans and the seas or to consider questions about the chicken and the egg, about what came before the start of time, and about where the edge of space is. Minsky adds: "And what of sentences like *'This statement is false,'* which can throw the mind into a spin? I don't know anyone who recalls such incidents as frightening. But then, as Freud might say, this very fact could be a hint that the area is subject to censorship."[14]

Minsky feels that the notions of "cognitive repression" and the "cognitive unconscious" will allow us to go beyond Freud. He uses Freud's discussion of jokes as an example. Freud's 1905 work on jokes explained that inner censors serve as barriers against forbidden thoughts. Most jokes are stories designed to fool the censors. It is a way to enjoy a prohibited wish. This is why so many jokes involve taboos concerning cruelty and sexuality. But it troubled Freud that this theory did not easily account for "nonsense jokes." One of Freud's hypotheses about the power of the nonsense joke was that senselessness reflects "a wish to return to carefree childhood, when one was permitted to think without any compulsion to be logical." The idea of the cognitive unconscious supports this view: paradox and senselessness need to be repressed in the process of developing emergent intelligence, whether in machines or in people. Absurd results of reasoning are taboo, as threatening as sex. The censors work as hard to suppress them; they have no innocence.

.   .   .   .   .   .

Psychoanalytic Culture and Computer Culture

Psychological cultures do not exist only in the world of professionals. Artificial intelligence and psychoanalysis set the context in which professional psychologists and the amateur psychologists we all are think about thinking. From a sociological perspective on this wider psychological culture, object theories make ideas in AI and psychoanalysis more "appropriable," easier for people to take up as ways of thinking about themselves, than theories about information or drive. In other words, object theories give psychoanalysis and AI a greater presence as philosophies in everyday life. Fairbairn's dense texts and the mathematical theory of connectionism might not be any more accessible to lay thinking than technical papers on information processing or on psychoanalytic drives. But when object-oriented theories are popularized and move out into the general culture, they have a special appeal. Ideas about objects and agents are more concrete than ideas about drives and flowcharts. They are seductive because it is easy to "play" with them. And they speak to a common problem. We all have the experience of not feeling completely "at one" with ourselves: inner voices offer conflicting advice, reassurance, and chastisement. These experiences are easily and satisfyingly translated into a drama of inner objects.

Freudian ideas about slips of the tongue became well known and gained wide acceptance for reasons that had little to do with positive assessments of their scientific validity. Freudian slips became part of the wider psychological culture because they made it easy to play with what might be hidden behind them. The slips are almost tangible ideas. They are manipulable. Slips are appealing as objects to think with. You can analyze your own slips and those of your friends. The theory of slips provided a way for psychoanalytic ideas to become part of everyday life. They helped to make psychoanalytic theory appropriable.

A Freudian perspective on the appropriability of psychoanalytic ideas might go further to suggest that the theory of significant slips is appealing because it puts us in immediate contact with the taboo. We are afraid of the sexual and aggressive sides of our natures, but we want to be in touch with them as well. Psychoanalytic ideas give us a way to play with what is forbidden. Similarly, we are afraid to think of ourselves as machines, yet we want to find a way to acknowledge this very real, if disturbing, part of our experience. Playing with AI, with the idea of the mind as computer, makes this possible. Now, playing with computational and psychoanalytic theories of objects and agents allows us to go even further. The idea of agents gives us a way to acknowledge the experience of fragmentation. The rational bias in our culture presents consistency and coherency as natural, but feelings of fragmentation abound. Indeed, it has been argued that they are a contemporary cultural malaise.[15] Theories within psychoanaysis and AI that speak simply and dramatically to the experience of a divided self have particular power.

In the past the computer culture and the psychoanalytic culture have been separate. In the main, psychoanalytic ideas for thinking about the self were congenial to people who had little contact with computational ones. If and when members of the psychoanalytic culture met computational models of mind, they were most likely to be information-processing models that seemed out of step with a psychoanalytic outlook. These models described sequences, not associations, and their model of determination was narrow rather than wide. But increasingly, the computational ideas put forward and reported in the popular, as well as the academic, press are not about rules and information but about agents, connections, and societies of mind. These new metaphors have a biological aesthetic—they are the kind of things that could be going on in a brain. They suggest broad determination and dynamic repression. They describe a system in conflict. And, most important, they resonate with the psychoanalytic ideas that are currently abroad, ideas not about drives and their vicissitudes but about objects and their interactions.

When the computer presence relegitimated the idea of memory, it was reinforcing an idea about psychology that predated computation. But ideas about recursion and agents are not precomputational. Dare one speculate what will pass between computation and our psychological culture if AI finds a voice finally divorced from what was static in logic and if psychoanalysis finds a voice finally divorced from the issues of nineteenth-century drive theory?

Notes

1. Many of the ideas in this paper emerged in a series of conversations with Seymour Papert, a collaborator in the development of my notion of the role of sustaining myths in the sociology of the sciences of mind.
2. Cited in Jonathan Miller, *States of Mind* (New York: Pantheon, 1983), p. 23.
3. Cited in Sherry Turkle, *The Second Self: Computers and the Human Spirit* (New York: Simon and Schuster, 1984), p. 256.
4. Cited in Jeremy Bernstein, *Science Observed* (New York: Basic Books, 1982), pp. 110–11.
5. Sigmund Freud, "Project for a Scientific Psychology," *The Standard Edition of the Complete Psychological Works of Sigmund Freud*, vol. 1, trans. and ed. James Strachey (London: Hogarth Press, 1960).
6. George Boole, *The Laws of Thought*, vol. 2 of *His Collected Works* (La Salle, Ill.: Open Court Publishing Company, 1952).
7. A suggestive effort to construct psychoanalytic algorithms was made by French psychoanalyst Jacques Lacan in his theory of the *mathèmes*. The power of this idea derives from its effort to legitimate systematicity and a closer relationship with science in psychoanalytic studies. See Sherry Turkle, *Psychoanalytic Politics: Freud's French Revolution* (New York: Basic Books, 1978).
8. This phrase is borrowed from Douglas R. Hofstadter, who discusses computation and the Boolean aesthetic in "Waking Up From the Boolean Dream, or Subcognition as Computation," in *Metamagical Themas: Questing for the Essence of Mind and Pattern* (New York: Basic Books, 1985).
9. Alan Kay, "Software's Second Act," *Science 85* (November 1985): 122.
10. Thomas H. Ogden, "The Concept of Internal Object Relations," *The International Journal of Psycho-Analysis* 64 (1983):227.

11. See Jay R. Greenberg and Stephen A. Mitchell, *Object Relations in Psychoanalytic Theory* (Cambridge: Harvard University Press, 1983).
12. Thomas Kuhn, *The Structure of Scientific Revolutions*, 2d ed. (Chicago: University of Chicago Press, 1970).
13. Marvin Minsky, *The Society of Mind* (New York: Simon and Schuster, 1987).
14. Ibid., 183. Fieldwork with children and computers is rich in examples of the kind of fright that Minsky expects. For example, an incident where it was evoked by a first contact with recursion is reported in Sherry Turkle, *The Second Self*. Interviews with adults on early experiences also reveal many such memories—fear of prisms, of mirrors reflecting mirrors, fear of questions such as "How far away are the stars?"
15. See, for example, Christopher Lasch, *The Culture of Narcissism* (New York: Norton, 1979).

— 14 —

Why People Think Computers Can't

Marvin Minsky

Most people are convinced computers cannot think. That is, *really* think. Everyone knows that computers already do many things that no person could do without "thinking." But when computers do such things, most people suspect that there is only an illusion of thoughtful behavior, and that the machine

- doesn't know what it's doing.
- is only doing what its programmer told it to.
- has no feelings. And so on.

The people who built the first computers were engineers concerned with huge numerical computations: that's why the things were *called* computers. So, when computers first appeared, their designers regarded them as nothing but machines for doing mindless calculations.

Yet even then a fringe of people envisioned what's now called "artificial intelligence"—or "AI" for short—because they realized that computers could manipulate not only number but also *symbols*. That meant that computers should be able to go beyond arithmetic, perhaps to imitate the information processes that happen inside minds. In the early 1950s, Turing began a chess program, Oettinger wrote a learning program, Kirsch and Selfridge wrote vision programs, all using the machines that were designed just for arithmetic.

Today, surrounded by so many automatic machines, industrial robots, and the R2-D2s of *Star Wars* movies, most people think AI is much more advanced than it is. But still, many "computer experts" don't believe that machines will ever "really think." I think those specialists are too used to explaining that there's nothing inside computers but little electric currents. This leads them to believe that there can't be room left for anything else—like minds, or selves. And there are many other reasons why so many experts still maintain that machines can never be creative, intuitive, or emotional, and will never really think, believe, or understand anything. This essay explains why they are wrong.

Can Computers Do Only What They're Told?

We naturally admire our Einsteins and Beethovens, and wonder if computers ever could create such wondrous theories or symphonies. Most people think that "creativity" requires some mysterious "gift" that simply cannot be explained. If so, then no computer can create—since, clearly, anything machines can do can be explained.

To see what's wrong with that, we'd better turn aside from those outstanding works our culture views as very best of all. Otherwise we'll fall into a silly trap. For, until we first have some good ideas of how we do the *ordinary* things—how ordinary people write ordinary symphonies—we simply can't expect to understand how great composers write great symphonies! And obviously, until we have some good ideas about *that,* we'd simply have no way to guess how difficult might be the problems in composing those most outstanding works—and then, with no idea at all of how they're made, of course they'll seem mysterious! (As Arthur Clarke has said, *any* technology sufficiently advanced seems like magic.) So first we'd better understand how people and computers might do the ordinary things that we all do. (Besides, those skeptics should be made to realize that their arguments imply that ordinary people can't think, either.) So let's ask if we can make computers that can use ordinary common sense; until we get a grip on that we hardly can expect to ask good questions about works of genius.

In a practical sense, computers already do much more than their programmers tell them to. I'll grant that the earliest and simplest programs were little more than simple lists and loops of commands like *"Do this. Do that. Do this and that and this again until that happens."* That made it hard to imagine how more could emerge from such programs than their programmers envisioned. But there's a big difference between "impossible" and "hard to imagine." The first is about *it;* the second is about *you!*

Most people still write programs in languages like Basic and FORTRAN, which make you write in that style—let's call it "do now" programming. This forces you to imagine all the details of how your program will move from one state to another, from one moment to the next. And once you're used to thinking that way, it is hard to see how a program could do anything its programmer didn't think of—because it is so hard to make that kind of program do *anything* very interesting. Hard, not impossible.

Then AI researchers developed new kinds of programming. For example, the "General Problem Solver" system of Newell, Shaw, and Simon lets you describe processes in terms of statements like "if you're on the wrong side of a door, go through it"—or, more technically, "if the difference between what you have and what you want is of kind D, then try to change that difference by using method M."[1] Let's call this "do whenever" programming. Such programs automatically apply each rule whenever it's applicable—so the programmer doesn't have to anticipate when that might happen. When you write in this style, you still have to say what should happen in each "state" the process gets into—but you don't have to know in advance when each state will occur.

You also could do such things with the early programming language COMIT, developed by Yngve at MIT, and the SNOBOL language that followed it. Today, that programming style is called "production systems."[2] The mathematical theory of such languages is explained in my book.[3]

That "General Problem Solver" program of Newell and Simon was also a landmark in research on artificial intelligence because it shows how to write a program to solve a problem that the programmer doesn't know how to solve. The trick is to tell the program what kinds of things to TRY; you need not know which one actually will work. Even earlier, in 1956, Newell, Shaw, and Simon developed a computer program that was good at finding proofs of theorems in mathematical logic—problems that college students found quite hard—and it even found some proofs that were rather novel. (It also showed that computers could do "logical reasoning"—but this was no surprise, and since then we've found even more powerful ways to make machines do such things.) Later, I'll discuss how this relates to the problem of making programs that can do "commonsense reasoning."

Now, you might reply, "Well, everyone knows that if you try enough different things at random, of course, eventually, you can do anything. But if it takes a million billion trillion years, like those monkeys hitting random typewriter keys, that's not intelligence at all. That's just evolution or something."

That's quite correct—except that the GPS system had a real difference—it didn't do things randomly. To use it, you also had to add another kind of knowledge—"advice" about when one problem-state is likely to be better than another. Then instead of wandering around at random, the program can seek the better states; it sort of feels around, the way you'd climb a hill in the dark, by always moving up the slope. This makes its "search" seem not random at all, but rather purposeful. The trouble—and it's very serious—is that it can get stuck on a little peak, and never make it to the real summit of the mountain.

Since then, much AI research has been aimed at finding more "global" ways to solve problems, to get around that problem of getting stuck on little peaks which are better than all the nearby spots, but worse than places that can't be reached without descending in between. We've discovered a variety of ways to do this, by making programs take larger views, plan further ahead, reformulate problems, use analogies, and so forth. No one has discovered a "completely general" way to always find the very highest peak. Well, that's too bad—but it doesn't mean there's any difference here between men and machines—since people, too, are almost always stuck on local peaks of every kind. That's life.

Today, most AI researchers use languages like LISP, that let a programmer use "general recursion." Such languages are even more expressive than "do whenever" languages, because their programmers don't have to foresee clearly either the kinds of states that might occur or when they will occur; the program just constrains how states and structures will relate to one another. We could call these "constraint languages."[4]

Even with such powerful tools, we're still just beginning to make programs that can learn and can reason by analogy. We're just starting to make systems that will learn to recognize which old experiences in memory are most analogous to present problems. I like to think of this as "do something sensible" programming. Such a program would remember a lot about its past so that, for each new problem, it would search for methods like the ones that worked best on similar problems in the past. When speaking about programs that have *that* much self-direction, it makes no sense at all to say "computers do only what they're told to do," because now the programmer knows so little of what situations the machine may encounter in its future—or what it will remember from its past.

A generation later, we should be experimenting on programs that *write better programs to replace themselves.* Then at last it will be clear how foolish was our first idea—that never, by their nature, could machines create new things. This essay tries to explain why so many people have guessed so wrongly about such things.

Could Computers Be Creative?

I plan to answer "no" by showing that there's no such thing as "creativity" in the first place. I don't believe there's any substantial difference between ordinary thought and creative thought. Then why do we think there's a difference? I'll argue that this is really not a matter of what's in the mind of the artist—but of what's in the mind of the critic; the less one understands an artist's mind the more creative seems the work the artist does.

I don't blame anyone for not being able to do the things creative people do. I don't blame them for not being able to explain it, either. (I don't even blame them for thinking that if creativity can't be explained, it can't be mechanized; in fact, I agree with that.) But I do blame them for thinking that, just because they can't explain it themselves, then no one *ever* could imagine how creativity works. After all, if you can't understand or imagine how something might be done at all, you certainly shouldn't expect to be able to imagine how a machine could do it!

What is the origin of all those skeptical beliefs? I'll argue first that we're unduly intimidated by admiration of our Beethovens and Einsteins. Consider first how hard we find it to express the ways we get our new ideas—not just "creative" ones but everyday ideas. The trouble is, when focusing on creativity, we're prone to notice it when others get ideas that we don't. But when we get our own ideas, we take them for granted, and don't ask where we "get" them from. Actually we know as little—maybe less—of how we think of ordinary things. We're simply so accustomed to the marvels of everyday thought that we never wonder—until unusual performances attract attention. (Of course, our superstitions about creativity serve other needs, e.g., to give our heroes special qualities that justify the things we ordinary losers cannot do.)

Should we suppose that outstanding minds are any different from ordinary minds at all, except in matters of degree? I'll argue both ways. I'll first say "No, there's nothing special in a genius, just some rare, unlikely combination of virtues—none very special by itself." Then, I'll say "Yes, but in order to *acquire* such a combination, you need at least a lucky accident—and maybe something else—to make you able, in the first place, to acquire those other skills."

I don't see any mystery about that mysterious combination itself. There must be an intense concern with some domain. There must be great proficiency in that domain (albeit not in any articulate, academic sense). And one must have enough self-confidence, immunity to peer pressure, to break the grip of standard paradigms. Without that one might solve problems just as hard—but in domains that wouldn't be called "creative" by one's peers. But none of those seems to demand a basic qualitative difference. As I see it, any ordinary person who can understand an ordinary conversation must have already in his head most of the mental power that our greatest thinkers have. In other words, I claim that "ordinary, common sense" already included the things it takes—when better balanced and more fiercely motivated—to make a genius. Then what makes those first-raters so much better at their work? Perhaps two kinds of difference-in-degree from ordinary minds. One is the way such people *learn* so many more and deeper skills.

The other is the way they learn to *manage* using what they learn. Perhaps beneath the surface of their surer mastery, creative people also have some special administrative skills that better knit their surface skills together. A good composer, for example, has to master many skills of phrase and theme—but those abilities are shared, to some degree, by everyone who *talks* coherently. An artist also has to master larger forms of form—but such skills, too, are shared by everyone who knows good ways to "tell a tale." A lot of people learn a lot of different skills—but few combine them well enough to reach that frontal rank. One minor artist masters fine detail but not the larger forms; another has the forms but lacks technique.[5]

We still don't know why those "creative masters" learn so much so well. The simplest hypothesis is that they've come across some better way to choose how and what to learn! What might the secret be? The simplest explanation: such a "gift" is just some "higher-order" kind of expertise—of knowing how to gain and use one's other skills. What might it take to learn *that*? Obvious: *one must learn to be better at learning!*

If that's not obvious, perhaps our culture doesn't teach how to think about learning. We tend to think of learning as something that just happens to us, like a sponge getting soaked. But learning really is a growing mass of skills: we start with some but have to learn the rest. Most people never get deeply concerned with acquiring increasingly more advanced learning skills. Why not? Because they don't pay off right away! When a child tries to spoon sand into a pail, the child is mostly concerned with filling pails and things like that. Suppose, though, by some accident, a child got interested in how that pail-filling activity itself improved over time, and how the mind's inner dispositions affected that improvement.

If only once a child became involved (even unconsciously) in how to learn better, then that could lead to exponential learning growth.

Each better way to learn to learn would lead to better ways to build more skills—until that little difference had magnified itself into an awesome, qualitative change. In this view, first-rank "creativity" could be just the consequence of childhood accidents in which a person's learning gets to be a little more "self-applied" than usual.[6] If this image is correct, then we might see creativity happen in machines, once we begin to travel down the road of making machines that learn—and learn to learn better.

Then why is genius so rare? Well, first of all, the question might be inessential, because the "tail" of every distribution must be small by definition. But in the case of self-directed human thought-improvement, it may well be that all of us are already "close to some edge" of safety in some sociobiological sense. Perhaps it's really relatively easy for certain genes to change our brains to make them focus even more on learning better ways to learn. But quantity is not the same as quality—and, possibly, no culture could survive in which each different person finds some wildly different, better way to think! It might be true, and rather sad, if there were genes for genius that weren't hard at all for evolution to come upon—but needed (instead of nurturing) a frequent, thorough weeding out, to help us keep our balance on some larger social scale.

. . . . .

Could A Computer Know What Something Means?

We can't think very well about meaning without thinking about the meaning of something. So let's discuss what numbers mean. And we can't think about what numbers mean very well without thinking about what some particular number means. Take Five. Now, no one would claim that Bobrow's algebra program could be said to understand what numbers "really" are, or even what Five really is. It obviously knows something of arithmetic, in the sense that it can find sums like "5 plus 7 is 12." The question is—does it understand numbers in any *other* sense—say, what are 5 or 7 or 12—or, for that matter, what are "plus" or "is"? Well, what would *you* say if I asked, "What is Five"? I'll argue that the secret lies in that little word *"other."*

Early this century, the philosophers Russell and Whitehead suggested a new way to define a number. "Five," they said, *is the set of all possible sets with five members.* This set includes every set of Five ballpoint pens, and every litter of Five kittens. The trouble was, this definition threatened also to include sets like "these Five words" and even "the Five things that you'd least expect." Sets like those led to so many curious inconsistencies and paradoxes that the theory had to be doctored so that these could not be expressed—and *that* made the theory, in its final form, too complicated for any practical use (except for formalizing mathematics, where it worked very well indeed). But, in my view, it offers little promise for capturing the meanings of everyday common sense. The trouble is with its basic goal: finding for each word some single rigid definition. That's fine for formalizing mathematics. But for real life, it ignores a basic fact of mind: what something means to me depends to some extent on everything else I know—and no one else knows just those things in just those ways.

But, you might complain, when you give up the idea of having rigid definitions, don't you get into hot water? Isn't ambiguity bad enough; what about the problems of "circular definitions," paradoxes, and inconsistencies? Relax! We shouldn't be *that* terrified of contradictions: let's face it, most of the things we people think we "know" are crocks already overflowing with contradictions; a little more won't kill us. The best we can do is just be reasonably careful—and make our machines careful, too—but still there are always chances of mistakes. That's life.

Another kind of thing we scientists tend to hate are circular dependencies. If every meaning depends on the mind it's in—that is, on all other meanings in that mind—then there's no place to start. We fear that when some meanings form such a circle, then there would be no way to break into the circle, and everything would be too subjective to make good science.

I don't think that we should fear the fact that our meanings and definitions run around in vicious circles, each depending on the others. There's still a scientific way to deal with this: just start making new kinds of theories—about those circles themselves! You don't *have to* break into them—you only need to have good theories *about* them. Of course, this is hard to do, and likely to get complicated. It was to avoid complication that all those old theories tried to suppress the ways that meanings depend on one another. The trouble is, that lost all the power and the richness of our wondrous meaning-webs! Let's face another fact: our minds really *are* complicated, perhaps more so than any other structure science ever contemplated. So we can't expect the old ideas to solve all the new problems.

Besides, speaking of breaking into the meaning-circle, many science-fiction writers have pointed out that no one ever really *wants* to get oneself inside another mind. No matter if that's the only hope of perfect communication—of being absolutely sure you understand exactly, at every level of nuance, what other people mean. The only way you could do that is by becoming exactly like that person but even then the game is lost, since then you couldn't understand any more (perfectly, that is) just what it was that your old self had tried to say.

What Is a Number, That a Mind Might Know It?

Now let's return to what numbers mean. This time, to make things easier, we'll think about Three. What could we mean by saying that Three hasn't any single, basic definition, but is a web of different processes that depend upon each other? Well, consider all the roles "Three" plays.

One way a person tells when there's a Three is to recite "One, Two, Three," while pointing to the different things. Of course, while doing that, you have to manage to (1) touch each thing once and (2) not touch any twice. One easy way to do *that* is to pick up one object, as you say each counting-word, and remove it. Soon, children learn to do that in their minds or, when it's too hard to keep track, to use some physical technique like finger-pointing.

Another way to tell a Three is to establish some Standard Set of Three things. Then you bring *your* set of things there and match them one-to-one: if all are matched and you have nothing left, then you had Three. And, again, that "Standard Three" would work quite well. To be sure, this might make it hard to tell which method you're using—"counting" or "matching"—at the moment. Good. It really doesn't matter, does it? (Except, perhaps, to philosophers.) For doers, it's really good to be able to shift and slip from one skill-process to another without even realizing it.

Another way to know a Three is by perceptual groups. One might think of Three in terms of arranging some objects into groups of One and Two. This, too, you can do mentally, without actually moving the objects, or you might lay them out on a table.

. . . . .

Which way is right—to count, or match, or group—which is the "real" meaning of a number? The very question shows how foolish is any such idea: each structure and its processes have both their own uses, and ways to support the others. This is what makes the whole into a powerful, versatile skill-system. Neither chicken nor egg need come first; they both evolve from something else.

It's too bad that so many scientists and philosophers despise such networks and only seek to construct simple "chains" of definitions in which each new thing depends only on other things that have been previously defined. That is what has given "reductionism" a bad name. The commonsense meaning of Three is not a single link in one long chain of definitions in the mind. Instead, we simply let the word activate some rather messy web of different ways to deal with Threes of things, to use them, to remember them, to compare them, and so forth. The result of this is great for solving problems since, when you get stuck with one sense of meaning, there are many other things to try and do. If your first idea about Three doesn't do some job, in some particular context, you can switch to another. But if you use the mathematician's way, then, when you get into the slightest trouble, you get completely stuck!

If this is so, then why *do* mathematicians prefer their single chains to our multiply-connected knowledge-nets? Why would anyone prefer each thing to depend upon as few other things as possible instead of as many as possible? The answer has a touch of irony: mathematicians *want* to get stuck! This is because, as mathematicians, we *want* to be sure above all that as soon as anything goes wrong, we'll be the first to notice it. And the best way to be sure of that is to make everything collapse at once! To mathematicians, that sort of fragility is *good,* not bad, because it helps us find out if any single thing that we believe is inconsistent with any of the others. This ensures absolute consistency—and that is fine in mathematics. It simply isn't good psychology.

Perfect consistency is not so relevant to real life because—let's face it—minds will *always* have beliefs that turn out to be wrong. That's why our teachers use a very wrong theory of how to understand things, when they shape our *children's* mathematics, not into robust networks of ideas, but into those long, thin, fragile chains or shaky towers of professional mathematics. A chain breaks whenever there's just one single weak link, just as a slender tower falls whenever we disturb it just a little. And this could happen to a child's mind, in mathematics class, who only takes a moment to watch a pretty cloud go by.

The purposes of children, and of other ordinary people, are not the same as those of mathematicians and philosophers. They need to have as few connections as can be, to simplify their careful, accurate analyses. In real life the best ideas are those robust ones that connect to as many other ideas as possible. And so, there is a conflict when the teachers start to consult those academic technicians about curricula. If my theory's right, they're not just bad at that; they're just about as bad at that *as possible*! Perhaps this helps explain how our society arranges to make most children terrified of mathematics. We think we're making things easier for them to find what's right, by managing to make things go all wrong almost all the time! So when our children learn about numbers (or about anything else) I would prefer that they build meshy networks in their minds, not slender chains or flimsy towers. Let's leave that for when they take their graduate degrees.

For learning about Two, a preschool child learns in terms of symmetry and congruence—two hands, two feet, two shoes—one doesn't need to count or refer to some standard ideal set. (It is only later that one learns that, every time you count, you get the same result.) We learn of Three in terms of rhymes and tales of Threes of Bears and Pigs and Turtle Doves (whatever those might be) that tell of many different *kinds* of Threes.

Note that those Bears are two and one, Parents and Child, while their famous bowls of porridge make a very different kind of Three—"too hot, too cold, just right"—that shows the fundamental dialectic compromise of two extremes. (So do the bears' forbidden beds—too hard, too soft, just right.) Just think of all the different kinds of Threes that confront real children in the real world, and the complex network of how they all relate to one another in so many different, interesting ways. There simply isn't any sense to choosing one of them to be "defined" so as to come before the rest.

Our culture tries to teach us that a meaning really ought to have only a single, central sense. But if you programmed a machine that way, then, of course it couldn't really understand. Nor would a person either, since when something has just one meaning then it doesn't really "mean" at all because such mental structures are so fragile and so easy to get stuck that they haven't any real use. A network, though, yields many different ways to work each problem. And then, when one way doesn't work and another does, you can try to figure out why. In other words, the network lets you *think,* and thinking lets you build more network. For only when you have several meanings in a network is there much to think about; then you can turn things around in your mind and look at them from different perspectives. When you get stuck, you can try another view. But when a thing has just one meaning, and you get stuck, there's no way out except to ask Authority. That's why networks are better than logical definitions. There never is much meaning until you join together many partial meanings; and if you have only one, you haven't any.

Could a Computer Know About the Real World?

Is there some paradox in this idea, that every meaning is built on other meanings, with no special place to start? If so, then isn't all a castle built on air? Well, yes and no. Contrary to common belief, *there's really nothing wrong at all with circular definitions.* Each part can give some meaning to the rest. There's nothing wrong with liking several different tunes, each one the more because it contrasts with the others. There's nothing wrong with ropes—or knots, or woven cloth—in which each strand helps hold the other strands together—or apart! There's nothing very wrong, in this strange sense, with having one's entire mind a castle in the air!

But then, how could such a mind have any contact with reality. Well, maybe this is something we must always face in any case, be we Machine or Man. In the human condition, our mental contact with the real world is really quite remote. The reason we don't notice this, and why it isn't even much of a practical problem, is that the sensory and motor mechanisms of the brain (that shape the contents of, at least, our infant minds) ensure enough developmental correspondence between the objects we perceive and those that lie out there in raw reality: and that's enough so that we hardly ever walk through walls or fall down stairs.

But in the final analysis, our idea of "reality" itself is rather network-y. Do triangles "exist" or are they only Three of Lines that share their vertices? What's real, anyway, about a Three—in view of all we've said; "reality" itself is also somewhat like a castle in the air. And don't forget how totally some minds, for better or usually for worse, *do* sometimes split away to build their own imaginary worlds. Finally, when we build intelligent machines we'll have a choice: either we can constrain them as we wish to match each and every concept to their outside-data instruments, or we can let them build their own inner networks and attain a solipistic isolation totally beyond anything we humans could conceive.

To summarize: of course computers couldn't understand a real world— or even what a number is—were they confined to any single way of dealing with them. But neither then could child or philosopher. It's not a question of computers at all, but only of our culture's foolish quest for meanings that can stand all by themselves, outside of any mental context. The puzzle comes from limitations of the way our culture teaches us to think. It gives us such shallow and simplistic concepts of what it means to "understand" that—probably—no entity could understand *that* way. The intuition that our public has—that if computers worked that way, they couldn't understand—is probably quite right! But this only means we musn't program our machines that way.

Can a Computer Be Aware of Itself?

> Even if computers do things that amaze us, they're just mechanical. They can't believe or think, feel pain or pleasure, sorrow, joy. A computer can't be conscious, or self-aware—because it simply has no self to feel things with.

Well. What do you suppose happens in your head when someone says a thing like that to *you?* Do you understand it? I'll demonstrate that this problem, too, isn't actually about computers at all. It isn't even about "understanding." This problem is about you. That is, it turns around that little word "you." For when we feel that when we understand something, we also seem to think there must be some agent in our heads that "does" the understanding. When we believe something, there must be someone in our heads to do the believing. To feel, someone must do the feeling.

Now, something must be wrong with that idea. One can't get anywhere by assuming there's someone inside oneself—since then there'll have to be another someone inside that one, to do *its* understanding for it, and so on. You'll either end up like those sets of nested Ukrainian Russian dolls, or else you'll end up with some "final" inner self. In either case, as far as I can see, that leaves you just exactly where you started.[7] So what's the answer? The answer is—we must be asking the wrong question: perhaps we never had anything like "self-awareness" in the first place—but only thought we had it! So now we have to ask, instead—why do we *think* we're self-aware?

My answer to this is that we are *not,* in fact, really self-aware. Our self-awareness is just illusion. I know that sounds ridiculous, so let me explain my argument very briefly. We build a network of half-true theories that gives us the illusion that we can see into our working minds. From those apparent visions, we think we learn what's really going on there. In other words, much of what we "discover" about ourselves, by these means, is just "made up." I don't mean to say, by the way, that those made-up ideas are *necessarily* better than or worse than theories we make about all other things that we don't understand very well. But I do mean to say that when we examine carefully the quality of the ideas most people have about their selves—ideas they got by using that alleged "self-awareness"—we don't find that quality very good at all.

By the way, I'm not saying that we aren't aware of sound and sights, or even of thoughts and ideas. I'm only saying that we aren't "self-aware." I'm also sure that the structures and processes that deserve to be called "self" and "awareness" are *very* complicated concept-networks. The trouble is that *those* are hardly at all like what we think they're like. The result is that in this area our networks don't fit together well enough to be useful for understanding our own psychology very well.

Now let's try to see what some of the meanings we attach to "self" are like. When you and I converse, it makes perfect sense for me to call you "you" and to call me "me." That's fine for ordinary social purposes, that is when neither of us cares about the fine details of what is going on inside our minds. But everything goes wrong at once as soon as one's concerned with that—because those you's and me's conceal most the of intricacy of what's inside our minds that really do the work. The very purpose of such words like "you" and "self" is to symbolize away what we don't know about those complex and enormous webs of stuff inside our head.

When people talk, the physics is quite clear: I shake some air, which makes your eardrums move, and some "computer" in your head converts vibrations into, say, little "phoneme" units. Next, oversimplifying, these go into strings of symbols representing words, so now somewhere in your head you have something that "represents" a sentence. The problem is, what happens next?

In the same way, when you see something, the waves of light excite your retinas, and this causes signals in your brain that correspond to texture fragments, bits of edges, color patches, or whatever. Then these, in turn, are put together (somehow) into a symbol-structure that "represents" a shape or outline, or whatever. What happens then?

We argued that it cannot help to have some inner self to hear or read the sentence, or little person, hiding there to watch that mental television screen, who then proceeds to understand what's going on. And yet that seems to be our culture's standard concept of the self. Call it the "Single Agent" theory: that inside every mind resides a certain special "self" that does the real mental work. Since this concept is so popular, we ought to have a theory of why we all believe such a ridiculous theory!

In fact, it isn't hard to see why we hold onto such ideas—once we look past the single self and out into society. For then we realize how valuable to us is this idea of Single Agent Self—no matter how simplistic, scientifically—in social matters of the greatest importance. It underlies, for instance, all the principles of all our moral systems; without it, we could have no canons of *responsibility,* no sense of blame or virtue, no sense of right or wrong. In short, without the idea of a Single Self, we'd scarcely have a culture to begin with. It also serves a crucial role in how we frame our plans and goals, and how we solve all larger problems—for, what *use* could solving problems be, without that idea of a self to savor and exploit their solutions.

And, furthermore, that image of a single self is central to the very ways we knit our personalities together—albeit though, as Freud has pointed out, it's not the image of us as we *are* that counts, but as we'd like to *be,* that makes us grow. That's why I didn't mean to say that it is bad to have illusions for our Selves. (Why, what could one prefer to that, anyway?) And so, in short, no matter that it bollixes up our thinking about thinking: I doubt if we could survive without that wonderful idea of Single Self.

To build good theories of the mind, we'll have to find a better way. We find that hard to do because the concept of the Single Self *is* so vitally important for those other reasons.[8] But, just as science forced us to accept the fact that what we think are single things—like rocks or mice or clouds—must sometimes be regarded as complicated other kinds of structures, we'll simply have to understand that Self, too, is no "elementary particle," but an extremely complicated construction.

We should be very used to this. There's nothing wrong with the idea of Single Houses, either. They keep us warm and dry, we buy them and sell them, they burn down or blow away; they're "things" all right but just up to a point. But when you really want to understand how Houses work, then you must understand that Houses aren't really "things" at all but constructions. They're made of beams and bricks and nails and stuff like that, and they're also made of forces and vectors and stresses and strains. And in the end, you can hardly understand them at all without understanding the intentions and purposes that underlie the ways they're designed.

So this wonderful but misleading Single Agent Self idea leads people to believe machines can't understand, because it makes us think that understanding doesn't need to be constructed or computed—only handed over to the Self—a thing that, you can plainly see, there isn't room for in machines.

Can a Computer Have a Self?

Now we can watch the problem change its character, before our eyes, the moment that we change our view. Usually, we say things like this:

> A computer can't do (xxx), because it has no self.

And such assertions often seem to make perfect sense—until we shed that Single Agent view. At once those sayings turn to foolishness, like this:

> A computer can't do (xxx), because all a computer can do is execute incredibly intricate processes, perhaps millions at a time, while constructing elaborately interactive structures on the basis of almost unimaginably ramified networks of interrelated fragments of knowledge.

It doesn't make so much sense any more, does it? Yet all we did was face one simple, complicated fact. The second version shows how some of our skepticism about computers emerges from our unwillingness to imagine what might happen in the computers of the future. The first version shows how some of our skepticism emerges from our disgracefully empty ideas about how *people* really work, or feel, or think.

Why are we so reluctant to admit this inadequacy? It clearly isn't just the ordinary way we sometimes repress problems that we find discouraging. I think it is a deeper thing that makes us hold to that belief in precious self-awareness, albeit it's too feeble to help us explain our thinking—intelligent or otherwise. It's closer to a childish excuse—like "something made me do it," and "I didn't really mean to"—that only denies Single Self when fault or blame comes close. And rightly so, for

questioning the Self is questioning the very notion of identity—and under-
neath I'm sure we're all aware of how too much analysis could shred the
fabrics of illusion that clothe our mental lives.

I think that's partly why most people still reject computational theories
of thinking, although they have no other worthy candidates. And that
leads to denying minds to machines. For me, this has a special irony
because it was only after trying to understand what computers—that is,
complicated mechanisms—*could* do, that I began to have some glimpses
of how a *mind* itself might work. Of course we're nowhere near a sharp
and complete theory of how human minds work—yet. But, when you
think about it, how could we ever have expected, in the first place, to
understand how minds work until after expertise with theories about very
complicated machines? (Unless, of course, you had the strange but popular idea that minds aren't complicated at all, only different from anything
else, so there's no use trying to understand them.)

I've mentioned what I think is wrong with popular ideas of self—but
what ought we to substitute for that? Socially, as I've hinted, I don't
recommend substituting anything—it's too risky. Technically, I have
some ideas but this is not the place for them. The "general idea" is to first
develop better theories of how to understand the webs of processes we
(or our machines) might use to represent our huge networks of fragments
of commonsense knowledge. Once we've some of those that seem to
work, we can begin work on other webs for presenting knowledge about
the first kind. Finally, we work on sub webs—within those larger webs—
that represent *simplified* theories of the entire mess! There's no paradox
at all in this, provided one doesn't become too greedy—i.e., by asking
that those simplified models be more than coarse approximations.

To do this will be quite complicated—but rightly so, for only such a
splendid thing would seem quite worthy as a theory of a Self. For just as
every child must connect a myriad of different ways to count and measure
and compare, in order to understand that simple "concept of number," so
each child must surely build an even more intricate such network, in
order that it understand itself (or even just a wishful image of itself)
enough to grow a full-fledged personality. No less will do.

Could a Computer Have Common Sense?

. . . Isn't it odd, when you think about it, that the very earliest AI
programs excelled at "advanced, adult" subjects. I mentioned that the
Newell-Simon program written in 1956 was quite good at certain kinds of
mathematical logic. Then, in 1961, James Slagle wrote a program that
could solve symbolic calculus problems at the level of college students (it

got an A on an MIT exam). Around 1965 Bobrow's program solved high school algebra problems. And only around 1970 did we have robot programs, like Terry Winograd's, which could deal with children's building blocks well enough to stack them up, take them down, rearrange them, and put them in boxes.

Why were we able to make AI programs do such grown-up things so long before we could make then do childish things? The answer was a somewhat unexpected paradox. It seems that "expert" adult thinking is often[9] somehow simpler than children's ordinary play! Apparently it can require more to be a novice than to be an expert, because (sometimes, anyway) the things an expert needs to know can be quite few and simple, however difficult they may be to discover or learn in the first place. Thus, Galileo was very smart indeed, yet when he saw the need for calculus, he couldn't manage to invent it. But any student can learn it today.

.   .   .   .   .

Today we know a lot about making that sort of "expert" program, but we still don't know nearly enough to build good commonsense problem-solving programs. Consider the kinds of things little children can do. Winograd's program needed ways to combine different kinds of knowledge: about shapes and colors, space and time, words and syntax, and others, just to do simple things inside that "children's world of building blocks"; in all it needed on the order of a thousand knowledge fragments, where Stagle needed only about a hundred—although the one just "played with toys" while the other could solve college-level problems. As I see it, "experts" often get by with deep but narrow bodies of knowledge—while common sense is almost always technically a lot more complicated.

Nor is it just a mere matter of quantity and quality of knowledge: Winograd needed more *different kinds* of ways for processes to control and exploit each other. It seems that commonsense thinking needs a greater variety of different *kinds* of knowledge, and needs different *kinds* of processes. And then, once there are more different kinds of processes, there will be more different kinds of interactions between them, so we need yet more knowledge.

To make our robots have just their teeny bit of common sense, and that was nothing to write home about, our laboratory had to develop new kinds of programming—we called it "heterarchy," as opposed to the "hierarchy" of older programs and theories. Less centralized, with more interaction and interruption between parts of the system, one part of Winograd's program might try to parse a phrase while another part would try to rectify the grammer with the meaning. If one program guessed that "pick" is a verb, in "Pick up the block," another program-part might check to see if "block" is really the kind of thing that *can* be picked up. Common sense requires a lot of that sort of switching from one viewpoint to another, engaging different kinds of ideas from one moment to another.

In order to get more common sense into our programs, I think we'll have to make them more reflective. The present systems seem to me a bit too active; they try too many things, with too little "thought." When anything goes wrong, most present programs just back up to previous decisions and try something else—and that's too crude a base for making more intelligent machines. A person tries, when anything goes wrong, to *understand* what's going wrong, instead of just attempting something else. We look for casual explanations and excuses and—when we find them—add them to our networks of belief and understanding—we do intelligent learning. We'll have to make our programs do more things like that.

Can Computers Make Mistakes?

To err is human, et cetera. I'll bet that when we try to make machines more sensible, we'll find that *knowing what causes mistakes* is nearly as important as knowing what is correct. That is, in order to succeed, it helps to know the most likely ways to fail. Freud talked about censors in our minds that serve to repress or suppress certain forbidden acts or thoughts; those censors were proposed to regulate much of our social activity. Similarly, I suspect that we accumulate censors for ordinary activities—not just for social taboos and repressions—and use them for ordinary problem solving, for knowing what *not* to do. We learn new ones, whenever anything goes wrong, by remembering some way to recognize those circumstances, in some "subconscious memory"—so, later, we won't make the same mistake.[10]

Because a "censor" can only *suppress* behavior, [its] activity is invisible on the surface—except in making fewer blunders. Perhaps that's why the idea of a repressive unconscious came so late in the history of psychology. But where Freud considered only emotional and social behavior, I'm proposing that they're equally important in commonsense thinking. But this would also be just as hard to observe. And when a person makes some good intellectual decision, we tend to ask what "line of thought" lay behind it—but never think to ask, "What thousand prohibitions warded off a thousand bad alternatives?"

This helps explain why we find it so hard to explain how our commonsense thinking works. We can't detect how our censors work to prevent mistakes, absurdities, bugs, and resemblances to other experiences. There are two reasons, in my theory, why we can't detect them. First, I suspect that thousands of them work at the same time, and if you had to take account of them, you'd never get anything else done. Second, they have to do their work in a rather special, funny way, because they have to *prevent* a bad idea before you "get" that idea. Otherwise you'd think too slowly to get anywhere.

Accordingly, much of our thinking has to be unconscious. We can only sense—that is, have enough information to make theories about—what's near the surface of our minds. I'm concerned that conscious thought is just one product of complex "adversary processes" that go on elsewhere in the mind, where parts of thoughts are always under trial with complicated presentations of the litigants, and lengthy deliberations of the juries.[11] And then, our "selves" hear just the final sentences of those unconscious judges.

How, after all, could it be otherwise? There's no way any part of our mind could keep track of all that happens in the rest of our mind, least of all that "self"—that sketchy little model of the mind inside the mind. Our famous "selves" are valuable only to the extent they simplify and condense things. Each attempt to give "self-consciousness" a much more comprehensive quality would be self-defeating; like executives of giant corporations, they can't be burdened with detail but only compact summaries transmitted from other agents that "know more and more about less and less." Let's look at this more carefully.

Could a Computer Be Conscious?

When people ask that question, they seem always to want the answer to be "no." Therefore, I'll try to shock you by explaining why machines might be capable, in principle, of even more and better consciousness than people have.

Of course, there is the problem that we can't agree on just what "conscious" means. Once I asked a student, "Can people be conscious?"

"Of course we can—because we are."

Then, I asked: "Do you mean that you can know everything that happens in your mind?"

"I certainly didn't mean *that*. I meant something different."

"Well," I continued, "what did you mean by 'conscious' if you didn't mean knowing what's happening in your mind?"

"I didn't mean conscious of what's *in* my mind, just *of* my mind."

Puzzled, I had to ask, "er, what do you mean?"

"Well, er, it's too hard to explain."

And so it goes. Why can we say so little about our alleged consciousness? Apparently because we can't agree on what we're talking about. So

I'll cheat and just go back to "self-awareness." I've already suggested that although it is very useful and important, it really doesn't do what we think it does. We assume we have a way to discover true facts about our minds but really, I claim, we only can make guesses about such matters. The arguments we see between psychologists show all too well that none of us have perfect windows that look out on mental truth.

If we're so imperfect at self-explanation, then I don't see any reason (in principle, at least) why we couldn't make machines much better than we are ourselves at finding out about themselves. We could give them better ways to watch the ways their mechanisms work to serve their purposes and goals. The hardest part, of course, would lie not in acquiring such inner information, but in making the machine able to understand it—that is, in building programs with the common sense they'd need in order to be able to use such "insight." Today's programs are just too specialized, too dumb—if you'll pardon the expression— to handle anything as complicated as a theory of thinking. But once we learn to make machines smart enough to understand such theories, then (and only then) I see no special problem in giving them more "self-insight."[12]

Of course, that might not be so wise to do—but maybe we will have to. For I suspect our skeptics have things upside down, who teach that self-awareness is a strange, metaphysical appendage beyond and outside mere intelligence, which somehow makes us human, yet hasn't any necessary use or function. Instead, it might turn out that, at some point, we *have* to make computers more self-conscious, just in order to make them smarter! It seems to me that no robot could safely undertake any very complex, long-range task, unless it had at least a little "insight" into its own dispositions and abilities. It ought not start a project without knowing enough about itself to be pretty sure that it will stay "interested" long enough to finish. Furthermore, if it is to be able to learn new ways to solve hard, new kinds of problems, it may need, again, at least a simplified idea of how it already solves easier, older problems. For this and other reasons, I suspect that any really robust problem solver, one that can adapt to major changes in its situation, must have some sort of model of itself.

On the other side, there are some minor theoretical limitations to the quality of self-insight. No interesting machine can, in general, predict ahead of time exactly what it will do, since it would have to compute faster than it can compute. So self-examination can yield only "general" descriptions, based on simplified principles. People, too, can tell us only fragments of details of how they think, and usually end up saying things like "It occurred to me." We often hear of "mystical experiences" and tales of total understanding of the self. But when we hear the things they say of what they learned—it seems they only learned to quench some question-asking portion of the mind.

So "consciousness" yields just a sketchy, simplified mind model, suitable only for practical and social uses, but not fine-grained enough for scientific work. Indeed, our models of ourselves seem so much weaker than they ought to be that one suspects that systematic mechanisms oppose (as Freud suggested) the making of too-realistic self-images. That could be to a purpose, for what would happen if you really could observe your "underlying" goals—and were to say, "Well, I don't *like* those goals" and change them in some willy-nilly way? Why, then, you'd throw away an eon's worth of weeding out of nonsurvivors—since almost every new invention has some fatal bug. For, as we noted earlier, a part of evolution's work is rationing the creativity of our brain-machines.

But when and if we chose to build more artfully intelligent machines, we'd have more options than there were in our own evolution—because biology must have constrained the wiring of our brains, while we can wire machines in almost any way we wish. So, in the end, those artificial creatures might have richer inner lives than people do. (Do I hear cries of "treason"?) Well, we'll just have to leave that up to future generations—who surely wouldn't want to build the things *that* well without good reasons to.

Can We Really Build Intelligent Machines?

It will be a long time before we learn enough about commonsense reasoning to make machines as smart as people are. We already know a lot about making useful, specialized, "expert" systems, but we don't yet know enough to make them able to improve themselves in interesting ways. Nevertheless, all those beliefs which set machine intelligence forever far beneath our own are only careless speculations, based on unsupported guesses on how human minds might work. The best uses for such arguments are to provide opportunities to see more ways that *human* minds can make mistakes! The more we know of why our minds do foolish things, the better we can figure out how we so often do things so well. In years to come, we'll learn new ways to make machines and minds both act more sensibly. We'll learn about more kinds of knowledge and processes, and how to make machines learn still more knowledge for themselves, while learning for ourselves to think of "thinking," "feeling," and "understanding" not as single, magic faculties, but as complex yet comprehensible webs of ways to represent and use ideas.

In turn, those new ideas will give us new ideas for new machines, and those, in turn, will further change our ideas on ideas. And though no one can tell where all of this may lead, one thing is certain, even now: there's something wrong with any claim to know, today, of differences of men and possible machines—because we simply do not know enough today, of either men or possible machines.

Notes

1. Of course, I'm greatly simplifying that history.
2. Allen Newell and Herbert Simon, *Human Problem Solving* (Englewood Cliffs, N.J.: Prentice-Hall, 1972).
3. Marvin Minsky, *Computation: Finite and Infinite Machines* (Englewood Cliffs, N.J.: Prentice-Hall, 1967).
4. This isn't quite true. LISP doesn't really have those "do whenevers" built into it, but programmers can learn to make such extensions, and most AI workers feel that the extra flexibility outweighs the inconvenience.
5. Of course each culture sets a threshold to award to just a few that rank of "first-class creativity"—however great or small the differences among contestants. This must make social sense, providing smallish clubs of ideal-setting idols, but shouldn't then burden our philosophy with talk of "inexplicability." There must be better ways to deal with feelings of regret at being "second rate."
6. Notice that there's no way a parent could notice—and then reward—a young child's reflective concern with learning. If anything, the kid would seem to be doing *less* rather than more—and might be urged to "snap out of it."
7. Actually, there might be value in imagining the Self as like those dolls— each a smaller "model" of the previous system, and vanishing completely after new stages.
8. Similarly, we find Einstein's space-time-integration very difficult because, no matter how it bollixes up our thinking about Special Relativity, I doubt if we could survive without that wonderful idea of Separate Space.
9. but certainly not always
10. More details of this theory are in my paper on *Jokes*.
11. Like the "skeptics" in Kornfeld's thesis.
12. I think that we are smart enough to understand the general principles of how we think, if they were told to us. Anyway, I sure hope so. But I tend to doubt that we have enough built-in self-information channels to figure it out by "introspection."

15

Intuition Versus Computer-Type Rationality

Herbert L. and Stuart E. Dreyfus with Tom Athanasiou

Despite what you may have read in magazines and newspapers, regardless of what your congressman was told when he voted on the Strategic Computing Plan, twenty-five years of artificial intelligence research has lived up to very few of its promises and has failed to yield any evidence that it ever will. The time has come to ask what has gone wrong and what we can reasonably expect from computer intelligence. How closely can computers processing facts and making inferences approach human intelligence? How can we profitably use the intelligence that can be given to them? What are the risks of enthusiastic and ambitious attempts to redefine our intelligence in their terms, of delegating to computers key decision-making powers, of adapting ourselves to the educational and business practices attuned to mechanized reason?

In short, we want to put the debate about the computer in perspective by clearing the air of false optimism and unrealistic expectations. The debate about what computers should do is properly about social values and institutional forms. But before we can profitably discuss what computers *should* do we have to be clear about what they *can* do. Our bottom line is that computers as reasoning machines can't match human intuition and expertise, so in determining what computers should do we have to contrast their capacities with the more generous gifts possessed by the human mind.

. . . . . .

Computers are certainly more precise and more predictable than we, but precision and predictability are not what human intelligence is about. Human beings have other strengths, and here we do not mean just the shifting moods and subtle empathy usually ceded to humanity by even the most hard-line technologists. Human emotional life remains unique, to be sure, but what is more important is our ability to recognize, to synthesize, to intuit.

. . . . . .

From Herbert L. and Stuart E. Dreyfus, *Mind Over Machine*, pp. xi–xiv, 31–36, 202–6. The Free Press, 1986. Reprinted by permission.

The acquisition of medical diagnosis skill using X-ray film has recently been studied. After a few years of training, radiologists seem to form diagnostic hypotheses and draw conclusions from sets of relevant features as described in our third stage of skill acquisition, competence. [The first two stages are novice and advanced beginner—Eds.] But do experts perform in that way? Our skill acquisition model suggests that after enough experience with the films of patients with a particular condition, the pattern of dark and light regions associated with that condition is stored in memory, and when a similar pattern is seen, the memory is triggered and the diagnosis comes to mind. There would be no decomposition of the patterns on the film into features, and no need for rules associating conditions with features. If you doubt that a dark and light pattern could look to the specialist like a collapsed lung lobe without need for detection of features and application of rules, imagine a patient with a glass chest. Even a novice doctor would see at a glance that a lung lobe was collapsed. Why should it be surprising that with enough experience an X-ray might look as familiar and informative to the expert as the actul chest looks to the novice doctor and that the expert should be able to "see" an abnormality through the X-ray as the novice doctor would see it through glass?

.　.　.　.　.

With enough experience in a variety of situations, all seen from the same perspective or with the same goal in mind but requiring different tactical decisions, the mind of the proficient performer seems to group together situations sharing not only the same goal or perspective but also the same decision, action, or tactic. At this point not only is a situation, when seen as similar to a prior one, understood, but the associated decision, action, or tactic simultaneously comes to mind.

An immense library of distinguishable situations is built up on the basis of experience. A chess master, it has been estimated, can recognize roughly 50,000 types of positions, and the same can probably be said of automobile driving. We doubtless store many more typical situations in our memories than words in our vocabularies. Consequently, such situations of reference bear no names and, in fact, seem to defy complete verbal description.

With expertise comes fluid performance. We seldom "choose our words" or "place our feet"—we simply talk and walk. The skilled outfielder doesn't take the time to figure out where a ball is going. Unlike the novice, he simply runs to the right spot.

.　.　.　.　.

The grandmaster chess player can recognize a large repertoire of types of position for which the desirable tactic or move immediately becomes obvious. Excellent chess players can play at the rate of five to ten seconds a move and even faster without serious degradation in performance. At that speed they must depend almost entirely on intuition and hardly at all on analysis and comparing alternatives.

．　．　．　．　．

The moral of [this discussion] is: there is more to intelligence than calculative rationality. Although irrational behavior—that is, behavior contrary to logic or reason—should generally be avoided, it does not follow that behaving rationally should be regarded as the ultimate goal. A vast area exists between irrational and rational that might be called *arational*. The word rational, deriving from the Latin word *ratio,* meaning to reckon or calculate, has come to be equivalent to calculative thought and so carries with it the connotation of "combining component parts to obtain a whole"; arational behavior, then, refers to action without conscious analytic decomposition and recombination. *Competent performance is rational; proficiency is transitional; experts act arationally.*

．　．　．　．　．

Socrates stands at the beginning of our tradition as the hero of critical, objective thought. There is something to be said for his sort of detached calculative rationality, but we have seen that it should be appealed to only by a beginner or an expert who, having left his domain of experience, can no longer trust his instincts. Nietzsche, who wrote at what he considered the end of our Western philosophical tradition, had a view of Socratic rationality similar to our own. For Nietzsche, Socrates was not the hero of our culture but its first degenerate, because Socrates had lost the ability of the nobles to trust intuition. "Honest men do not carry their reasons exposed in this fashion," Nietzsche maintained.

Of course, Socrates' "rationality" was not a personal sickness. Athenian society was coping with monumental changes, not the least of which was the transformation of Athens into an imperial power. Deliberative reflection no doubt served as a device for evaluating the continued relevance of traditional ways. But Socrates seems to have overreacted and tried to call all traditional wisdom into question. As Nietzsche saw it, Socrates was symptomatic of a whole culture that, having lost its intuitive sense, desperately sought rules and principles to guide its actions:

> Rationality was at the time divined as a *saviour;* neither Socrates nor his "invalids" were free to be rational or not, as they wished—it was *de*

rigueur, it was their *last* expedient. The fanaticism with which the whole
of Greek thought throws itself at rationality betrays a state of emergency: one was in peril, one had only *one* choice: either to perish or—be
absurdly rational.[1]

Aristotle, living a generation after Socrates, occupied an ambiguous
position as the opponent of Socrates and Plato. He realized that even if,
as Socrates and Plato had believed, people were continuously following
rules, they needed wisdom or judgment in order to apply those rules to
particular cases. But Aristotle nonetheless seems to have thought that
before one could act, one had to deduce one's actions from one's desires
and beliefs. The basis of action was, for Aristotle, the practical syllogism:
If I desire S and I believe that A will bring about S, then I should do A.
Both Aristotle's sense of the importance of judgment and his problem-
solving view of intelligence were compatible with his definition of man as
zōion logon echon, the animal equipped with *logos,* for when Aristotle
thought of man as an animal equipped with *logos,* the word *logos* could
still mean speaking, or the grasping of whole situations, as well as logical
thought. But when *logos* was translated into Latin as *ratio,* meaning
"reckoning," its field of meaning was decisively narrowed. It was a fateful
turn for our Western tradition: man, the logical animal, was now he who
counted, he who measured.

All that was necessary to complete the degeneration of reason into
calculation was to equate concepts with collections of objective features,
e.g., house = object, shelter, for man; man = thing, living, thinking. By
the time Hobbes wrote, around 1600, it was possible to claim not only
that reasoning meant reckoning, but that reckoning was nothing more
than "the addition of parcels." Four centuries later we so consider reckoning our essence that, trying to create machines in our own images, we see
only the problem of creating machines that can make millions of inferences per second.

We have gone farther than Aristotle and Hobbes could have imagined,
generalizing Aristotle's model of intelligence to all skills, even physical
skills, so that even the animal part of man, which Aristotle understood as
animated, that is, self-moving, is thought to function by unconscious
calculation. At Wright State University Dr. Roger Glaser and Dr. Jerrold
Petrofsky have performed ground-breaking research using computer-
controlled electrical impulses to exercise the paralyzed limbs of spinal-
cord-injured individuals. Yet that amazing therapy is surrounded by
heated debate. Dr. Petrofsky, who apparently believes that man's animal-
ity is rational, has begun to make extraordinary claims, predicting that
the new techniques will eventually lead to computer-controlled free walking for the paralyzed—a strikingly literal example of the first-step fallacy

that has buttressed faith in AI for years. He has begun to search for facts about muscle condition, limb position, and terrain that can be combined by rule to produce flexible walking. In a conversation with us, Dr. Glaser opposed that optimism as unrealistic and as cruelly raising false hopes: "We have no idea how the subconscious process that replaces the conscious step-by-step procedure used by beginners works," he said. "We might walk by using sophisticated subconscious rules, but how can we find them? Or walking might involve some process of direct pattern recognition followed by a learned response."

We sometimes work out solutions to problems in our heads, but we rarely "figure out" how to move our bodies. Thus thinking looks like a better candidate for computerization than walking. And if thinking *is* reckoning, then it is reasonable to expect that, as futurists have been telli_____ __ _____ the computer as logic machine is the next stage of evo_____ ___ _____ _lace us.

T_____ ___ _____ is no disputing
com_____ ___ _____ formed our rela-
tior_____ ___ _____ world of extraordi-
nar_____ ___ _____ main our servants,
hel_____ ___ _____ our humanity, and
cas_____

___ _____ element of anxiety
has_____ ___ _____ drives us in several
des_____ ___ _____ not against the im-
pro_____ ___ _____ general, offering, as
Jer_____ ___ _____ of the *Whole Earth
Re_____ ___ _____ omputers altogether.
M_____ ___ _____ tting us off from our
in_____ ___ _____ inds and with nature.
Fi_____ ___ _____ y did in the days of
G_____ ___ _____ rything."

___ _____ reactions is not that
th___ _____ ted. The enemies of
technology focus valuable attention on the fact that computers are no panaceas, but such opposition can at best slow their proliferation. A real victory over the improper application of advanced technology—a victory of mind over machine—can come only with recognition that technology has many proper as well as improper uses and with a widely cultivated ability to tell the difference. Computers are perhaps the most powerful, and certainly the most flexible, devices we have yet built. They have many positive, indeed many wonderful uses. The question is not how to eliminate them but how to make the most of their powers.

Likewise, nostalgia for what is being lost is a healthy reaction to the glorification of the "hacker culture." The back-to-nature mystics, however, confuse the supposed dangers of technological devices with the real danger of the technological mentality. In opposing computers they miss the real problem: total dependence upon calculative thinking and a loss of respect for the less formalizable powers of the mind. They fail to see that computers properly used need not alienate us from our everyday experience-based intuition or whatever other intuitive powers we may possess.

Of the computer opponents only the romantics are on the right track. They oppose not technology but technological rationality. But by rejecting *all* rationality, they fail to see that calculative rationality is appropriate for beginners and in novel situations and that deliberative rationality is not opposed to intuition but based upon it. Put in its proper place rational deliberation sharpens intuition.

The question is whether we are going to accept the view of man as an information processing device, or whether we are still enough in touch with our pre-Platonic essence to realize the limits of the computer metaphor. With our mechanical contrivances now able to solve certain problems more effectively than we can, we are being forced to rethink some very old and by now very basic elements of our self-image. It is our hope that the rethinking will lead to a new definition of what we are, one that values our capacity for involved intuition more than our ability to be rational animals.

What we do now will determine what sort of society and what sort of human beings we are to become. We can make such a decision wisely only if we have some understanding of what sort of human beings we already are. If we think of ourselves only as repositories of factual knowledge and of information processing procedures, then we understand ourselves as someday to be surpassed by bigger and faster machines running bigger and more sophisticated programs. Those who embrace that limited conception of intelligence welcome the change with enthusiasm.[2]

Should we become servants of expert systems and, demanding of our experts their rules and facts, become careless of the intuitive powers that fall outside our stunted vision, we will in one generation lose our professional expertise and confirm those expectations. Our children brought up on Logo and our *competent* specialists crammed with procedures will indeed be inferior to the systems they have been trained to imitate.

But fortunately there are other possibilities. We can use computers to track the vast array of facts and law-governed relationships of our modern technological world, yet continue to nurture the human expertise that inference engines cannot share by encouraging learners to pass from rule following to experience-based intuition. If we do so, our experts will be empowered by their computer aids to make better use of

their wisdom in grappling with the still unresolved problems of technological society.

The chips are down; the choice is being made right now. And at all
levels of society computer-type rationality is winning out. Experts are an
endangered species. If we fail to put logic machines in their proper place,
as aids to human beings with expert intuition, then we shall end up
servants supplying data to our competent machines. Should calculative
rationality triumph, no one will notice that something is missing, but
now, while we still know what expert judgment is, let us use that expert
judgment to preserve it.

Notes

1. Nietzsche, *Twilight of the Idols,* Aphorism #10 (Penguin Classics, 1968),
   p. 33 (italics in original).
2. As Roger Schank wrote: "Ultimately, AI will be assimilated into every
   other discipline. . . . The ability to create better and better knowledge
   systems will allow people in every field to develop new ideas and to find
   new approaches to their oldest problems. AI will encourage a renais-
   sance in practically every area that it touches." Roget Schank, *The Cognitive Computer* (Addison-Wesley, 1984), pp. 221–22.

Quality of Work Life

Vincent Guiliano (reading 16) takes us through the history of office automation, starting with Morse's telegraph, Bell's telephone, Edison's dictating machine, and the typewriter, and going all the way to video terminals and electronic mail. He distinguishes between the preindustrial office (totally dependent on the performance of individuals), the industrial office (organized as a rigid production system), and the information-age office, which has the potential to combine the two systems for the benefit of individual workers and clients. Guiliano holds great hope that the intelligent use of new information technologies will yield considerable benefits to individuals and organizations.

Margrethe Olson and Sophia Primps (reading 17) describe their research on employees who "telecommute": work at home using telecommunications devices. Their evaluations are mixed. Some workers, particularly professionals, have benefited from this arrangement; others, especially women with low skills and child-rearing responsibilities, have fared less well.

Jan Zimmerman (reading 18) focuses directly on the impact of telecommuting on women. In an essay that traces the history of women working at home from the early stages of the industrial revolution to the global assembly line of the future, she persuasively argues that telecommuting enables employers to exploit and discriminate against female workers. She also observes that telecommuting, which keeps women at home, has been a boon to the pro-family movement.

Rob Kling and Suzanne Iacono (reading 19) emphasize the role of non-economic factors in making decisions to computerize. Rather than computerization being solely a natural by-product of progress and efficiency, they say it is also a result of "computerization movements" that seek to use computers to reform society. Among the movements the authors examine are artificial intelligence, office automation, and personal computing. Kling and Iacono also discuss the existence—and the limited impact—of counter-computerization movements.

—— **16** ————————————————————

The Mechanization of Office Work

Vincent E. Guiliano

Mechanization was applied first to the processing of tangible goods: crops in agriculture, raw materials in mining, industrial products in manufacturing. The kind of work that is benefiting most from new technology today, however, is above all the processing of an intangible commodity: information. As machines based mainly on the digital computer and other microelectronic devices become less expensive and more powerful, they are being introduced for gathering, storing, manipulating, and communicating information. At the same time information-related activities are becoming ever more important in American society and the American economy; the majority of workers are already engaged in such activities, and the proportion of them is increasing. The changes can be expected to alter profoundly the nature of the primary locus of information work: the office.

An office is a place where people read, think, write, and communicate; where proposals are considered and plans are made; where money is collected and spent; where businesses and other organizations are managed. The technology for doing all these things is changing with the accelerating introduction of new information-processing machines, programs for operating them, and communications systems for interconnecting them. The transformation entails not only a shift from paper to electronics but also a fundamental change in the nature and organization of office work, in uses

From Vincent E. Guiliano, "The Mechanization of Office Work," *Scientific American*, Sept. 1982. Reprinted by permission.

of information and communications, and even in the meaning of the office as a particular place occupied during certain hours.

Office mechanization started in the second half of the nineteenth century. In 1850 the quill pen had not yet been fully replaced by the steel nib, and taking pen to paper was still the main technology of office work. By 1900 a number of mechanical devices had established a place in the office, notably Morse's telegraph, Bell's telephone, Edison's dictating machine, and the typewriter.

In 1850 there were at most a few dozen "writing machines" in existence, and each of them was a unique, handmade creation. Typewriters were among the high-technology items of the era; they could be made in large numbers and at a reasonable cost only with the adoption and further development of the techniques of precision manufacturing with interchangeable parts developed by Colt and Remington for the production of pistols and rifles during the Civil War. By the late 1890s dozens of companies were manufacturing typewriters of diverse designs, with a variety of layouts for the keyboard and with ingenious mechanical arrangements. (Some even had the type arrayed on a moving, cylindrical element and thus were 70 years ahead of their time.) By 1900 more than 100,000 typewriters had been sold and more than 20,000 new machines were being built each year. As precision in the casting, machining, and assembly of metal parts improved and the cost of these processes was lowered, typewriters became generally affordable in offices and homes. The evolution of typewriter usage was comparable to what is now taking place—in only about a decade—in the usage of office computers and small personal computers.

With the typewriter came an increase in the size of offices and in their number, in the number of people employed in them, and in the variety of their jobs. There were also changes in the social structure of the office. For example, office work had remained a male occupation even after some women had been recruited into factories. (Consider the staffing of Scrooge's office in Charles Dickens' *A Christmas Carol*.) Office mechanization was a force powerful enough to overcome a long-standing reluctance to have women work in a male environment. Large numbers of women were employed in offices as a direct result of the introduction of the typewriter.

The first half of the twentieth century saw a further refinement of existing office technologies and the introduction of a number of new ones. Among the developments were the teletypewriter, automatic telephone switching, ticker tape, the electric typewriter, duplicating machines and copiers, adding machines and calculators, tape recorders for dictation, offset printing presses small enough for office use, and data-processing equipment operated with punched paper cards. The new devices were accompanied by a rapid expansion in the volume of office communications and in the number of people engaged in white-collar work.

The first computers in offices were crude and very expensive by today's standards. By the mid-1960s most large businesses had turned to computers to facilitate such routine "back office" tasks as storing payroll data and issuing checks, controlling inventory, and monitoring the payment of bills. With advances in solid-state circuit components and then with microelectronics the computer became much smaller and cheaper. Remote terminals, consisting of either a teletypewriter or a keyboard and a video display, began to appear, generally tapping the central processing and storage facilities of a mainframe computer. There was steady improvement in the cost effectiveness of data-processing equipment. All of this was reflected in a remarkable expansion of the computer industry. The late 1960s and the 1970s also saw the advent of inexpensive copiers, minicomputers, small and affordable private automated branch exchanges (electronic switchboards), the word processor (the typewriter's successor) and then, toward the end of the 1970s, the microcomputer.

An anthropologist visiting an office today would see much that he would have seen 25 years ago. He would see people reading, writing on paper, handling mail, talking with one another face to face and on the telephone, typing, operating calculators, dictating, filing, and retrieving files from metal cabinets. He would observe some new behavior too. He would see a surprising number of people working with devices that have a typewriterlike keyboard but also have a video screen or an automatic printing element. In 1955 the odds were overwhelming that someone working at an alphabetic keyboard device was female and either a typist or a keypunch operator. No longer. The keyboard workers are both female and male, and the typewriterlike devices now accomplish an astonishing variety of tasks.

Some of the keyboard workers are indeed secretaries preparing or correcting conventional correspondence on word processors. Other workers are at similar keyboards that serve as computer terminals. In one office they are managers checking the latest information on production performance, which is stored in a corporate data base in the company's mainframe computer. Economists are doing econometric modeling, perhaps calling on programs and data in a commercial service bureau across the continent. Librarians are working at terminals connected to a national network that merges the catalogs of thousands of participating libraries. Attorneys and law clerks are at terminals linked to a company whose files can be searched to retrieve the full text of court decisions made anywhere in the country. Airline personnel and travel agents make reservations at terminals of a nationwide network. Some of the devices are self-contained personal computers that engineers and scientists, business executives, and many other people depend on for computation, data analysis, scheduling, and other tasks.

Many of the users of terminals and small computers can communicate with one another and with their home offices through one of the half-dozen "electronic mail" networks now in existence in the United States. A surprising number of people are doing these things not only in the office but also at home, on the factory floor, and while traveling. This article was written with a portable personal computer at home, in a hotel in Puerto Rico, and at a cottage in New Hampshire. I have drawn on information from personal files in my company's mainframe computer and have also checked parts of the text with colleagues by electronic mail.

What all of this adds up to is a shift from traditional ways of doing office work based mainly on paper to reliance on a variety of keyboard-and-display devices, or personal work stations. A workstation may or may not have its own internal computer, but it is ultimately linked to a computer (or to several of them) and to data bases, communications systems, and any of thousands of support services. Today the workstations in widest service handle written and numerical information. In less than a decade machines will be generally available that also handle color graphics and store and transmit voice messages, as the most advanced work stations do today.

.   .   .   .   .

Until [the early 1980s] most workstations and their supporting devices and data-base resources were designed to serve a single purpose: to prepare text, access stock-market data, or make air travel reservations, for example. The stockbroker's terminal started out as a replacement for the ticker tape, the word processor as a replacement for the typewriter. The first terminals therefore served as complete workstations only for people who were engaged in a more or less repetitive task.

Now the capabilities of the workstation have been extended by developments in the technology of information processing, in communications, and in enhancements of the "software," or programs, essential to the operation of any computer system. A variety of resources and functions have become accessible from a single workstation. The stockbroker can not only check current prices with his terminal but also retrieve from his company's data base a customer's portfolio and retrieve from a distant data base information on stock-price trends over many years. Millions of current and historical news items can also be called up on the screen. He can issue orders to buy or sell stock, send messages to other brokers, and generate charts and tables, which can then be incorporated into a newsletter addressed to customers.

It is not only in large corporations that such tools are found. Low-cost personal computers and telecommunications-based services available to individuals make it possible for them to enjoy a highly mechanized work environment; indeed, many professionals and many office workers in small businesses have workstation resources superior to those in large corporations where the pace of office mechanization has been slow.

By the year 2000 there will surely be new technology for information handling, some of which cannot now be foreseen. What can be predicted is that more capable machinery will be available at lower cost. Already a personal computer the size of a briefcase has the power and information storage capacity of a mainframe computer of 1955. For a small computer an approximate measure of performance is the "width" of the data path, that is, the number of bits, or binary digits, processed at a time. Computational speed can be represented roughly by the frequency in megahertz of the electronic clock that synchronizes all operations in the central processor. Memory capacity is expressed in bytes; a byte is a group of eight bits. The customary unit is the kilobyte, which is not 1,000 bytes but rather 2^{10}, or 1024. Only three years ago a powerful personal computer had 48 kilobytes of working memory and an 8-bit processor running at a rate of one megahertz.

Today about the same amount of money buys a machine with 256 kilobytes of working memory and a 16-bit processor chip that runs at 4 megahertz or more. Storage capacity and processing power will continue to increase—and their costs will continue to decrease—geometrically. By the year 2000 memory and processing power should be so cheap that they will no longer be limiting factors in the cost of information handling; they will be available as needed anywhere in an organization. The next 20 years will also see the continuing extension of high-capacity communications, of networks for the exchange of information between workstations and other computers, and of centralized data banks. Together these developments will provide access to information, to processing capacity, and to communications facilities no matter where the worker is or what time it is.

New technology inevitably affects the organization of work. One can define three evolutionary stages of office organization, which I shall designate preindustrial, industrial, and information-age. Each stage is characterized not only by its technology but also by its style of management, personnel policies, hierarchy of supervisory and managerial staff, standards of performance, and human relations among office workers and between the workers and their clients or customers.

The first two stages correspond to the well-understood artisan and industrial models of production; the nature of the third stage is only now becoming clear. The operation of a preindustrial office depends largely on the performance of individuals, without much benefit from either systematic work organization or machines. The industrial office organizes people to serve the needs of a rigid production system and its machines. The information-age office has the potential of combining systems and machines to the benefit of both individual workers and their clients.

Most small business, professional, general management, and executive offices are still at the preindustrial stage. In a preindustrial office little conscious attention if any is paid to such things as a systematic flow of work, the efficiency or productivity of work methods, or modern information technologies. What information-handling devices are present (telephones, copiers, and even word processors) may be central to the operation, but there is no deliberate effort to get the maximum advantage from them. Good human relations often develop among the employees; loyalty, understanding, and mutual respect have major roles in holding the organization together. An employee is expected to learn his job, to do what is wanted and needed, and to ask for help when it is necessary. Varied personal styles of work shape the style of the operation and contribute to its success.

Preindustrial office organization generally works well only as long as the operation remains small in scale and fairly simple. It is inefficient for handling either a large volume of transactions or complex procedures requiring the coordination of a variety of data sources. If the work load increases in such an office, or if business conditions get more complex, the typical response is to ask people to work harder and then to hire more employees. Such steps are likely to be of only temporary benefit, however. Without the help of additional systems or technology, effectiveness and morale may soon begin to break down.

One response to the limitations of preindustrial office organization has been to bring to bear in the office the principles of work simplification, specialization, and time-and-motion efficiency articulated for factory work some 70 years ago by Fredrick W. Taylor. The result is the industrial-stage office, which is essentially a production line. Work (in the form of paper documents or a folder of papers related to one customer) moves from desk to desk just as parts move from station to station along as assembly line. Each worker gets a sheaf of papers in an "in" box; his job is to perform one or two incremental steps in their processing and then to pass the paper through an "out" box to the next person, who performs the next steps. Jobs are simple, repetitive, and unsatisfying. A worker may do no more than staple or file or copy, or perhaps check and confirm or correct one element of data. And of course everyone has to

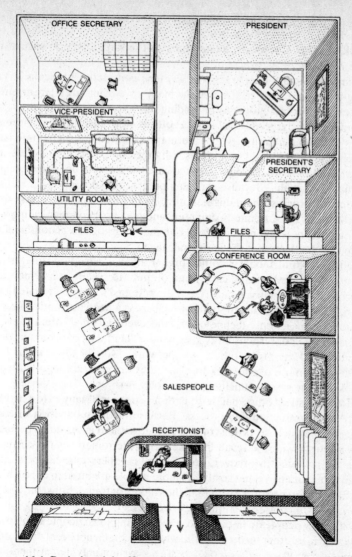

Figure 16.1 Preindustrial office organization dates back to the mid-nineteenth century but is still typical of most professional, small-business, and even corporate management offices today. It is represented here by a hypothetical real estate brokerage. There is little systematic organization. Each person does his job more or less independently, moving about as necessary to retrieve a file, take a client to see a property, or attend a meeting where the sale of a house is made final. Individuals can have different styles of work, and human relations are important. The preindustrial model of office organization can still be effective for some small operations. Conversion to information-age methods is fairly easy.

work together during the same hours in the same office to sustain the flow of paper.

The production-line approach has been considered particularly suitable for office activities in which the main job is handling a large volume of customer transactions, as in sending out bills or processing insurance claims. Many large production-line offices were instituted in the early days of computerization, when information had to be gathered into large batches before it could be processed by the computer; input to the machine then took the form of punched cards and output consisted of large books of printouts. Because early computers could do only a few steps of a complex process, the industrial office had to shape people's tasks to fit the needs of the machine. Computers and means of communicating with them have now been improved, but many large transaction-handling offices are still stuck at the industrial stage.

The industrial model of office organization is based on a deliberate endeavor to maximize efficiency and output. To create an assembly line the flow of work must by analyzed, discrete tasks must be isolated, and work must be measured in some way. There is a need for standardization of jobs, transactions, technologies, and even personal interactions. A fragmentation of responsibility goes hand in hand with bureaucratic organization and the proliferation of paperwork. Most of the workers have little sense of the overall task to which they are contributing their work, or of how the system functions as a whole.

The industrial office has serious disadvantages. Many errors tend to arise in a production-line process. Because of the subdivision of tasks, efforts to correct errors must often be made without access to all pertinent information, with the result that the errors are sometimes not corrected but compounded. Moreover, production-line operations can be surprisingly labor-intensive and costly. As more people are hired to cope with an error rate that increases faster than the volume of transactions, the cost per transaction increases and efficiency declines.

Effective people do not want to stay in boring jobs; people who do stay often lack interest in their work, which becomes apparent to the customer. Even if workers do their best, the system may defeat them, and customer service is likely to be poor. Because a given item can take weeks to flow through the pipeline, it is often difficult to answer customer inquiries about the current status of an account, and even harder to take corrective action quickly. For example, a clerk may be able to check a sales slip and agree that a customer's bill is incorrect; in many instances, however, the clerk is able to change the account only by feeding a new input into the production line, with little assurance that it will have the desired effect. As a result, the billing error can be adjusted incorrectly or can be repeated for several months.

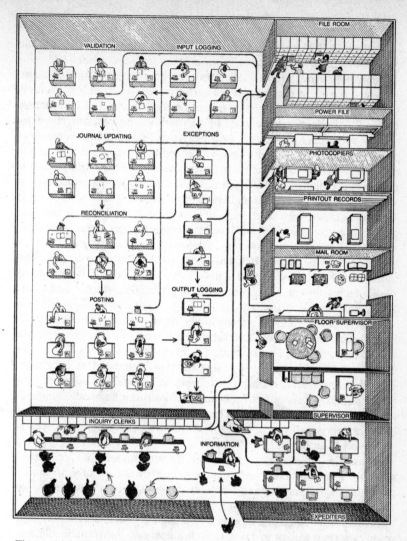

Figure 16.2 The industrial office, essentially a production line, has been favored for operations handling a large number of transactions, as in this, the claims adjustment department of an insurance company. Tasks are fragmented and standardized. Documents are carried from the mail room to the beginning of the production line and eventually emerge at the other end; the flow is indicated by the arrows. Successive groups of clerks carry out incremental steps in the processing of a claim; in general, they leave their desks only to retrieve files or examine computer printouts. If clients make inquiries, they are dealt with by clerks who may be able in time to answer a specific question but can seldom follow through to solve a problem. The work is usually dull. The flow of information is slow, and service is poor.

183

In the mid-1970s the recognition of these limitations, combined with the availability of new workstation information systems, motivated a few progressive banks and other service organizations with a heavy load of transactions to take the next step: they converted certain departments to a mode of operation more appropriate to the information age. The information-age office exploits new technology to preserve the best aspects of the earlier stages and avoid their failings. At its best it combines terminal-based workstations, a continuously updated data base, and communications to attain high efficiency along with a return to people-centered work rather than machine-centered work. In the information-age office the machine is paced to the needs and abilities of the person who works with it. Instead of executing a small number of steps repetitively for a large number of accounts, one individual handles all customer-related activities for a much smaller number of accounts. Each worker has a terminal linked to a computer that maintains a data base of all customer-related records, which are updated as information is entered into the system. The worker becomes an account manager, works directly with the customer, and is fully accountable to the customer.

Information is added incrementally to the master data base. The stored data are under the control of the worker, who can therefore be made responsible for correcting any errors that arise as well as for handling all transactions. Since information is updated as it becomes available, there is no such thing as "work in process," with its attendant uncertainties. An inquiry or a change in status can be handled immediately over the telephone: the sales slip can be inspected, the customer's account can be adjusted, and the bill that is about to be mailed can be corrected accordingly.

The design of effective systems and the measurement of productivity are still important in the information-age office with a large volume of transactions, but the context is different from that of the industrial office. Productivity is no longer measured by hours of work or number of items processed; it is judged by how well customers are served. Are they satisfied? Are they willing to bring their business back? Are they willing to pay a premium for a high level of service?

To the extent that the answers are "yes," the company gains an important competitive advantage. Even if cost cutting is not the only objective, the company can expect dramatic savings in personnel costs. Staff reductions of as much as 50 percent have been common in departments making the changeover to a workstation system. Those employees who remain benefit from a marked improvement in the quality of their working life.

The benefits of the information-age office are not limited to the transaction-intensive office. A similar transformation can enhance productivity, effectiveness, and job satisfaction in offices concerned with

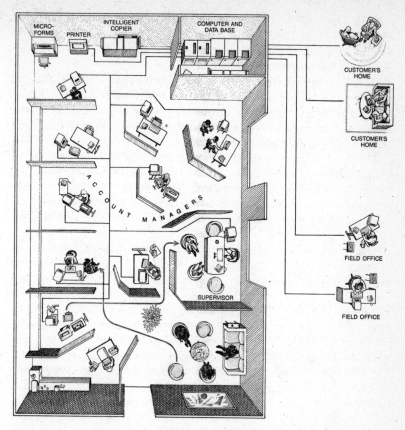

Figure 16.3 The information-age office exploits new technology to preserve the values of the preindustrial office while handling a large volume of complex information. The drawing shows an information-age claims adjustment department. Each adjuster mans a workstation, which is linked to a computer that maintains and continually updates all client records. Each adjuster can therefore operate as an account manager, handling all operations for a few clients rather than one repetitive operation for a large number of clients. Necessary action can be taken immediately. Forms are updated and letters are written at the same workstation that gives access to stored data, and the forms and letters can be printed automatically. The same facilities are available to adjusters visiting a client's home or working in one of the company's field offices (right). The work is more interesting, service to clients is improved, and costs are reduced.

management, general administration, and research. Most such offices are still in the preindustrial stage. They can be transformed to the information-age stage by the introduction of such person-centered technologies as the workstation and electronic mail.

Once most of the activities of a job are centered on the workstation the nature of the office can be transformed in still another way: there is no longer any need to assemble all workers at the same place and time. Portable terminals and computers, equipped with appropriate software and facilities for communication (including the telephone), create a "virtual" office, which is essentially anywhere the worker happens to be: at home, visiting a client or customer, in a hotel, or even in an airplane. The remote workstation can communicate electronically with the central office, and so it extends the range of places where written and numerical material can be generated, stored, retrieved, manipulated, or communicated.

The effects of small-computer technology on the locale of work are analogous to those of the telephone. Because of the almost universal distribution of telephones it is not necessary to go to the office to call a customer or a co-worker, but until now it has been necessary to go there to write or dictate a letter, to read mail, or to find something in a file. Now the workstations and ancillary electronic devices of an automated office can be linked to external terminals and personal computers. The job is no longer tied to the flow of paper across a designated desk; it is tied to the worker himself. The individual can therefore organize his own time and decide where and when he wants to do his work. Individuals who work best early in the morning or late at night can do so.

.   .   .   .   .   .

What will happen to the physical office? It has its virtues, after all. The office provides a home for organizations, a place for people to come together face to face, and a work-oriented environment away from home. Many people need the structure of an office schedule; they like (or at least they are accustomed to) compartmentalization of the day and the week into time for work and time for other activities. Another role for the office is to house centralized forms of communications technology, such as facilities for video conferences, that are too expensive for the home. For these reasons and others I think the physical office will remain a part of working life, at least for as long as I am working. There will be continuing change, however, in how often some workers go to the office and in why they go there.

Many powerful factors are operating together to propel the transformation of office work. A complex set of feedback loops links economic and social change, new developments in information technology, the widespread adoption of the technology, and the introduction of the new office organization the technology makes possible. The large number of information workers, for example, stimulates interest in enhancing their productivity. The concern for productivity serves to increase demand for technologies that can reduce the cost of handling information. Thus several trends reinforce one another to generate an ever stronger market for information products and services. The infiltration of the new devices into the workplace in turn creates an environment in which working electronically is the normal expectation of the worker.

Economics is a major factor. It is becoming far cheaper to communicate electronically than it is to communicate on paper. The transition to word processing from multidraft secretarial typing can reduce secretarial costs from more than $7 per letter to less than $2. Even more dramatic savings are associated with electronic mail, which can bring the cost of sending a message down to 30 cents or less. Electronic filing, in which a "document" is stored and indexed in a computer memory, brings further savings. (The highest-cost activites in manual correspondence are making multiple copies, filing them, and retrieving them.) Such obvious reductions in cost are overshadowed by the savings in the time of managers and executives, the largest element by far in the cost of running an office.

The savings are becoming more significant each year as the cost of the electronic technology is reduced. For example, fast semiconductor memory is a tenth as expensive now as it was in 1975; the cost will drop by another factor of 10 by 1995. The result has been to bring into the individual consumer's price range information-handling capabilities that only a few years ago called for very expensive equipment.

As the market for mechanized workstations expands, more money is invested in research and development for communications, electronics, software, office mechanization systems, and the like. The time span between the development, introduction, and obsolescence of a product becomes shorter. Each year brings a new generation of semiconductor devices; each generation makes possible a new set of applications. The dramatic improvement in products in turn builds demand for them and strengthens the trend toward office mechanization.

Whether a company's business is in farming, mining, manufacturing, transporation, or retailing, its management, marketing, distribution, and other operating controls are basically office-centered, information-handling activities. As the number of blue-collar workers decreases, the proportion of white-collar workers even in manufacturing organizations continues to increase. In virtually all commercial enterprises one finds executives, managers, clerks, and secretaries; in most organizations there are also more specialized information workers, such as engineers and scientists, attorneys, salesmen, librarians, computer programmers, and word processors. These people constitute the human capital resources that can make an information-intensive economy viable.

Yet a tendency to think of white-collar workers in offices as support personnel, outside the economic mainstream, has tended to inhibit the transformation of office work. Physical activities that produce food, minerals, and manufactured goods have been regarded as the only truly productive ones, whereas the handling of information has been considered necessary but essentially nonproductive. This way of looking at things (which may have been appropriate in an industrial society) persists today, even in the minds of economists who call for the "reindustrialization of America." It deeply affects the thinking of corporate management.

Even though most work in American society is information work and most such work is done in offices, the benefits of an increase in the productivity of office workers are not always within the field of view of managers. For those who retain a preindustrial view of office organization, the very concept of productivity seems irrelevant or inappropriate in the context of offices or information work. Those who have an industrial office orientation tend to focus on labor-saving measurements; the installation of new technology and a system for exploiting it is evaluated only in the context of cutting visible office costs.

It is in offices that the basic decisions are made that determine the cost effectiveness of an entire organization. The office is the place where the timeliness of a decision or of a response can have immense consequences. If the office is ineffective, the organization must be ineffective. As it happens, moreover, a high degree of mechanization of the kind described in this article is much less expensive in the office than analogous mechanization is in the factory or on the farm.

The mechanization of office work is an essential element of the transformation of American society to one in which information work is the chief economic activity. If new information technology is properly employed, it can enable organizations to attain the following objectivies:

1. a reduction of information "float," that is, a decrease in the delay and uncertainty occasioned by the inaccessibility of information that is being typed, is in the mail, has been misfiled, or is simply in an office that is closed for the weekend;

2. the elimination of redundant work and unnecessary tasks such as retyping and laborious manual filing and retrieval;

3. better utilization of human resources for tasks that require judgment, initiative, and rapid communication;

4. faster, better decision making that takes into account multiple, complex factors; and

5. full exploitation of the virtual office through expansion of the workplace in space and time.

— 17 —————————————————

Working at Home with Computers

Margrethe H. Olson and Sophia B. Primps

Recently there has been a surge of interest in work at home, explained in part by a fascination with the notion of the "electronic cottage" as popularized by Alvin Toffler (1980) and the fact that many home personal computers are frequently employed for work-related tasks. Technological developments are such that in the near future many office jobs will only be dependent on access to a computer terminal with telecommunications access to a company computer; all other materials required (including communications to other people) will be available through the computer. It has been estimated that as much as 50 percent of all office work could be performed at or near employees' homes rather than at a central office location (Harkness, 1977).

A common term for this work arrangement is *telecommuting*, which means the substitution of telecommunications for physical travel to work; the work is brought to the individual rather than vice versa. Work performed through telecommuting is a specific case of a more general trend called *remote work*, which refers to work performed away from a central work site.

. . . . .

The Role of Technology

The role of computer and communications technology in the telecommuting phenomenon is broader than home information-processing gim-

From Margrethe Olson and Sophia Primps, "Working at Home with Computers: Work and Nonwork Issues," *JSI*, Vol. 40, No. 3, pp. 97–112 (1984). Reprinted by permission.

micks and fads that currently fascinate the public. Since the constraints of work location and of work scheduling are removed, the relationship between employer and employee may be altered. This change is expected with regular telecommuting from home, as well as in other instances where the employee works off-site (as in a single-office satellite workcenter).

Technology can play three potential roles in remote work: (a) as a substitute for other forms of communications with supervisors and co-workers, (b) as a management tool for performance evaluation and monitoring, and (c) as a mechanism for obtaining information necessary to perform one's job (Olson, 1982)

Technology also plays an important role in determining the nature of the work performed in the home. It can have an impact on the number of different tasks in a job and the complexity of each task. Further, it can determine the amount of autonomy of control the job incumbent has over rate of production, work order, performance monitoring, or other related processes. Variations in these two facets of work design, influenced by technology, complexity, and autonomy, characterize two schools in the management literature. The "job enrichment" school of work design is characterized by the desirability of high complexity and autonomy, whereas the "scientific management" school is typified by low complexity, routine work, and little autonomy or control. These two approaches to work design, together with some of the outcomes with each approach, are summarized in Table 17.1 and are referred to again later.

In our research we have found remote work relatively easy to accommodate under either the scientific management or the job enrichment approach to work design. On one hand, many routine clerical jobs using computers, such as full-time word processing, can be reorganized to minimize autonomy and complexity. Pervasive performance controls can be created through technology so that the employee can "do little damage" despite being out of the direct sight of the supervisor. On the other hand, for an employee in a job with high autonomy and self-control over the work performed, performance monitoring is of less concern to management and consequently work location is less critical. Computer programmers working independently on long-term projects characterize this group. Thus, in both cases, remote work is accommodated by the specific characteristics of performance control and management or supervisory style. In cases between the two extremes, the issues of control and supervision make the implementation or remote work arrangements more complex.

General Issues Related to Work at Home

The remainder of this paper is concerned with people using computer and communications technology to work in the home at least part of the time

Table 17.1. Computer Technology and Work Design

High autonomy and job complexity	Low autonomy and job complexity
Impact of technology	
Employee sets own work pace	Work is paced by machine
Performance is monitored by output, delivered by employee	Performance is monitored by process, electronically
Large, complex unit of work	Small, simple unit of work
Long turnaround on deliverables	Short turnaround on deliverables
Relation between employee and employer	
Employee has unique skills	Employee has substitutable skills
More informal control procedures	More formal control procedures
Traditional attendance behavior not relevant	Attendance procedures may be required
Improved relationship with supervisor	Deterioration in relationship with supervisor
Focus on job commitment rather than organizational commitment	Decreased promotability
	Loss of compensation status benefits
Relation between work and nonwork	
Control over work schedule	Potential lack of separation between work and nonwork
Increased opportunity for leisure/time with family	Reduced time for leisure, if assuming child-care responsibilities
Reduced stress	Increased stress
Reduced distractions from the workplace	Potential exploitation
Increased job involvement—possible "workaholism"	
Potential integration of work and family	

while employed full-time in large organizations. It attempts to demonstrate, on the basis of exploratory data collected from company experiments, that a gain or loss of individual autonomy and to a lesser degree, job complexity, associated with work at home has broader implications for the individual than does the nature of the work being performed. Two specific issues are examined: (a) the redefinition of the affiliation between the employee and the organization, and (b) the change in the relation between work and nonwork domains of life experienced by the employee, in terms of time use, family responsibilities, personal freedom, and leisure.

Research Background

The authors [investigated] company experiments on work at home [from 1981 to 1984]. The observations reported here are based on unstructured interviews with employees who work at home on a regular basis, as well as interviews with management to understand the circumstances behind the arrangements. The interviews of employees were constructed to investigate three areas: (a) the specific work habits of the employee at home, including hours of work and extent of routine; (b) impact on the work content itself through any changes in actual job characteristics, types of control or monitoring mechanisms, and communications patterns with the supervisor and with co-workers; and (c) impact on the nonwork domain, including changes in leisure activities, family responsibility, volunteer or community activities, and any changes in personal habits. The interviews with management explored the reasons for considering work at home, and aimed to record perceived changes in work content or leadership style associated with work at home.

The organizations selected fell into two categories: (a) 14 companies running formal pilot programs, and (b) 6 high-technology companies in which work at home is common, referred to here as "informal programs."

Pilot Programs

The 14 work-at-home programs we surveyed are typical of the several dozen that have taken place in the United States during the last few years. The industries represent the types of companies that have expressed the greatest amount of interest in work at home to date. Nine are members of the financial and insurance industries, which tend to be heavily dependent on information processing and are constantly searching for ways to improve the productivity of this labor-intensive component of their work. Four are computer and communication firms which, while seeking ways to improve productivity, are also aware of the market that would be created for their products if work at home were to become widely accepted. One company is a manufacturing firm.

Most of the jobs involved extensive use of computers—programming, word processing, and data entry were most common. Four pilot programs were open to any professional-level job that met these characteristics and were approved by management; these included course developers, internal auditors, and product managers. Most required computer support for at least a portion of the work. Table 17.2 summarizes the job classifications of employees in formal work-at-home pilot programs. In all cases, work-

Table 17.2. Job Classifications of Formal Work-at-Home Pilot Programs

Type of job	No. of pilot programs	No of employees interviewed
Computer professional	5	21
General professional	4	5
Word processing clerical	5	9

at-home employees volunteered for the plan and were selected based on interest and appropriateness of the characteristics of their jobs, as defined by the organization. In a few instances, volunteer employees assumed new assignments or changed their job descriptions in order to accommodate the organization's criteria.

The total number of employees typically involved in a pilot program was rather small. Over half of the programs involved only three or four people. Three involved between 10 and 15 people, and one had approximately 30 people working at home part-time. In four programs, employees worked at home 2 and 3 days per week. In the others, employees worked at home full-time and occasionally (1 day per week or every other week) went to the office to pick up and deliver work or to attend staff meetings.

Informal Work-at-Home Arrangements

There are fewer people working at home on a regular basis than the news media would lead one to suspect. But there does appear to be a growing trend toward informal work-at-home arrangements in certain companies in specific industries, such as California's high-technology computer industry. These are really extensions of the occasional arrangements described earlier, except that such arrangements have become an accepted work norm. In the firms to which we had access, management sanctioned work at home implicitly, if not explicitly.

Six companies, all located in California, were surveyed. Four were computer firms, while two were research and development firms closely associated with the computer and communications industries. In all of these companies, professional employees had extensive computer support for their work and used it intensively. All companies had flexible work hours and made equipment readily available to employees working at home outside regular hours. The employees who worked at home regularly were professionals receiving full salary and benefits. Few worked the 9-to-5 schedule, and it was common to work substantial amounts of overtime on a sporadic basis.

Our objective in comparing the results from the formal pilot programs to the informal arrangements was to understand the role of computer and communications technology in facilitating work at home in differing organizational cultures. We wished to investigate whether the opportunity to work at home is more a function of the technological capability, or whether it is more directly related to the culture or work environment of the organization.

Results

In this section we summarize the reasons employees choose to work at home and the trade-offs involved, based on our data. This summary is divided into the two areas of concern defined earlier: the connection between the employee and the employer, and the relation between work and nonwork domains. Table 17.1 summarizes the observed effects of work at home for each of these two areas in terms of increases or decreases in autonomy.

Relationships Between Employees and the Organization

The degree of autonomy and self-supervision of the jobs being performed at home is of particular importance to the relationship between the employee and the organization. Job status (i.e., clerical versus professional) tends to determine whether the employee experienced increased or decreased autonomy in the performance of his or her job. For example, it was expected that a supervisor faced with a loss of opportunity for face-to-face control would respond by establishing more formal and stringent performance control procedures for the remote employee. In fact, policies concerning attendance and absenteeism, tardiness, and personal time off were redefined for clerical work-at-home employees in a way that decreased their autonomy. In the cases we examined, these changes were implemented through a change in the employee's status from full-time to permanent part-time. Under this classification, the employee was paid either for actual time worked (hourly) or amount of work done (piece rate). The employee was not credited with sick days or vacation time. In the case of piece-rate payment (e.g., number of insurance claims processed), absenteeism and tardiness policies became irrelevant to the organization. The piece-rate system itself became a means of enhancing control (payment based solely on output).

In contrast, highly skilled professionals, who were accustomed to a high degree of autonomy in their work, had that autonomy reinforced by working at home. Of course, these effects on the supervisory process were highly dependent on the existing personal style of the supervisor, as well as on the culture of the organization.

Employees' perceptions of the changes in their relationships to their supervisors reflected the changes in supervision consistent with increased or decreased autonomy. Our findings indicated that the majority of clerical employees believed that their relationship with supervisors had deteriorated. In contrast, the majority with informal arrangements believed that their relationship with the supervisor was enhanced, suggesting that they had achieved a desirable situation of increased autonomy. This improvement was apparently possible, however, only because these employees were in a position to dictate their terms of employment, given the demand for their particular skills.

Employees in the formal pilot programs believed that working at home would have a negative effect on their prospects for promotion. This was noted in particular by the professionals. It was apparent that the lack of visibility or contact with co-workers and supervisors experienced by work-at-home employees was viewed as a serious impediment to promotion opportunities. Clerical employees in the formal programs were not as concerned about promotion as were the professionals, primarily because they assumed that their jobs had little potential for promotion regardless of work location.

Employees in the informal programs were more mixed in their views about the impact of remote work on promotion. For some, promotion to management was a dubious distinction at best. For this group, the increase in autonomy afforded by the opportunity to work at home was more important than the possibility of promotion within the organization.

Another manifestation of the contrast between the two extremes associated with remote work is the way the organization addressed the compensation of work-at-home employees (see Table 17.3). The pilot programs

Table 17.3. Employment Status and Payment Method For Work-at-Home Employees

No. of pilot programs	Employment status	Payment method	Benefits
8	Permanent full-time	Salary	Yes
3	Permanent part-time	Hourly	Prorated
1	Permanent part-time	Hourly	No
1	Permanent part-time	Output	No
1	Contact	Hourly	No

with clerical employees were typically devised to pay hourly or piece rates, with prorated or no benefits. Employees in these programs typically had family responsibilities and viewed work at home as a trade-off: it was one of few options available to them for earning income and simultaneously providing child care. These employees generally believed that if they were to work outside the home, the additional costs required for commuting and child care would outweigh the advantages of higher income. Some were also unwilling to delegate child care to others.

In one company, employees forfeited substantial benefits, such as paid vacation, in order to work at home. In another program, the compensation was changed from a salaried to a piece-work basis, yet the employees had no control over the fluctuations in work volume. At times, no work at all was forwarded to them, with the result that they earned no pay. The on-site employees doing the same work continued to be paid full-times wages with benefits.

Professional employees in both the formal and informal programs remained on full salary with benefits, with one exception. In this case, the employee's status was changed to a contract basis (no benefits) and work became part-time. Some individuals did notice a change in net pay because of the reduced costs of commuting, clothing, and lunches.

Finally, an important facet of the employment relationship is the impact of physical separation on commitment to the organization. In interviews with the at-home professionals in the high-technology firms, we found relatively low commitment to the employees' particular organization. Rather, commitment was strong to their particular product, project, or work group. In those cases where the employee felt that his or her commitment to the organization increased, it was because the employee attributed the organization with the opportunity to create work arrangements to suit personal preference. Several employees joined a company specifically because they knew they would be able to work at home. Several also said that if the company removed the privilege they might seek employment elsewhere. Where commitment increased for those in the formal pilot programs, it was generally because the company allowed the employees to resolve conflicting responsibilities such as child care and the need to work.

To generalize from the experiences of a small group of organizations, it appears that work at home can either result in increased autonomy and freedom from traditional organizational constraints such as required attendance, *or* it can result in a loss of autonomy through more formal control procedures, the loss of promotion opportunities, and a change in compensation or work status. The critical variable appears to be the extent to which the organization views the employee as an irreplaceable resource. The value of the employee's skills to the organization determines the degree of change in autonomy experienced by the employee.

Only those individuals with skills in demand are in a position to require certain conditions necessary for creating a high quality of work life. Some of these employees choose to work at home.

The professionals with the informal arrangements in high-technology firms are typical of those whose skills are in demand. These individuals, like many of their peers who do not work at home, have moved frequently from one company to another within the industry and the geographical area. They exhibit strong loyalty to their work group rather than to their company, and tend to value individual autonomy over organizational commitment. Companies tolerate these attitudes as well as demands to work at home because of their need for the highly specialized skills the employees have to offer. These employees have managed to maximize their autonomy.

Organizations that have viewed work at home as a means for employees to exercise additional control over their work have enhanced the quality of work life for these employees, and they have realized benefits such as decreased overhead costs and an ability to attract and retain certain highly valued employees. For the clerical group, the opportunity to work at home was associated with higher commitment. This suggests that work at home *can* benefit both employer and employee if implemented in this manner even if the employee's skills are not in great demand. Unfortunately, work-at-home programs for clerical employees, whose skills are not in great demand and whose options are limited, have not been concerned with improved quality of work life.

The Connection Between Work and Nonwork Domains

In addition to influencing changes in the traditional notions of organizational membership, work at home can alter the traditional relationship between the employee's work and nonwork life. Nonwork life includes family, leisure, and "maintenance" activities—i.e., all roles outside the context of work itself. Perhaps the most obvious reason for changes in the link between work and nonwork is the work-at-home employee's increased control over his or her time. If the employee is not required to work a specified schedule or number of hours by the organization, he or she can decide what hours are the most efficient regarding work style, the demands of other family members, and scheduling of leisure activities. Also, the elimination of regular commuting can add substantially to the amount of time available.

Our observations also suggest that the lack of physical separation between the work and nonwork domains and the constant presence of the "work setting" may aggravate tendencies toward "workaholism" for those employees who are highly involved in their jobs. However, this appeared to be more of a problem for the professional group. The design of the home workplace further affects the degree to which work intrudes on one's nonwork life. The ideal situation, though costly, is a separate area used *only* for work. Many employees needed to usurp important living space (e.g., the dining room table) for elaborate computer equipment that could not be moved easily.

Implications for Male Professionals

Those we interviewed who were motivated by personal preference to work at home were men who did not have responsibility for child care. Typical reasons these men gave for choosing to work at home included reduced distractions and an improved work environment. Although the reasons tended to be work-related, many of these employees also reported that relationships with their children improved as a result of working as home. However, it should be noted that their spouses were also at home to keep the children from "bothering Daddy while he's working." In general, the male professional employees reported that working at home helped reduce stress. This was attributed to lack of interruptions and avoidance of office politics. Another important factor was the elimination of the stress of commuting.

Work at home also had a positive impact on leisure for this group. These employees reported improvements in both the amount of time available for leisure activities and the types of leisure activities in which they engaged. In particular, they took advantage of free time during the weekdays to engage in sports such as skiing, hang gliding, and wind surfing, avoiding weekend crowds.

Implications for Females with Child-Care Responsibilities

It appeared that the relation between work and nonwork domains altered substantially when the employee had child-care responsibilities. All employees interviewed who chose to work at home for reasons of child care were women. This group was also characterized by jobs with low autonomy, for two reasons. First, the female population in our sample tended to occupy low-status clerical jobs. Second, they perceived the work-at-home arrangement as their only means to provide an income while fulfilling child-care responsibilities. In this situation, where the employee's skills are easily substitutable and the employee is dependent on the organization for the special arrangement, the result can be exploitation: the employee perceived that she must accept reduced work status, the imposition of controls (e.g., through piece-rate work), lower compensation, and fewer benefits relative to those of her counterparts performing the same work on site.

In addition to the negative impact on the employee's job status and work content, the lack of separation between work and family demands placed a great deal of pressure on the employee and allowed little time for leisure. From the anecdotal evidence it was apparent that nonprofessional work-at-home mothers, typically with preschool children and no additional help, juggled work and family responsibilities throughout the day. Descriptions of the typical workday included playpens next to the work area, interrupting work for car pools, and lunchtime, and returning to work during children's naps. The one woman in our sample in a professional job who also assumed child-care responsibility had supplemental care. Because her job required long periods of concentration with few distractions, child-care arrangements were essential and a fairly rigid schedule of work hours was set up around other family responsibilities.

Not surprisingly, these women consistently reported stress associated with work at home (cf. Nine to Five National Association of Working Women, 1984). They reported that they constantly juggled a complex schedule of activities and experienced conflict over the simultaneous demands of work and family roles. The sources of stress were not clear and remain an important subject for further exploration. However, this exploratory analysis does raise the important question of whether work at home is beneficial for the employee with primary child care, or whether it is disruptive to both work and child care, given the stress levels reported.

For males and females without primary child-care responsibilities, work at home can enhance the integration of work and family through increased control over time, decreased commuting, and availability during the day. For women with primary child care, work at home may require an additional set of responsibilities—work plus family. Further, the demands of family responsibilities may tie the female employee to the home setting physically and psychologically, isolating her from other organization members. The social isolation and decreased visibility may reduce the sharing of job knowledge, mutual support from colleagues, and, ultimately, opportunities for advancement in the organization. Members of this group can be exploited if they work at home unless the organization views them as valued resources, whose commitment and contributions are irreplaceable. Overall, though, the technology that allows work at home tends to have a negative effect on women who lack unique skills and have primary child-care responsibilities.

The Role of Technology Revisited

At the outset, we described three potential roles of technology in remote work: as a substitute for communications with supervisors and co-workers, as a tool for performance monitoring, and as a means for obtaining job-related information. We found that, in the formal pilot programs, employees used the available computer and communications technology in traditional ways, and in the same way as their on-site counterparts. For this group, computer technology was utilized primarily for information or word processing, and only minimally to substitute for other forms of communication. Thus, the technology did not serve as a substitute for communication with supervisors and co-workers. Electronic mail was used extensively for intraorganizational communications in only one of the companies studied. In that organization, many people worked in locations separated from the central office by large geographical distances. Among the professionals, several had extensive networks of contacts in other companies accessed through electronic mail. However, the use of this facility was not contingent upon working at home.

Although the capability exists, none of the organizations utilized telecommunications for monitoring the work performance of at-home employees. In situations where the organization saw the need for greater control, this was accomplished through changes in pay status to a piece rate or hourly wage. Finally, the role of technology in providing a mechanism for obtaining work-related information was not substantially different among either the nonprofessionals or the professionals.

It is apparent that companies and employees did not fully utilize the available technological capabilities. Consequently, the potential opportunities for adapting the computer technology to improve employee-employer relationships were not realized.

References

Applegath, J. 1982. *Working free.* Washington, DC: World Future Society Press.

Gregory, J., and K. Nussbaum. 1982. Race against time: Automation of the office. *Office: Technology and People.* 1(2, 3): 197–236.

Harkness, R. C. 1977. *Technology assessment of telecommunications/transportation interactions.* Menlo Park, CA: SRI International.

Hewes, J. J. 1981. Worksteads: *Worksteads: Living and working in the same place.* New York: Doubleday.

Nilles, J. M., F. R. Carlson. P. Gray., and G. G. Hanneman. 1976. *The telecommunications–transportation tradeoff.* New York: Wiley.

Nine to Five National Association of Working Women. 1984. *The nine to five national survey on women and stress.* Cleveland, OH: Author.

Olson, M. H. 1982. New information technology and organization culture. *Management Information Systems Quarterly* 6(5): 71–92.

Olson, M. H. 1983. Remote office work: Changing work pattern in space and time. *Communications of the ACM* 26(3): 182–187.

Short J., E. Williams and B. Christie. 1976. *The social psychology of telecommunications.* London: Wiley.

Sproull. L., S. Kiesler, and D. Zubrow, 1984. Encountering an alien culture. *Journal of Social Issues* 40(3): 31–48.

Toffler, A. 1980. *The third wave.* New York: Morrow.

Turner, J. A., and R. A. Karasek. 1984 Software ergonomics: Effects of computer application design parameters on operator task performance and health. *Economics* 27(6): 663–90.

Unions battle against jobs in the home. 1984. *New York Times.* 1, 32.

Zuboff, S. 1982. New worlds of computer-mediated work. *Harvard Business Review,* 60(5): 142–52.

—— **18** ——————————————————————

Some Effects of the New Technology on Women

Jan Zimmerman

Caretaking functions, like day care for children of working mothers, adult day care for the dependent segment of the senior population, and rehabilitative services for the physically or mentally handicapped, are being reinterpreted as chores that should be handled in a "loving family atmosphere"—and that means women—mothers, grandmothers, daughters or wives—at home. Very, very rarely have men undertaken such unpaid, labor-intensive activities.

·   ·   ·   ·   ·

With this twist of the prism, the disadvantages of telecommuting—working by computer from the home—suddenly appear. Telecommuting, lauded as a means of saving gasoline and saving time, encourages the transmutation of societal obligations into women's obligations, while simultaneously serving the ends of the pro-family movement, which seeks to re-create the nuclear family. With electronic homework, the elderly can become self-supporting and "more desirable to have around," says a National Science Foundation Report.[1] With electronic homework, says futurist Alvin Toffler, citing an Institute for the Future prediction, "married secretaries caring for small children at home [can] continue to work."[2] (Nowhere is there any mention of telecommuting fathers caring for small children or the elderly.)

The values behind these images are trebly discriminatory. First, the assumption that women should be the ones to stay home and care for the young, the old, and the infirm reinforces socialization patterns that link gender and nurturance. Second, the statement implies that neither women's worklife nor their caretaking function is important enough to demand full-time attention. (A women cannot concentrate on meaningful work with an infant on her lap. Nor can a women working on a computer at home keep a toddler away from the stove or read to an elderly parent; day care is still essential.) And third, it presumes that women will continue to hold routine, low-paying data entry jobs, thus completing the vicious circle that makes those jobs the only ones caretaking mothers can hold. These assumptions run counter to two recent social changes: women's surge into the labor force (women composed nearly 45 percent of the U.S. and over 40 percent of the U.K. work force in 1983, and contributed to the financial support of 97 percent of the U.S. families and 88 percent of British ones); and the increasing variety of family forms (less than 7 percent of American households in 1980 were composed of the traditional homemaker Mom, breadwinner Dad, two or more kids, and a dog; in the U.K. the figure is about 11 percent.)[3] By reinforcing these outdated assumptions, however, electronic homework is manipulated to re-create the past.

Telecommuting promises two very different types of work experiences for those at the upper and lower ends of the occupational scale: data entry clerks and secretaries will handle routine tasks under continuous computer scrutiny of their performance and hours, while professionals will have discretionary working hours and unrestricted freedom to use the computer for personal tasks, such as home accounting and data base access.[4] As long as job segregation persists (women make up more than 95 percent of all secretaries and data entry personnel in the U.S. but less than 30 percent of all managers and administrators[5]), telecommuting will be much less tantalizing for women than it is for men in managerial positions, to whom it offers greater independence.

For immigrant, sometimes undocumented, third world women, who constitute the majority of electronics assemblers, computer homework has a different meaning. Those minimally skilled Asian women and Latinas who hope to earn a living in high-tech environments like "Silicon Valley" in California find themselves packing circuit boards on a piecework basis in their kitchens and garages, often pressing their children into service, even if dangerous chemicals are involved. Manufacturers, who generally hire subcontractors to manage such "electronics sweatshops," appreciate this system: they can pay below minimum wages, eliminate benefits, and avoid occupational health and safety regulations. Labor contractors frequently collect kickbacks from the women, who are grateful for the work, willing to accept overhead costs as their own, and too terrified of immigration authorities to report abuses.[6]

If these obviously exploitative conditions could be eliminated, home-based assembly work *could* be of real benefit. Since low-paid assembly workers often cannot afford day care, second cars, or housing near their jobs, homework offers these women a chance to reduce the cost and time of commuting long distances by public transit, and an escape from hostile, often racially discriminatory work environments. Because they do not address those needs, the efforts of some labor unions to outlaw homework meet with resistance from female workers.[7]

There are alternatives that would both avoid the abuses of piecework practices and meet justifiable labor union concerns. Unions, for instance, could act as hiring halls, monitoring pay scales to assure that fair wages and benefits are paid. Or assembly workers could form collectives to establish decentralized, neighborhood-based, mini-worksites, with day care on the premises and occupational safety and health conditions regulated. Electronics assembly or computer workstation sites could be shared by people working for many different companies. As independent contractors, these women could set their own hours, work without supervision, and elect their own facilities coordinator.

. . . . .

Rose-Colored Glasses, Rose-Covered Cottages

Cottage industry, or "putting out," first became popular during the early stages of the Industrial Revolution. Prior to that time, each household manufactured fabric and clothing from start to finish, with all members of the household sharing both the work and the money made by selling surplus goods. Spinning wheels and looms [were] owned by the family, which controlled the tempo and quality of its work; housework [was] a visible part of the seamless web of tasks whose completion allowed the family to survive. But by the mid-eighteenth century, entrepreneurs who owned the new carding machines and water-powered spinning frames that had been developed to speed up textile manufacture took control of the production process.[8]

The entrepreneurs rationalized production by "putting out" the separate tasks of cleaning, spinning, drying, weaving, and sewing to different households, which then competed to provide the greatest output at the lowest prices. Initially, the cash income of the cottage industry household rose, but at the cost of reduced activity in the barter economy.[9] As the Industrial Revolution continued, new inventions like water-and steam-powered looms provided the impetus to centralize production in factories. By 1790 the brief age of "putting out" was over. Centralized factories increased output, as expected, but they also further divided textile work into the specific, repetitive tasks done by speeders, drawers, dressers, and warpers. In the process, workers lost money and entrepreneurs and investors made it.

Under the factory system, the family wage disappeared and the net purchasing power of the household decreased.[10] Employers paid low wages for relatively unskilled tasks, continually reduced piecework pay rates to keep earnings down, and at least one former cash-earner, often the wife, had to remain at home to provide child care and support services for all the family members employed outside the home.[11] (Even at that time, single women working in the mills earned, on the average, only 57 cents for every male dollar. The lowest factory wage for men was almost invariably greater than the highest wage for women.)[12]

On the other hand, the investors who put up capital to buy equipment earned enormous profits. For one textile firm, the Boston Manufacturing Company, sales skyrocketed from $3,000 annually in 1814 to more than 100 times that figure eight years later. Between 1814 and 1823 the company's assets grew almost twentyfold, increasing from $39,000 to $771,000.[13] The company paid its investors 20 percent dividends every year between 1817 and 1825, but laborers' wages—and household income—did not keep pace.[14] Textile technology spread its impact throughout the entire society, but its rewards were distributed more narrowly.

Although the cottage industry mentality compared rather favorably to factory-based, mass-production techniques, it still separated housework from "productive" labor. In many cases, housework that had been shared by all members of the artisan family devolved solely upon women, while other family members concentrated on producing goods that brought in cash. Since cash payment was associated with textile labor inside the home, but not with the effort women expended on housework, cottage industries contributed to the devaluation of women's work. The much-acclaimed computerized retreat to the electronic cottage thus appears less a vision of "progress" than a backward glance through a rearview mirror. Rather than a stop en route to an even better artisan age, the electronic cottage heralds another form of owner-controlled, supervised, and rationalized labor. Perhaps poet Paul Valery, who said that "the future is not what it used to be," had only the half of it; the past doesn't seem to be what it used to be, either.

The Global Assembly Line

In the decentralized factory of the future, the worker will buy her/his own tools (i.e. a computer and peripherals) and absorb all overhead costs—rent, heat, equipment insurance, telephone connection charges and instal-lation fees (if one line is continually tied up for computer transmission, the worker must install another line for calls), electrical usage, and instal-lation of a power supply that is more stable than the one electrical utilities usually provide to residences. In addition, employees are likely to lose opportunities for promotion (promotion usually means becoming a super-visor, but there will be fewer workers to supervise), to receive lower earnings because of competitive, piecework payment or flat-rate, con-tract wage schemes, and forfeit benefits (for example, will a company pay worker's compensation for an employee working at home? Will home workers become self-employed, independent subcontractors liable for all social security taxes, instead of half? Revisions in the 1983 U.S. tax code cite these very consequences as a definition of the term "independent contractor").

If workers don't accept the wonderful way in which new technology decreases employers' costs, they will be out of luck. Already, "offshore offices" are following runaway factories to the cheap labor of the Sunbelt and the third world. George R. Simpson, chairman of Satellite Data Corporation, says his company relays printed materials by satellite from New York to Barbados, where data entry clerks are paid only $1.50 an hour. As he told *Business Week* magazine, "We can do the work in Barbados for less than it costs in New York to pay for floor space . . . The economics are so compelling that a company could take a whole building in Hartford, Connecticut, and transfer the whole function to India or Pakistan."[15]

Already, similar economic arguments have been used to justify transfer-ring electronics assembly jobs to the export processing zone (EPZs) of developing countries. Eighty-five to 90 percent of the workers on the global assembly line are female; since most are single and under the age of 25, they are expected to submit docilely to patterns of domination within the factory that mimic traditional patriarchal family relationships.[16] For the manufacturer, the global assembly line offers great savings. Hourly wages that average $4.50 in the United States are typically reduced to hourly rates of $1.15 in Hong Kong, 90 cents in Mexico, 48 cents in Malaysia, or 19 cents in Indonesia.[17] Shipping costs on lightweight components are low, and the components are taxed only on the value added in the overseas assembly process (for example, soldering fine gold wires to the microscopic integrated circuits etched on a silicon chip). For large production runs, "packing" various chips onto a board is usually done overseas, but small runs are often "jobbed out" to minority women in industrial nations. The seemingly endless supply of cheap, female labor in Asia and Latin America obviates any need to provide decent benefits, good working conditions, protection for workers' health, promotional opportunities, or job enrichment; the resulting high job turnover rate is welcomed as a means of discouraging unionization.[18]

As long as the profits of new technology return primarily to those who finance it, rather than to those who produce it, international capital will continue to flow to third world countries so those profits can be maximized. Manufacturers argue that cheap overseas labor is the only way to produce competitively priced goods, a red herring that distracts attention from the millions of dollars spent on advertising, marketing, packaging, promotion, executive salaries, bonuses, "golden parachutes" (benefits for executives who lose their jobs), and dividends. There are, in fact, many other ways to produce competitively priced goods; the question is: from whose pocket will the competitive edge be picked?

. . . . .

Dividing the World: Hands

Robots for factory automation . . . computers for office automation . . . techniques for manufacturing silicon chips and solar cells . . . gene splicing . . . nuclear power plants. New jobs in new technologies, all right, but not for women. By 1979 these fields had opened up more than 13 million jobs in the United States,[19] but women hold very few of them, especially at the technical level. (Depending on the field, women's participation ranges roughly from 5–25 percent of all engineers, scientists and technicians.)[20] As in more traditional job categories, women's jobs are concentrated at the lower end of the pay scale. Whether a women assembles circuit boards or plucks chickens, she still makes only about $4.50 or £3 an hour, and she still makes less than men doing the same job.[21]

Neither women's increased participation in the labor force nor the loss of high-paying, unionized, male jobs in the old industrial fields of coal mining, steel production, and car manufacture has appreciably narrowed the wage gap; U.S. women earned 64 cents to the male dollar in 1955, 59 cents in 1981, and 62 cents in 1982. In the U.K. women have not fared much better. In 1983, women working full-time earned 64.5 percent of what men were paid.[22] On a broad social level, this situation will not improve in the short term. Although high-technology industries forecast phenomenal growth over the next ten years, the expansion of low-paying, predominantly female, service jobs will be more than five times the projected growth in high-technology employment.[23] Unless the concept of equal pay for work of comparable worth becomes a reality, new technology will not erase the boundary between women's work and men's work, but rather will redraw it in indelible ink.

For instance, three-quarters of women's jobs in the computer field are in the low-paying categories of keypunch, data entry, and computer operations, while three-quarters of men's jobs are in the much more remunerative fields of machine repair, programming, systems analysis, and other computer specialties. Female computer professionals in the U.S. earn only 75 percent of men's wages; the average wage disparity over all computer occupations is $5,000 to $7,000 each year—enough for a small car, a year's rent, or an annual college tuition bill.[24] Only one percent of female computer professionals earn more than $50,000 a year, which is generally considered the dividing line between middle management and the executive suite.

Although Ada Byron Lovelace programmed the world's first computer (Charles Babbage's differential engine) in 1840 and women like Captain Grace Hopper (U.S. Naval Reserves) programmed the Mark I, one of the first modern computers, during and after World War II, their contributions were considered exceptional. At first, programming was considered a routine, clerical task (female), but later, as a computer technology matured, programming was elevated to an esoteric realm for "wizards" (male). Partly because there were few formal opportunities to learn computer science in schools (learning took place on an *ad hoc* basis in science and engineering centers where computers were plentiful and women were scarce), women's participation was set back nearly twenty years. When colleges started to offer training in computer science in the late 1960s, women began to enter the field in greater numbers. Now that women constitute nearly 30 percent of all programmers, programming is once again becoming a routine clerical job through the use of automatic program generators, structured programming techniques, and packaged applications software.[25] A July 1980 issue of *Computerworld* unintentionally headlined the irony. One story read, "Programmers seen needing fewer skills," while another in the same issue announced, "Project opens computer science jobs to women."[26]

Computing is not alone, of course. Even the young, nontraditional practitioners of appropriate technologies in the Aquarian age can't avoid pervasive job segregation. Anecdotal evidence shows women in charge of gardening, food preservation, aquaculture, and worm farming, men in charge of alternative energy sources—solar, wind, hydro, and biomass.[27] The biological sciences have always been friendlier to women than the physical sciences, but even here there is segregation. Genetic engineering and molecular biology sport tweedy "masculine" labels, whereas physiology and botany flirt with "feminine" tags. (This fact makes 1983 Nobel laureate Barbara McClintock's achievement in elucidating the role of "jumping genes" of Indian corn all the more spectacular; her recognition was thirty years late.)

Partially as a consequence of discriminatory employment in many fields, new technology encourages its own discriminatory application. The ubiquitous computer enables management to record keystroke and error rates for typists and data entry personnel, to pay piecework rates for information processing, and to implement work speedups. In 1981 the National Institute for Occupational Safety and Health found that clerical workers using video display terminals (VDTs) exhibit higher levels of job stress than any other category of workers ever studied, including air traffic controllers. By comparison, professional writers and editors using VDTs, who control the pace of their own work and receive recognition for it, have the lowest stress.[28] Women, of course, hold 95 percent of those high-stress clerical VDT job slots.[29]

Similarly, women suffer disproportionately from the work rationalization or deskilling effects of most office automation schemes. Rationalization is the division of labor into ever finer, more specialized tasks. In many automated offices secretarial work is now divided so that one person does word processing all day, another does electronic filing, and a third answers the phone.[30] Women who have recently entered middle management are equally vulnerable. Writer Barbara Garson quotes an executive who describes the future corporate structure quite graphically:[31]

> We are moving from the pyramid shape to the Mae West. The employment chart of the future will still show those swellings on the top, and we'll never completely get rid of those big bulges of clerks on the bottom. What we're trying to do right now is pull in that waistline (expensive middle management and skilled secretaries).

Computers and other new forms of office technology—electronic mail, telefacsimile, voice store-and-forward message systems, voice and optical character recognizers—offer management new opportunities to control a female work force that has just started to organize; the technology itself does not demand that those opportunities be seized. A sensitive company could implement office automation in humanitarian ways to increase productivity and reduce error rates, but still offer job diversification, create career ladders, and save jobs. In reality, though, the men at the top who control the process are increasingly distant from the women controlled by it.

Other technologies find different razors to slice labor into similar gender divisions. In the case of hazardous new manufacturing processes, for example, women of childbearing age may be excluded from employment unless they agree to be sterilized. In 1978 five women needed their jobs at American Cyanamid's Willow Bend, West Virginia, pigment plant so badly that they underwent sterilization rather than be demoted to lower-paying jobs.[32] Focusing on women's unsuitability as a class whether or not they intend to have children diverts attention from the fact that jobs hazardous to women's reproductive systems may be just as hazardous to men's; efforts should be directed to make the jobs safe for both.

Genetic screening offers another means of excluding women from the workplace. Seventeen companies in the United States have already begun using this technique (and forty-two more intend to), ostensibly to identify workers who may be susceptible to occupational diseases caused by environmental pollutants. This rather inaccurate screening method can be used to discriminate against women with "faulty genes."[33]

Robotics, which is seen as a threat to "male" jobs on assembly lines in heavy industry, is concurrently identified as a new career opportunity for displaced male workers. Robotics will also displace the predominantly female workers who assemble electronics, manufacture textiles, inspect and test products, or pack and wrap goods,[34] but no one talks about retraining women to produce or supervise their robotic replacements. In a typical catch-22, women are offered no training opportunities, yet they are excluded from this new field because they lack appropriate skills.

The New Automation: Who Puts the Coffee in the Coffee Pot?

"I thought we spent all that money on the machine because they said the operators never had to sit around and do nothing." Typist Ellen Levy still shudders when she thinks of that remark, made by her boss's wife while Ellen waited for the computer to finish printing out a job.[35] The temptation is so great to identify the worker with her equipment that the person behind the work disappears in a fog of words: "When can your machine do this?" Human beings become the machines they operate: word processors, keypunchers, spinners, stampers. The identification becomes so pervasive that supervisors wonder, "If the machine is doing all the work, why is it taking so long?"

When computers, electric weed clippers, robots, microwave ovens, and digitally controlled coffee makers are taken for granted, the labor involved becomes invisible. But someone still has to run the weed clipper, prepare the meal, enter data, put coffee in the coffee maker, and throw away the old grounds. Particularly when that someone is a woman, whose labor is likely to be either unpaid or underpaid, we forget that work is even required. Such invisibility is particularly destructive to women since new technologies disperse the work done by paid female labor to unpaid female consumers.

Banking or paying bills by computer, touch-tone phone, or videotex, for instance, will result in extensive job loss for bank tellers, bookkeepers, record keepers, and clerical workers, all female jobs. We know that the work these women do is valuable; Citibank in New York charges customers who have less than $500 in their accounts twenty-five cents for every transaction with a human being, instead of with an automatic teller machine (ATM). (The bank tried to limit "live" teller service to customers whose minimum balance was $5,000, but public outrage forced them to devise a less offensive, but equally income-discriminatory, formula. Can you imagine waiting outside on a cold, rainy day to use an ATM?)[36]

Companies whose accounts will be electronically credited when clients "dial-their-dollars-direct" from home will collectively save millions of dollars that they now spend on clerical and bookkeeping services. They ought to *pay* their customers for the savings they will realize, but of course they won't. Hey, we're supposed to be grateful for the convenience; after all, we're not standing in the rain waiting in line, are we?

When people start checking out their food, shoes, and jogging suits on home TV screens, retail clerks and grocery checkers will join the unemployment line, right behind the tellers and billing clerks. Hard-working librarians will be next, since videotex will offer a handy substitute for the library services cut during an era of tight budgets. And don't forget the airplane reservation clerk and ticket sellers, whose jobs will also fall prey to the green screen. The telephone company already has replaced live voices and real people with speech synthesizers that provide forwarding numbers, time, weather, and directory assistance.

Unpaid consumers, primarily women, will be doing the work women used to be paid to do. In 1982 females in the U.S. made up 60 percent of all retail sales clerks, 85 percent each of librarians, billing clerks, and cashiers,

92 percent of telephone operators, 94 percent of bank tellers, and 98 percent of telephone receptionists—all occupations that will become obsolete in the Information Age. On the other half of the great gender divide, men will come out ahead with new technology. In 1982 in the U.S. they held the overwhelming percentage of jobs for which demand will rise as home information services proliferate: 78 percent of all shipping clerks; 90 percent of electronic technicians; 94 percent of delivery services and route workers, 95 percent each of TV, computer, and home appliance repair and installation technicians.[37] Look at it this way: all those unpaid female consumers will have been really well trained!

Notes

1. R. Rheinhold, "Study Says Technology Could Transform Society," *New York Times,* June 14, 1982, p. A16.
2. A. Toffler, *The Third Wave* (New York: William Morrow, 1980), p. 215; (London: Pan Books, 1981).
3. "Women and the Family" and "Women and the Economy," *WEAL Washington Report,* 12 (June/July 1983), p. 1, 3; and Matrix, *Making Space: Women and the Man-made Environment* (London: Pluto Press, 1984), p. 5: "About one in nine of all households consist of a man *with* a paid job, a woman *without* one and children under the age of 16." M. Harris, *America Now: The Anthropology of a Changing Culture* (New York: Simon & Schuster, 1981), p. 97.
4. B.A. Gutek, "Women's Work in the Office of the Future," in J. Zimmerman, *The Technological Woman: Interfacing with Tomorrow* (New York: Praeger, 1983), pp. 159–68.
5. N.F. Rytina, "Earnings of Men and Women: A Look at Specific Occupations," *Monthly Labor Review,* April 1983, pp. 25–31.
6. For this and the following paragraph, see: R. Morales, "Cold Shoulder on a Hot Stove," in Zimmerman, pp. 169–80. P. Mattera, "Home Computer Sweatshop," *The Nation,* 2 (April 1983): pp. 390–92.
7. Morales, op. cit., pp. 173–79. "Worksteaders' Clean Up," *Newsweek,* January 9, 1984, pp. 86–87. N. Katz, "Join the Future Now: Women and Work in the Electronics Industry," San Francisco State University, 1981, photocopy. Ursula Huws discusses union participation in the U.K. in her book *The New Homeworkers* (London: Low Pay Unit, 1984).
8. T. Dublin, *Women at Work* (New York: Columbia University Press, 1979), pp. 5, 14–16. R. Baxautall *et al.,* eds., *America's Working Women* (New York: Random House, 1976), pp. 13–15, 20–21. A. Oakley, *Woman's Work: The Housewife Past and Present* (New York: Vintage Books, 1976), pp. 32–36.
9. Ibid., pp. 35–36.

10. J. Pinchbeck, *Women Workers and the Industrial Revolution* (New York: August M. Kelley, reprinted 1969), p. 4; (London: Virago, 1981).

11. Dublin, op. cit., p. 109. E. Abbott, *Women in Industry: A study in American Economic History* (New York: Arno Press, reprinted 1969), p. 269.

12. Ibid., 279. Dublin, op. cit., p. 66. See also A. Clark, *The Working Life of Women in the Seventeenth Century* (London: Routledge & Kegan Paul, 1982; first published 1919).

13. Dublin, op. cit., p. 18.

14. According to Dublin, op. cit., p. 137, "Output per worker averaged across these firms rose by almost 49% from 1836 to 1850, while daily wages increased only 4%."

15. "The Instant Offshore Office," *Business Week,* March 15, 1982, p. 136E.

16. M.P. Fernadez-Kelly, "Gender and Industry on Mexico's New Frontier," in J. Zimmerman, ed., *The Technological Woman: Interfacing with Tomorrow* (New York: Praeger, 1983), pp. 18–29.

17. Hourly wages from *Semiconductor International,* February 1982, chart shown in: A. Fuentes and B. Enrenreich, *Women in the Global Factory,* INC pamphlet no. 2, Institute for New Communications, (New York: South End Press, 1983), p. 9.

18. R. Morales, "Cold Shoulder on a Hot Stove," in Zimmerman, ed., op. cit., pp. 169–80.

19. "America Rushes to High Tech for Growth," *Business Week,* March 28, 1983, pp. 84–87, 90.

20. N.F. Rytina, "Earnings of Men and Women: A Look at Specific Occupations," *Monthly Labor Review,* April 1982, p. 25–31.

21. Rytina, op. cit., pp. 26–29.

22. "Consumer income report," *Current Population Reports,* P-60, U.S. Department of Commerce, Bureau of the Census, Washington, D.C. 1983. *Employment Gazette,* U.K., October 1983.

23. "America Rushes to High Tech Jobs for Growth," op. cit. E. Rothschild, "Reagan and the Real America," *The New York Review,* February 15, 1981, pp. 12–17.

24. J. Zimmerman, "Women in Computing: Meeting the Challenge in an Automated Industry," *Interface Age* 12 (December 1983): 79, 86–88.

25. P. Kraft, *Programmers and Managers* (New York: Springer Verlag, 1977).

26. B. Schulz, "Programmers Seen Needing Fewer Skills," *Computerworld,* July 28, 1980, pp. 1, 6. "Funded by National Science Foundation: Project Opens Computer Science Jobs to Women," *Computerworld,* July 28, 1980, p. 17.

27 J. Smith, *Something Old, Something New, Something Borrowed, Something Due: Women and Appropriate Technology* (Missoula, Montana: Women and Technology Project, 1978).

28. U.S. Department of Health and Human Services, "Potential Health Haz-

ards of Video Display Terminals," DHSS (NIOSH), no. 81-129, Cincinnati, Ohio, 1981.

29. J. Gregory, "The Next Move: Organizing Women in the Office," in J. Zimmerman, ed., *The Technological Woman: Interfacing with Tomorrow* (New York: Praeger, 1983), p. 260.

30. Ibid., pp. 260–72.

31. Quoted in B. Garson, "The Electronic Sweatshop: Scanning the Office of the Future," *Mother Jones* 6 (July 1981): 41.

32. A. Mereson, "The New Fetal Protectionism: Women Workers Are Sterilized or Lose Their Jobs," *Civil Liberties,* July 1982, pp. 6–7. See also: R. Petchesky, "Workers, Reproductive Hazards, and the Politics of Protection: An Introduction," *Feminist Studies* 5 (Summer 1979): pp. 233–45. W. Chavkin, "Occupational Hazards to Reproduction: A Review Essay and Annotated Bibliography, *Feminist Studies* 5 (Summer 1979): 311–25.

33. "Genes on the Job," *Science for the People,* November/December 1982, pp. 7–8.

34. B. Cornish, "Robots See, Hear, Feel," *The Futurist,* August 1981, pp. 11–13. O. Friedrich, "The Robot Revolution," *Time,* December 8, 1980, pp. 72–83. Rytina, op. cit.

35. S. Otos and E. Levy, "Word Processing: This Is Not a Final Draft," in Zimmerman, ed., *The Technological Woman,* op. cit., p. 152.

36. Citibank, New York, telephone conversation with author's research associate to confirm fees, February 24, 1984. "Citibank relents on rule limiting live-teller access," *Los Angles Times,* May 31, 1983, p. IV-2.

37. All statistics from Rytina, op. cit.

19

Computerization Movements

Rob Kling and Suzanne Iacono

Why is the United States rapidly computerizing? One common answer is that computer-based technologies are adopted because they are efficient economic substitutes for labor or other technologies (Simon, 1977).

From Rob Kling and Suzanne Iacono, "The Mobilization of Support for Computerization: The Role of Computerization Movements," *Social Problems,* Vol. 35, No. 3, June 1988. Reprinted by permission.

Rapid computerization is simply a by-product of cost-effective computing technologies. A variant of this answer views computerization as an efficient tool through which monopoly capitalists control their suppliers and markets and by which managers tighten their control over workers and the labor process (Braverman, 1975; Mowshowitz, 1976; Shaiken, 1986).

A second answer focuses on major epochal social transformations and argues that the United States is shifting from a society where industrial activity dominates to one in which information processing dominates (Bell, 1979). Computer-based technologies are simply "power tools" for "information workers" or "knowledge workers" as drill presses are power tools for machinists (Strassman, 1985).

The market assumptions of these common answers have also shaped the majority of social studies of computerization (for a detailed review of the empirical studies of computerization, see Kling, 1980, 1987). These studies focus on computerization in particular social settings that range in scale from small groups and workplaces (Shaiken, 1986) through single organizations (Kling, 1978; Kling and Iacono, 1984) to comparative multiorganizational studies (Laudon, 1974, 1986). These studies of computerization usually ignore the ways that participants in the settings under study develop beliefs about what computing technologies are good for and how they should organize and use them (see, for example, Attewell and Rule, 1984).

[Since 1973] we have conducted systematic studies of computerization in diverse organizations: banks (Kling, 1978, 1983), engineering firms and insurance companies (Kling and Scacchi, 1982), manufacturing firms (Kling and Iacono, 1984), public agencies (Kling, 1978), and schools (Kling, 1983, 1986). We have been participant observers of four specific computerization movements: artificial intelligence (1966–1974), computer-based education (1974–1988), office automation (1975–1988), and personal computing (1983–1988). And we have observed several computerization efforts at our home university as participants, firsthand observers, or coordinators of an assessment team.

On the basis of this research and experience, we believe computerization in the United States is not simply a product of economic forces directed toward progress and efficiency. A wide variety of noneconomic factors must be considered to fully understand the process of rapid computerization that has occurred in Western industrialized societies. The adoption, acquisition, installation, and operation of computer-based systems is often much more socially charged than the adoption and operation of other equipment such as telephone systems, photocopiers, air conditioners, or elevators. We have observed that participants are often highly mobilized to adopt and adapt particular computing arrangements through collective activities. We have recently begun to see that these collective activities take place both outside and within computerizing organizations and that they share important similarities with various other social, professional, intellectual, and scientific movements. We have observed that strongly committed advocates often drive computerization projects. They develop and encourage ideologies that interpret what computing is good for and how people in these projects should manage and organize access to computing. They usually import these ideologies from discourses about computerization external to the computerizing organization (Kling and Iacono, 1984).

In this paper we examine how specialized "computerization movements" (CMs) advance computerization in ways that go beyond the effects of promotion by the industries that produce and sell computer-based technologies and services. Our main thesis is that computerization movements communicate key ideological beliefs about the links between computerization and a preferred social order which help legitimize computerization for many potential adopters. These ideologies also set adopters' expectations about what they should use computing for and how they should organize access to it. In this paper we focus our attention primarily on the character of the computerization movements and their organizing ideologies.

We describe five specific computing technologies as the focus of computerization movements: urban information systems, artificial intelligence, computer-based education, office automation, and personal computing. These specific CMs, along with movements organized around other computing technologies, form a general computerization movement. There is a core ideology that supports the general CM and there are groups whose world views balance pursuit of the most advanced computing technologies with alternative social values. Computerization is a process deeply embedded in social worlds that extend beyond the micro realities of any particular organization or setting. In this paper we examine the character of CMs in the hope that our analysis will stimulate new lines of inquiry and more complete analyses.

Computerization Movements

Sociologists have used the concept of "movement" to study many different kinds of collective phenomena. The most common term found in this literature is "social movement," often used in a generic way to refer to movements in general. But sociologists also have written about professional movements (Bucher and Strauss, 1961), artistic movements, and scientific movements (Aronson, 1984; Star, forthcoming). From the diverse ways sociologists have used the concept, we found Blumer's

(1969:8) general definition most helpful for our analysis: social movements are "collective enterprises to establish a new order of life." This is an inclusive definition that allows us to consider elements relevant to computerization that other, narrower conceptions would rule out. We also found the work of Zald and his colleagues (McCarthy and Zald, 1977; Zald and Berger, 1978) helpful in that it allows us to consider social forms that are both highly and loosely organized as part of computerization movements.

Computerization movements (CMs) are a kind of movement whose advocates focus on computer-based systems as instruments to bring about a new social order. The mobilizing ideologies of [counter-computerization] movements (CCMs) oppose certain modes of computerization that their advocates view as bringing about an inappropriate social order. We will examine five CMs, some CCMs, and their mobilizing ideologies in the next section. . . .

Five Specific Computerization Movements

Below we describe five specific computerization movements to illustrate how the collective activities of each CM include the development of a mobilizing ideology and organized activity to promote it. These, along with other CMs we do not examine, form a general CM.[1] We examine key ideological elements of these CMs in the next section.

Urban Information Systems

Urban information systems process data about the activities of people in local government jurisdictions, the services they receive, and the internal operations of the local governments. These include systems to support tax collection, police operations, municipal libraries, and urban planning. Support for urban information systems was continuous with progressive reform movements that sought to professionalize local government administration in the early part of the century (Laudon, 1974). Local governments that have pursued these reforms are disproportionately likely to adopt computer-based applications (Danziger et al., 1982). The urban information systems CM is a relatively low profile movement that forms one segment of a larger professional reform movement to "take politics out of government operations." Its adherents conceive of computerization as fostering government by skilled professionals instead of political appointees and bringing the purported rationality and efficiency of business to governmental operations.

Urban information systems activists have employed associations for local government officials and professionals as movement organizations. Professional organizations for tax assessors, finance officers, and other administrators have been strong supporters of computerization within their areas of expertise. Urban planners, social service professionals, and computer specialists have also found organized support for their computer interests within the Urban and Regional Information Systems Association, which has held annual national conferences since the 1960s. Other professional associations, such as the International City Managers Association, provide information and staff support to foster automation in local governments.

During the 1960s the federal government stimulated the growth of urban information systems through a wide variety of grants: the U.S. Department of Housing and Urban Development supported planning and social services (Kling, 1978), the Department of Transportation for transportation planning, the Law Enforcement Assistance Administration for police systems, and Housing and Urban Development for several massive demonstration projects (Kraemer and King, 1978). Though federal funding for specific urban projects has been substantially reduced since the late 1960s, urban goverments now support their own information systems, which have become embedded in the operations of many departments, especially tax collection, finance, police, welfare, and planning (Danziger et al., 1982). The urban information systems movement is today highly institutionalized.

Artificial Intelligence

The belief that computer-based technologies can be programmed to "think" about complex cognitive tasks is the central tenet held by enthusiasts of artificial intelligence (Turkle, 1984). Early conceptions of artificial intelligence (AI) were framed within an abstract scientific discourse emphasizing the formal modeling of different domains of knowledge or the study and simulation of human cognitive processes. More recently, some AI enthusiasts have replaced abstract models of universal cognitive processes with the social construct of "expert" systems (Feigenbaum and McCorduck, 1983).

In one of the most enthusiastic accounts, Feigenbaum and McCorduck (1983) assert that almost no meaningful intellectual work will be possible in "the world of our children" that does not depend upon knowledge-based (e.g., AI) systems. This idea, a controversial one within academic computer science, argues for a vision of the world its proponents hope others will help them construct. Public proselytizing for AI, so conceived, has transformed it from a technological-scientific movement into a CM.

AI spread as a scientific movement within the computer science profession itself. In the 1960s and 1970s interest in AI was focused around the *Artificial Intelligence Journal* and the triennial International Joint AI Conference, originally organized by the secret Artificial Intelligence Council. More recently, a "special interest group" on Artificial Intelligence (SIGART) within the Association for Computing Machinery and the American Association of Artificial Intelligence (with its monthly publication, *The AI Magazine*) has been established.

Large-scale AI research began in the early 1960s under sponsorship of the Advanced Research Projects Agency of the Department of Defense, although other military agencies and the National Science Foundation also supported some of this initial research. This funding was used to create major laboratories at a handful of universities and research institutes. In the late 1970s AI became commercialized as established industrial firms such as Schlumberger and General Motors made substantial investments (Winston and Prendergast, 1984). AI specialty firms emerged along with expert systems and practical natural language processing. In the 1980s artificial intelligence was transformed into a CM as the mainstream business magazines (e.g., *Fortune, Business Week*) began to aggressively promote a fantasy definition of powerful and accessible AI being "here today."

Media accounts usually describe the value of current technologies in terms of their potential capabilities. They often exaggerate their level of development, the quality of product engineering, and the extent to which they have been adopted. They have also exaggerated the attributes of the few commercial applications that do exist. They suggest that these systems "think" like people rather than simply performing symbolic manipulations on formal representations of the tasks usually performed by people. The American press often promotes AI in sensationalist terms (cf. *Business Week,* 1984; Applegate and Day, 1984). A recent and representative example appears on the cover of the February 15, 1988, issue of *Insight on the News.* It heralds its cover story about AI with: "See machines think. Think, machine, think. The world of artificial intelligence." More accurate stories sometimes appear in magazines and newspapers, but they are less common than the sensationalized stories. Despite sensational claims in the press, none of the higher performance AI systems, with the exception of a particular checkers-playing program, has "learned" to modify its behavior, let alone learned the way that humans do.

Computer-Based Education

Computer-based education includes both computer-assisted instruction programs that interact with students in a dialogue and a broader array of educational computer applications such as simulations or instruction in computer programming (see, for example, Taylor, 1980). There is a major national push for extended application of computer-based education at all educational levels. For example, in the mid-1980s several private colleges and universities required all of their freshmen students to buy a specific kind of microcomputer, and others invested heavily in visions of a "wired campus." There was also a major push to establish computer literacy and computer science as required topics in the nation's elementary and secondary schools.

Computer-based education has been promoted with two different underlying ideologies in primary and secondary schools (Kling, 1983). Some educators argue that computer-based instructional approaches can help fulfill the traditional values of progressive education: the stimulation of intellectual curiosity, initiative, and democratic experiences. For example, Cyert (1984) has argued that computerized universities are qualitatively different than traditional universities; college students with microcomputers in their dorm rooms will be more stimulated to learn because they will have easy access to instructional materials and more interesting problems to solve. Papert (1979) argues that in a new computer-based school culture, children will no longer simply be taught mathematics; rather they will learn to be mathematicians. These visions portray an enchanted social order transformed by advanced computing technologies. Other advocates are a bit less romantic, but not less enthusiastic. For example, Cole (1972:143) argues:

> Because of . . . the insatiable desire of students for more and more information at a higher level of complexity and more sophisticated level of utilization . . . more effective means of communication must be used . . . computers can provide a unique vehicle for this transmission.

Others emphasize a labor-market pragmatism that we label "vocational matching" (Kling, 1983). In this view, people will need computer skills, such as programming, to compete in future labor markets and to participate in a highly automated society; a responsible school will teach some of these skills today. Advocates of computer-based education promote a utopian image of computer-using schools as places where students learn in a cheerful, cooperative setting and where all teachers can be supportive, enthusiastic mentors (Kling, 1986).

The computer-based education movement is not a well-organized national movement. It is far more diffuse and localized than other CMs. In some regions, such as the San Francisco Bay Area, consortia of teachers who are interested in computer-based instruction have formed local movement organizations. At the university level, the Apple Computer Company has formed a consortium of faculty from schools that have adopted

Apple computers on a major scale. While this movement organization is linked to a particular vendor, its participants advance a more general vision of computer-based education. Some magazines, such as *The Computer Teacher,* also promote the movement. Regional conferences, usually hosted by schools of education in state universities, are held for school teachers and administrators interested in computer-based education.

Academic researchers have been developing instructional courseware for primary and secondary schools since the 1960s. However, such products only became viable when computing equipment costs declined with the advent of microcomputers. By the 1980s, the promotion of computer-based education surged substantially. On one hand, there was growing public belief that public schools were having chronic problems in educating children. On the other, the personal computer industry was reducing the prices of basic equipment and promoting educational applications. Steve Jobs, co-founder of Apple Computer Corporation, lobbied hard and visibly to receive tax advantages by giving a free Apple IIe microcomputer to each public school district.

One microcomputer for a school of 500 students has little educational value, but the industry's marketing efforts and press promotion stimulated public interest in educational computing. Popular literature stresses the capabilities of equipment ignoring the lack of high quality courseware and inadequate teacher training in computer use (Kling, 1983). Many parents want their children exposed to computers in school although they know little at all about the details of computerized education.

The computer-based education movement received a symbolic boost in the spring of 1983 when the President's Commission on Excellence in Education released an urgent report that recommended that one semester of "computer science" be added to high school graduation requirements. This report is simply one high profile example of many local activities throughout the nation, much as the mobilization for the Equal Rights Amendment was one national level activity that captured the sentiments of many local variants of the woman's movement. In the 1980s there has been continuous ferment at state and local levels with coalitions of administrators, teachers, and parents banding together to lobby for various computer-based education programs in the public schools.

Office Automation

In the 1950s and early 1960s, "office automation" was synonymous with the introduction of computer-based technologies in offices—batch information systems in that period. Today office automation (OA) is a diffuse term that usually connotes the use of text-oriented computer-based technologies in offices.

OA technologies have two different kinds of roots. Stand-alone word processors evolved from magnetic card and magnetic tape typewriters. The organizational side of these office technologies evolved from kinds of secretarial work and the administrative services departments that commonly controlled organization-wide secretarial pools. The second root is computer systems that tied together text processing and electronic mail along with general purpose computing capabilities (e.g., workstations). These differences in technology are becoming less significant since specialized word processors are increasingly marketed with a wide array of information-handling functions. However, the social worlds that support these technologies still differ, even though they overlap.

Visionaries of automated offices used the term "office of the future" (Uhlig et al., 1979) as one of their major rallying points. An "office of the future" could never be built since there is no fixed future as a reference point for these technologies. However, a more prosaic conception lies beneath this vision: a terminal on every desk which provides text processing, mail, calendar, and file handling, communications, and other computing capabilities with a flexible interface. The scenarios emphasize the deployment of equipment, while OA advocates portray social relations as cheerful, cooperative, relaxed, and efficient—better jobs in better environments (Giuliano, 1982; Strassman, 1985).

These scenarios gloss the realities of work life in a highly automated office. Some office automation has led to deskilling and highly pressured jobs while work has been upskilled in other offices (Iacono and Kling, 1987). More seriously, the clerical work force is likely to retain jobs near the bottom of the American occupational structure in terms of pay, prestige, and control of working conditions. This is true even if the content of clerical jobs requires vastly more complex computer-related skills because most clerical jobs specialize in less discretionary delegated work (Iacono and Kling, 1987). The changes in professional work in automated offices are less clear; but it is clear that professionals cannot count on working with ample resources in a cooperative, cheerful environment (Kling, 1987). Computerization is a complex social and technical intervention into the operations of an organization. Conventional CM ideology emphasizes the power of new equipment and downplays the kinds of social choices that can allow powerful equipment to facilitate better jobs. When journalists criticize OA, it is often because the equipment does not deliver the miracles promised by the more enthusiastic advocates (Salerno, 1985).

Several national professional organizations promote OA. Associations of professional administrators and computer specialists have expanded their activities to include OA within their domains while the more specialized Association of Information Systems Professionals has a direct interest in the diffusion of OA. A large number of trade magazines and several academic journals, such as *Office* and *ACM Transactions on Office Information Systems*, support the OA movement. A strong office products industry also has developed and promotes OA equipment. This industry includes major vendors of mainframes, microcomputers, and specialized office equipment. Since the late 1970s the American Federation of Information Processing Societies has sponsored an annual trade show and conference of academic and professional OA activists.

Personal Computing

Personal computers (PCs), like video recorders, have become middle-class luxury appliances of the early 1980s. The PC movement began in the early 1970s as groups of hobbyists built their own primitive computers (Levy, 1984). Apple Computer Company and Tandy Corporation grew quickly in the late 1970s by providing two of the first microcomputers for which the purchaser did not have to solder parts and continually fiddle. The PC movement is one of the few CMs that has a distinctly "mass public" audience, although an elite audience of wealthy organizations that have also adopted and institutionalized PCs in some of their operations has developed as well.

Some of the early writings about PCs emphasized personal and social transformations that would accompany widespread PC use (Kay, 1977; Osborne, 1979). Today, the mobilizing ideologies of the PC movement have shifted to an instrumental pragmatism based on the capabilities of PCs for professionals. However, many PC enthusiasts believe that "almost everybody" should have a PC. For example, Jim Warren, the founder of a popular computer magazine and a series of immensely popular regional PC fairs, recently commented:

> I continue to feel that computers in the hands of the general public are crucial tools for positive social change. The only hope we have of regaining control over our society and our future is by extracting the information we need to make informed, competent decisions. And that's what computers do (quoted in Goodwin, 1988:114).

While PC applications such as word processing, financial analysis, and project scheduling are useful for professionals, "home applications" such as checkbook balancing, recipe storage, and home inventory are perhaps too marginal to justify a PC on instrumental grounds. The assumption that "almost everybody" should have a PC reveals the ideology of the PC movement.

The PC movement is national in scope and is popularized by high circulation national publications. By the mid-1980s, about 12 PC magazines were widely sold on newsstands in the United States. Many were specialized for audiences who own microcomputers made by particular vendors such as Apple, Tandy, and IBM. Others were specialized for business applications or some particular kind of technology (e.g., programming languages, UNIX). One very popular multivendor magazine, *Byte,* circulated approximately 500,000 copies per monthly issue.

The main national level activities are the trade fairs that travel from region to region. The software and hardware vendors in the microcomputer industry now play a major role in these ventures. The PC movement has numerous local movement organizations. The computer clubs in metropolitan areas are usually segmented along vendor lines (e.g., Apple, IBM, Atari) and still are dominated by hobbyists. In addition, computerized "bulletin boards" are operated by commercial firms and hobbyist amateurs. The two major commercial firms, Compuserve and The Source, serve business, professionals, and hobbyists nationwide. The operators of amateur-run bulletin boards are relatively transient and aim their services at computer hobbyists in their metropolitan areas. In the mid-1980s there were over 200 amateur-run bulletin boards in the United States. The PC movement overlaps with the office automation movement and the instructional computing movement since microcomputers have come to play a significant role as focal technologies for each. Some specialized organizations link these movements; for example, in the Los Angeles area, the PC Professional Association holds monthly meetings for PC coordinators who work in business organizations.

The mass media has become a major promoter of the PC movement. Newspapers such as the *Los Angeles Times* run a weekly personal computing column. It is difficult to separate the effects of journalistic promotion from industry advertising. As the PC industry has grown, IBM, Apple, and Commodore, among others, have placed advertisements on prime-time television. The PC industry's advertising is significantly enhanced by CM promotion. In the case of the PC movement, the mass media have played a major role. It is relatively rare to find the mass media investigating PC use. Dissatisfaction with the service and support provided by major computer sales chains is commonplace among PC enthusiasts. Journalists are much more apt to write about poor service in auto sales than in home computer sales.

Ideological Elements of CMs

Ideologies of computerization are especially developed in two kinds of writings. Some accounts focus on the coming of an "information society" and treat computerization as one key element in that transformation. In this view, the computerization of America will be an apolitical, bloodless "revolution." All will gain, with the possible exception of a few million workers, such as telephone operators and assemblyline welders, who will be temporarily displaced from "old-technology" jobs. Prophets of a new "information age" such as Bell (1979), Toffler (1980), Dizard (1982), Naisbitt (1983), and Strassman (1985) argue that this transformation is an inevitable and straightforward social process. Human intention, pluralities of interest, or large scale conflict play a minor role in these predictions of substantial social transformation.

A second source of computerization ideologies focuses on specific CMs and is of greater interest to us. Few CM activists, including those who publish their arguments, assert their key ideological themes directly. They can be located in books and articles only through a "symptomatic reading" in which the absence of relevant themes is as important or more important than those stated explicitly. For example, if an author argues that every household should have PCs for all family members, she is arguing in effect that families should restructure their budgets such that computing investments are not compromised.

We have identified five key and related ideas from such symptomatic reading. In addition, as students of and participants in several CMs, we have found these ideas recurring in continuing discussions with movement activists. As a system of beliefs, these themes help advance computerization on many fronts.

Computer-Based Technologies Are Central for a Reformed World

Many CM activists assert (or imply) that computer technology provides a historically unique opportunity for important social changes. For example, Papert (1979:74) advocates a special mode of computer-based education (e.g., Logo programming) and strongly criticizes other modes of computer-based education as reinforcing "traditional educational structures and thus play[ing] a reactionary role." Papert clearly articulates the belief that computing is special and different from all other educational innovations of the 1960s in its potential to redress serious widespread social inequities. Papert's analysis is largely individualistic and focuses upon children's cognitive abilities. He makes no attempt to ask whether social inequities are tied to a structured economic system rather than simply to the distribution of skills in society.

Similarly simple assertions about the special role of computing can be found in literature for each of the CMs. In the case of urban information systems, Evans and Knisley (1970) argue that a wide array of prosocial services can be specially supported by urban information systems. Giuliano (1982) suggests that office automation provides a unique historical opportunity to organize offices without tedious clerical work. Hayes-Roth (1984) suggests that AI-based expert systems will radically transform the professions by providing a unique capability in taking over cognitive expertise previously monopolized by people. In each case, these analysts imply that the computer-based technologies they advocate have a historically powerful and unique role.

CM activists often argue that computers are a central medium for creating the world they prefer. This belief gives proposals for computerization a peculiarly technocentric character: computer technologies are central to all socially valuable behavior. This belief is exemplified in characterizations of computer-based education in which any meaningful learning is computer-mediated (e.g., Papert, 1980).

One variant of this argument holds that computing is essential for modern organizations to compete effectively through increased productivity. Productivity, an economic conception, is linked to social progress through economic advance; and computerization, like prayer, will always have this desired result if it is "done properly." In addition, advocates attribute "productivity gains" primarily to new computing technologies, even though other elements, such as work organization and reward systems, play a critical role.

Concerns for productivity are closely linked to concerns for reducing costs. These concerns and the relatively high cost of "state of the art" computing, which helps set expectations, give the practice of computerization a relatively conservative political character. New systems of socially significant scale require the approval of higher level managers who are unlikely to approve arrangements which threaten their own interests (Danziger et al., 1982).

Improved Computer-Based Technologies
Can Further Reform Society

Advocates often portray the routine use of ordinary computer equipment as insufficient to reap the hoped-for benefits. Rather, the most advanced equipment is essential. This ideological theme is most clear when CM activists emphasize the continual acquisition of advanced equipment. They give minor attention to how to organize access to it or how to use it for social change. Nationally, the belief is reflected in funding priorities for computing research and development: the overwhelming support is for the development of new equipment. In the organizations we have studied, we have found relatively little money and attention spent on learning how to humanely integrate new computer-based technologies into routine social life.

CM activists often define the computing capabilities as those of future technologies, not the limits of presently available technologies. Many promotional accounts for computing reflect this future orientation. For example, Kay (1977) advocates book-sized personal computers that handle graphics, play music, and store vast amounts of data. Hiltz and Turoff (1978) advocate nationwide computer conferencing systems that would connect every household and office, much like telephones. During the last 20 years, computer-based technologies have become substantially cheaper, faster, smaller, and more flexible. Doubtless, computer-based systems will improve technologically during the next two decades just as cars, airplanes, typewriters, and telephones improved between 1910 and 1980. However, CM activists usually dismiss contemporary technologies, except to the extent that they foreshadow more interesting future technologies.

In any year, only a few organizations can purchase the state of the art equipment that CM activists recommend. With rapid changes in technological capabilities, today's technological leaders are surpassed by tomorrow's, unless they recycle their equipment so rapidly that they never "fall behind." Like people who purchase a new car or stereo with every model change, the heroes of this vision invest heavily and endlessly.

Computer hardware has become faster, cheaper, larger in scale, and more reliable in the last 30 years. Software support systems (like programming languages) have become more powerful and flexible, albeit at a much slower pace. Finally, computer applications have also improved technologically, but at a still slower pace. But a focus on future technologies helps deflect attention from the problems of using today's technologies effectively while offering the hope of salvation soon.

More Computing Is Better Than Less,
and There Are No Conceptual Limits to the Scope
of Appropriate Computerization

This theme goes beyond the previous one that "more computing can help reform society." CM activists usually push hard on two fronts: people and organizations ought to use state of the art computing equipment, and state of the art computing should become universal. In their writing and talks, CM activists usually emphasize the claim that certain groups use the forms of computing they prefer rather than explaining carefully how people alter their social lives to use or accommodate new technologies. For example, it is common for enthusiasts of computer-based education in schools to report that there are now several hundred thousand micro-computers in use but to spend relatively little time examining how children actually partake in computer-oriented classes.

CM activists imply that there are no limits to meaningful computing by downplaying the limits of the relevent technologies and failing to balance computing activity against competing social values. They portray computing technologies as mediating the most meaningful activities—the only real learning, the most important communications, or the most meaningful work—whether now or in the future that they prefer. Other media for learning, communicating, or working are treated as less important. Real life is life on-line (e.g., Feigenbaum and McCorduck, 1983; Papert, 1980; Hiltz and Turoff, 1978).

No One Loses from Computerization

Computer-based technologies are portrayed as inherently apolitical. While they are said to be consistent with any social order, CM advocates usually portray their use in a cheerful, cooperative, flexible, individualistic, and efficient world. This allows computer-based technologies to be shown as consistent with the most cherished social values. Computerization can enable long-term societal goals such as a stronger national economy and military. Any short-term sacrifices that might accompany these goals, such as displaced workers, are portrayed as minor unavoidable consequences.

CM activists rarely acknowledge that systematic conflicts might follow from computerizing major social institutions. Those that are acknowledged are defined as solvable by "rational discourse" and appropriate communication technologies (see, for example, Hiltz and Turoff, 1978). Many activists ignore social conflicts in their discussions of computerization and thus imply that computerization will reduce them. Some authors explicitly claim that computerized organizations will be less authoritarian and more cooperative than their less automated counterparts (Simon, 1977). In most of the accounts of office automation, staff are cheerfully efficient and conflicts are minor (see, for example, Strassman, 1985; Guiliano, 1982).

Similarly, in the literature on computer-based education, cheerful students and teachers who are invariably helpful and understanding populate computerized classrooms (see, for example, Taylor, 1980). Spitballs, paper planes, and secret paper messages that pass under the desks do not appear in this literature (Kling, 1983). Teachers who are puzzled by new technologies, concerned with maintaining order in their classrooms, or faced with broken equipment, competitive students, and condescending consultants are also ignored in the published discourse. Occasionally, an advocate of one mode of computer-based education may criticize advocates of others (e.g., Papert, 1979), but these are only sectarian battles within a CM.

Most seriously, the theme that computing fosters cooperation and rationality allows CM activists to gloss deep social and value conflicts that social change can precipitate (Kling, 1983; Kraemer, Dickhoven, Tierney, and King, 1987). In practice, organizational participants can have major battles about what kind of computing equipment to acquire, how to organize access to it, and what standards should regulate its use (Kling and Scacchi, 1982; Kling, 1987).

Uncooperative People Are the Main Barriers to Social Reform Through Computing

In many social settings we have found CM advocates arguing that poorly trained or undisciplined users undermine good technologies (Kling and Iacono, 1984). Even when they are making their procedures more complex or automating more exceptions and special contingencies, these advocates argue that the limitations of their co-workers rather than problems in their strategy of automation are the major impediments to nearly perfect computer-based systems. People place "unnecessary" limits on the complexity of desirable computer-based technologies and must be properly trained and taught to reorganize their activities and institutions accordingly.

These central themes of computerization movement ideology emphasize technological progress and deflect competing social values. They are foundation for social visions that include the extensive use of relatively advanced computer systems. In this vision, computer users should actively seek and acquire "the best" computer technologies and adapt to new technologies that become available, regardless of their cost. In this moral order, the users of the most advanced technologies are the most virtuous. And, as in melodramas where the good triumphs in the end (Cawelti, 1976:262), only developers and users of advanced computing technologies find the good life.

These ideological themes shape public images of computers and computerization. We do not claim that computer systems are "useless." We claim that these ideological positions help activists build commitment and mobilize resources for extensive computerization in organizations that adopt computer technologies beyond the value that mere utility justifies. It is ironic that activists often employ the imagery of science and objectivity (e.g., "knowledge") to advance their CMs. They deflect attention from what other analysts claim are the social problems raised by computerization—problems of consumer control, protection of personal privacy, quality of jobs, employment, and social equity, among others.

Counter-Computerization Movements

CMs generally advance the interests of richer groups in the society because of the relatively high costs of developing, using, and maintaining computer-based technologies. This pattern leads us to ask whether CMs could advance the interests of poorer groups or whether there are countermovements that oppose the general CM. Many CM activists bridle at these suggestions. They do not always value "helping the rich." In our fieldwork we have found that CM advocates sometimes see themselves as fighting existing institutional arrangements and working with inadequate resources. We do not suggest that "relative deprivation" triggers CMs. We are suggesting that CM participants can argue that they have nonelite positions even though they work with elite organizations. CM activists often develop coalitions with elite groups that can provide the necessary financial and social resources. The elite orientation of the general CM is sufficiently strong that one might expect some systematic "progressive" alternative to the major CMs.

There is no well-organized opposition or substantial alternative to the general CM. A general counter-computerization movement (CCM) might well be stigmatized as "Luddite." CM activists portray computing simply as a means to reform a limited set of social settings. Therefore, a specific CCM would have to oppose all or most computer-based education developments, or most office automation, or most PC applications, and so on. A successful ideological base for such opposition probably would have to be anchored in an alternate conception of society and the place of technology in it. In practice, there is no movement to counter computerization in general, though some writers are clearly hostile to whole modalities of computerization (Braverman, 1975: Mowshowitz, 1976; Reinecke, 1984; Weizenbaum, 1976). These writers differ substantially in their bases of criticism, from Frankfurt School critical theory (Weizenbaum) to Marxism (Mowshowitz, Braverman).

Criticism of computerization and CMs comes from social movements and organizations that focus on specific changes due to computerization, which they see as problematic. For example, civil libertarians are critical of those computing applications that most threaten personal privacy but are mute about other kinds of computing applications (Burnham, 1983). Consumer advocates may be highly critical of strategies of computerization that place consumers at a disadvantage in dealing with supermarkets on pricing or that increase consumers' liability for bank card problems.

But neither civil libertarians nor consumer advocates typically focus on problems of computerization in employment. Union spokesmen are especially concerned about how computerization affects the number and quality of jobs (Shaiken, 1986) but are mute about consumer issues. Antiwar activists criticize computer technologies that they think make war more likely but rarely speak to other computerization issues. Consequently, some analysts try to envision computer use shaped by other social values—such as consumer control over electronic payments or improvements in working life.

Major policy initiatives in these directions have come from other social movements. For example, consumer groups rather than CMs have been the main advocates of allowing debit card users protections such as ceilings on liability when cards are stolen, clear procedures for correcting errors, reverse payments, and stop payments (Kling, 1983). Civil liberties groups such as the American Civil Liberties Union (ACLU) have played a stronger role than CMs in pressing for the protection of privacy in automated personal record systems. Labor unions rather than CMs have been most insistent in exploring the conditions under which work with video display terminals (VDTs) has possible adverse health consequences. Each of these reform movements is relatively weak and specialized. Moreover, each initiative by CCMs to place "humane" constraints on laissez-faire computerization may be met by well-funded opposition within parts of the computer industry or computer-using industries. Consequently, the general drift is toward increased and intensive computerization with equipment costs, technological capabilities, and local organizational politics playing major enabling/limiting roles.

Discussion

We have argued that the computerization of many facets of life in the United States has been stimulated by a set of loosely linked CMs guided by mobilizing ideologies offered by CM activists who are not directly employed in the computer and related industries (Kling, 1983). We have characterized CMs by their ideological content, shown how five computer-based technologies are the focus of specific CMs, characterized core beliefs in their ideologies, and examined the fragmentary character of CMs. Our analysis differs from most organizational analyses of computerization by considering CMs that cut across the society as important sources of mobilizing ideologies for computing advocates. Their publications and meetings provide channels of communication for computing enthusiasts outside of the organizations that employ them and which they try to aggressively computerize.

But much more should be done in examining particular CMs. We need to learn in more detail about their participants and social organization and to better understand their relations with computerizing organizations, interest groups, the media, other CMs, and different segments of the computer industry. We hope that this analysis will encourage scholars to examine the CMs in other specific social settings and the activists who push them. These activists play a critical role in setting expectations about what a particular mode of computing is good for, how it can be organized, and how costly or difficult it will be to implement. These expectations can shape participants' attempts to computerize in a specific social setting such as a school, public agency, hospital, or business.

CM activists and participants play a role in trying to persuade their audiences to accept an ideology that favors everybody adopting "state of the art" computer equipment in specific social sectors. There are many ways to computerize, and each emphasizes different social values (Kling, 1983). While computerization is rife with value conflicts, CM activists rarely explain the value and resource commitments that accompany their dreams. And they encourage people and organizations to invest in computer-based equipment rather than paying equal or greater attention to the ways that social life can and should be organized around whatever equipment is acquired. CM activists provide few useful guiding ideas about ways to computerize humanely.

During the last 20 years, CMs have helped set the stage on which the computer industry expanded. As this industry expands, vendor organizations (like IBM) also become powerful participants in persuading people to automate. Some computer vendors and their trade associations can be powerful participants in specific decisions about equipment purchased by a particular company. They also can help weaken legislation that could protect consumers from trade abuses related to computing (Kling, 1983). But vendor actions alone cannot account for the widespread mobilization of computing in the United States. They feed and participate in it; they have not driven it. Part of the drive is economic, and part is ideological. The ideological flames have been fanned as much by CM advocates as by marketing specialists from the computer industry. Popular writers like Alvin Toffler and John Naisbitt and academics like Daniel Bell have stimulated enthusiasm for the general computerization movement and provided organizing rationales (e.g., transition to a new "information society") for unbounded computerization. Much of the enthusiasm to computerize is a by-product of this writing and other ideological themes advanced by CMs.

Most computer-based technologies are purchased by organizations. The advocates of computerization within specific organizations often form coalitions with higher level managers to help gain support and resources for their innovations, which they present as professional reform movements with limited scope. Office automation "only" influences general office practices. Advanced computerized accounting systems "only" influence the finance department and those who manage revenues and expenses. Computer-based inventory control systems "only" influence materials handling in manufacturing firms. In each administrative sector that is professionalized, some related group has taken on the mantle of computerization as a subject of reform. But these professionals do not identify with the computer industry; they identify themselves as accountants, doctors, teachers, or urban planners with an interest in certain computer applications.

By attempting to alter the character of social life across the society, the general computerization movement is basically revolutionary. A few computing promoters exploit a slick "revolutionary" image. As Langdon Winner (1984) aptly observes, Americans have been marketed "revolutionary" toothpastes, home entertainment centers, and plastics. None of these consumer items has revolutionized the social order. According to Winner, the sensible observer would treat claims about a "computer revolution" as marketing hype. We agree with Winner that shallow promotional claims dominate this public "revolutionary" discourse. Moreover, the social changes that one can attribute to computerization are often socially "conservative." But we suspect that the pervasiveness of computerization is having quiet cumulative effects in American life. Certain technologies are relatively plastic and substantially extend people's range of action. Basic technologies for transportation, energy, and communications, such as automobiles, electricity, and telephones, have become central elements of social life in advanced industrial societies. We suspect that computer-based technologies will be as socially important as these other technologies rather than peripheral, like plastics, processed foods, and hair spray.

There is not likely to be a general CCM. More seriously, it is unlikely that humanistic elements will be central in the mobilization of computing in the United States. Central humanistic beliefs are "laid onto" computerization schemes by advocates of other social movements: the labor movement (Shaiken, 1986), the peace movement, the consumer-rights movement (Kling, 1983), the civil liberties movement (Burnham, 1983). Advocates of the other movements primarily care about the way some schemes for computerization intersect their special social interest. They advocate limited alternatives to particular CMs but no comprehensive, humanistic alternative to the general computerization movement. In its most likely form, our "computer revolution" will be a conservative revolution that will reinforce the patterns of an elite-dominated, stratified society.

Note

1. Other computer technologies, such as electronic funds transfer (Kling, 1978; Kling, 1984), robotics, and supercomputing are also advanced by CMs. Not all applications of computer technologies, however, are the focus of CMs; for example, payroll systems are not promoted through CMs.

References

Applegate, J., and K. Day. 1984. Computer firms aim: artificial intelligence. *Los Angeles Times,,* Part IV August 7:1, 17.

Aronson, Naomi. 1984. Science as a claims-making activity: implications for social problems research. In Joseph Schneider and John Kitsuse, eds., 1–30, *Studies in the sociology of social problems.* Norwood, NJ: Ablex Publishing Co.

Attewell, Paul and James Rule. 1984. Computing and organizations: what we know and what we don't know. Communications of the ACM December 27:1184–92.

Bell, Daniel. 1979. The social framework of the information society. In Michael Dertouzos and Joel Moses, eds., 163–211. *The computer age: A twenty-year view.* Cambridge, MA: MIT Press.

Blumer, Herbert. 1969. Social movements. In Barry McLaughlin, eds., 8–29. *Studies in social movements: A social psychological perspective.* New York: Free Press.

Braverman, Harry. 1975. *Labor and monopoly capital: The degradation of work in the twenthieth century.* New York: Monthly Review Press.

Bucher, Rue, and Anselm Strauss. 1961. Professions in process. *American Journal of Sociology* 75:325–34.

Burnham, David. 1983. *The rise of the computer state.* New York: Random House.

Business Week. 1984. Artificial intelligence is here: computers that mimic human reasoning are already at work. July 9:54–57, 60–62.

Cawelti, John. 1976 *Adventure, mystery, and romance: Formula stories as art and popular culture.* Chicago: University of Chicago Press.

Cole, Ralph I. 1972. Some reflections concerning the future of society, computers and education. In Robert Lee Chartrand, ed., 135–145. *Computers in the service of society.* New York: Pergamon Press.

Cyert, Richard. 1984. New teacher's pet: the computer. *IEEE Spectrum* 21:120–22.

Danziger, James, William Dutton, Rob Kling, and Kenneth Kraemer. 1982. *Computers and politics: High technology in American local governments.* New York: Columbia University Press.

Dizard, Wilson. 1982. *The coming information age*. New York: Longman.

Evans, John W., and Richard A. Knisley. 1970. *Integrated municipal information systems: Some potential impacts*. Washington, DC: U.S. Department of Housing and Urban Development.

Feigenbaum, Edward, and Pamela McCorduck. 1983. *Fifth generation: Artificial intelligence and Japan's challenge to the world*. Reading, MA: Addison-Wesley.

Giuliano, Vincent. 1982. The mechanization of office work. *Scientific American* 247:148–64.

Goodwin, Michael. 1988. Wild-man Warren. *PC World,* January 6:108–09, 114.

Hayes-Roth, Frederick. 1984. The machine as partner of the new professional. *IEEE Spectrum,* 21:28–31.

Hiltz, Starr Roxanne, and Murray Turoff. 1978. *The network nation: Human communication via computer*. Reading, MA: Addison-Wesley.

Iacono, Suzanne, and Rob Kling. 1987. Changing office technologies and transformations of clerical work: a historical perspective. In Robert Kraut, ed., pp 53–75. *Technology and the transformation of white collar work*. Hillsdale, NJ: Lawrence Erlbaum.

Kaplan, Bonnie. 1983. Computers in medicine, 1950–1980: the relationship between history and policy. Ph.D. diss., University of Chicago.

Kay, Alan. 1977. Microelectronics and the personal computer. *Scientific American,* 237:230–44.

Kling, Rob. 1978. Automated welfare client-tracking and service integration: The political economy of computing. *Communications of the ACM* 21:484–93.

———. 1980. Computer abuse and computer crime as organizational activities. *Computers and Law Journal* 2:403–27.

———. 1983. Value conflicts in the deployment of computing applications: cases in developed and developing countries. *Telecommunications Policy* March:12–34.

———. 1986. The new wave of academic computing in colleges and universities. *Outlook* 19, Spring/Summer:8–14.

———. 1987. Defining the boundaries of computing across complex organizations. In Richard Boland and Rudy Hirschheim, eds., 307–62. *Critical issues in information systems*. London: John Wiley.

Kling, Rob, and Suzanne Iacono. 1984. The control of information systems development after implementation. *Communications of the ACM* December 27:1218–26.

Kling, Rob, and Walt Scacchi. 1982. The web of computing: computer technology as social organization. *Advances in Computers* 21. New York: Academic Press.

Kraemer, Kenneth, and John Kling. 1978. Requiem for USAC. *Policy Analysis* 5:313–49.

Kraemer, Kenneth, Siegfried Dickhoven, Susan Fallows Tierney, and John King. 1987. *Datawars: The politics of modeling in federal policymaking.* New York: Columbia University Press.

Laudon, Kenneth C. 1974. *Computers and bureaucratic reform.* New York: John Wiley and Sons.

————. 1986. *Dossier society: Value choices in the design of national information systems.* New York: Columbia University Press.

Levy, Steven. 1984. *Hackers: Heroes of the computer revolution.* Garden City, NY: Anchor/Doubleday.

McCarthy, John, and Mayer Zald. 1977. Resource mobilization and social movements: a partial theory. *American Journal of Sociology* 82:1212–41.

Mowshowitz, Abbe. 1976. *The conquest of will: Information processing in human affairs.* Reading, MA: Addison-Wesley.

Naisbitt, John. 1983. *Megatrends.* New York: Warner Books.

Osborne, Anthony. 1979. *Running wild: The next industrial revolution.* Berkeley, CA: Osborne-McGraw Hill.

Papert, Seymour. 1979. Computers and learning. In Michael L. Dertouzos and Joel Moses, eds., 73–86. *The Computer age: A twenty-year view.* Cambridge, MA: MIT Press.

————. 1980. *Mindstorms: Children, computers and powerful ideas.* New York: Basic Books.

Reinecke, Ian. 1984. *Electronic Illusions.* New York: Penguin Books.

Salerno, Lynne. 1985. What ever happened to the computer revolution? *Harvard Business Review* 85:129–38.

Shaiken, Harley. 1986. *Work transformed: Automation and labor in the computer age.* Lexington, MA: Lexington Books.

Simon, Herbert. 1977. *The new science of management decision.* Englewood Cliffs, NJ: Prentice-Hall.

Star, Susan Leigh. Forthcoming. *Regions of the mind: British brain research, 1870-1906.* Stanford: Stanford University Press.

————. 1988. Personal communication. February 15, Irvine, California.

Strassman, Paul. 1985. *Information payoff: The transformation of work in the electronic age.* New York: Free Press.

Taylor, Robert, ed. 1980. *The computer in the school: Tutor, tutee, tool.* New York: Teachers College Press, Columbia University.

Time. 1982. Machine of the year. *Time* January 3:13–39.

Toffler, Alvin. 1980. *The third wave.* New York: William Morrow.

Turkle, Sherry. 1984. *The second self: Computers and the human spirit.* New York: Simon and Schuster.

Uhlig, Ronald, David Farber, and James Bair. 1979. *The office of the future: Communication and computers.* New York: North-Holland.

Weizenbaum, Joseph. 1976. *Computer power and human reason.* New York: W.H. Freeman and Co.

Winner, Langdon. 1984. Myth information in the high-tech era. *IEEE Spectrum* 21:90–96.

Winston, Patrick and Karen Prendergast, eds. 1984. *The AI business: Commercial uses of artificial intelligence.* Cambridge, MA: MIT Press.

Yourdon, Edward. 1986. *Nations at risk: The impact of the computer revolution.* New York: Yourdon Press.

Zald, Mayer, and Michel Berger. 1978. Social movements in organizations: Coup d'etat, insurgency, and mass movements. *American Journal of Sociology* 83:823–61.

Justice: The Nature
and Distribution of Work

A U.S. government study (reading 20) addresses possible employment problems resulting from the use of computers, with special emphasis on the difficulty of making predictions, the lessons of history, the interaction of computers with economic forces, and the ways human decisions affect computer applications.

Nils Nilsson (reading 21) examines the potential economic effects of artificial intelligence, and concludes that AI probably will cause massive reductions in the amount of human labor needed to produce the world's goods and services. He claims that this development will be a blessing rather than a curse if we are able to dissociate income from employment and make changes in employment gradual and nondisruptive.

Juan Rada (reading 22) considers the plight of developing countries in the face of the computer revolution. His conclusions, drawn from empirical studies of the economies of third world countries, are mixed. For instance, he recognizes that some factors, like the lower barrier to entry in the international service market, may benefit developing nations, but he suggests that the general trend seems to be toward widening gaps between the wealthy and the poor.

Philip Bereano (reading 23), in his article on social relations within industrialized countries, argues that political, cultural, and economic forces determine how new technologies are used or abused. He is particularly concerned about how power differences in our society are linked to the use of computers and other technological things.

Computers and Employment

National Commission for Employment Policy

Of the equipment, *Scientific American* heralded:

> We believe that it is one of the most important inventions of the age. We will yet live to see it forming part of every household's furniture, for it is undoubtedly a family labor-saving machine.

Of the industry, the same periodical asserted:

> The rapid rise of the . . . business constitutes one of the wonders of this enterprising age. No industrial revolution can equal that which has been produced by it within the short space of sixteen years.

The first statement appeared in 1851; the second in 1862. The innovation was the sewing machine.[1] This [essay] begins with these 125-year-old quotes because of their resemblance to statements being made today. For example:

> The new microtechnology . . . is bringing change which is perhaps even more revolutionary than the industrial revolution . . .[2]

Predictions about specific ways that computers will affect the workplace are also being made. It is said that computers will:

- Decrease the skills required to perform jobs. Such "deskilling" implies an increasing routinization of work and an overeducated and bored work force.[3]

- Increase the skill requirements of jobs. This "upskilling" implies that many workers will be unable to qualify for available jobs unless they acquire additional training, and that the least educated segments of the population will have greater difficulty finding work than they do now.[4]

- Dramatically increase the mismatches between the skills that jobs require and those available in the work force, implying that the skills workers now possess will become largely obsolete.[5]

From National Commission for Employment Policy, *Computers in the Workplace: Selected Issues,* pp. 23–24, 29–30, 39–43, 65–66. Washington, D.C., March 1986.

- Mandate that workers be "computer" or "technology" literate.[6]
- Eliminate many clerical jobs as the "paperless office of the future" becomes a reality.[7]
- Eliminate many operative jobs as the "factory of the future" becomes a reality.[8]

These statements, which are being made in the United States and a number of Western European countries, carry with them the [implication] that computers pose a threat to workers and to young people preparing for work. The fears are generalized to include most workers and most young people and are also personalized: computers will replace me in my job; my job is going to change so much that I will not be able to perform it; my job will become boring; when I finish school I will not be qualified to get a good job.

The purpose of this [essay] is to assess the employment consequences of the "computer revolution." The [essay] challenges or modifies the predictions reported earlier and attempts to allay some of the fears they have produced.

Examining the impact of computers is a difficult undertaking for several reasons. First, there are many types of computers and computer-based equipment and they are used in a variety of ways. There are mainframe computers, minicomputers, and microcomputers, differentiated on the basis of computing power and price. Equipment which uses these computers, termed here computer-based equipment, includes, for example, industrial robotic and computer-assisted design systems, automatic teller machines, and computerized cash registers. For the purposes of this essay, the terms computers and computer-based equipment are used interchangeably.

Second, there is a paucity of information on the number of computers in place, on the ways they are being used, and on the occupations and industries affected. Where there is information on particular types of computer-based equipment, it is used [in this essay]; in other cases the discussion relies upon whatever reasonable evidence is available.

Finally, assessing the impact of computers is a difficult undertaking because they are not being adopted in a vacuum. Other forces are operating simultaneously and affecting the number and types of employment opportunities. For example, economic growth is stimulating increases in employment and can be offsetting decreases in employment due to the introduction of computers. Also, changes in both consumer preferences and patterns of world trade are affecting the number of jobs that exist and in which industries [these jobs] are found. . . .

A Historical Perspective

While the computer is one of the most recent innovations society has witnessed, it is not the first to have a major impact on employment or to be described as revolutionary. Although the computer is unique in many respects, experiences with previous innovations provide several insights into what we might reasonably expect.

First, whether an innovation is perceived as beneficial or detrimental to a nation's employment opportunities depends largely upon the overall well-being of the national economy during the period of diffusion. The introduction of the power loom into the English economy in the nineteenth century serves as an example. Because it was introduced during a period of—and contributed to—massive growth in factory jobs, it has generally been viewed as beneficial, despite the fact that its adoption displaced the hand-loom weavers.

Attitudes toward computers have changed over the decades as general economic conditions have changed. When there was high unemployment in the early 1960s and again in the early 1980s, there was fear that computers would cause massive nationwide unemployment; many of these fears dissipated as unemployment rates declined. History's judgment of computers will in part reflect the health of the economy during the period of diffusion.

Second, the full range of changes that will ensue from an innovation cannot be known in advance. For example, the sewing machine's greatest contribution to economic growth was in the method used to produce it. The machine tools developed to manufacture sewing machines could be used to produce a number of items in addition to the sewing machine; they were subsequently used to mass produce such things as other machine tools, steam engines, firearms, bicycles, and locomotives.

As a further example, while electric power was developed as an alternative to gas lighting and steam power, one of its greatest impacts was the flexibility it offered manufacturing operations. Machines driven by steam derived their power from a central shaft which ran the length of the factory, and they had to be clustered near the central shaft in order to run. With electric power, the location of the power source no longer constrained the location of the machinery. Equipment could be placed in a more efficient sequence, thereby increasing workers' productivity. Indeed, even the location of factories was no longer determined by their power source.

When computers were first developed, observers of the day could not foresee the full range of ways in which they are already used. Although the world is rife with predictions about computers of the future, it is similarly unlikely they will have captured either the full range of computer uses or the full range of changes they will bring about.

Third, major innovations tend to be adopted over many years, allowing people time to adjust to the ensuing changes. The spread of electric power, the "Second Industrial Revolution," spanned decades. It took about 50 years after Thomas Edison opened his first power station for electric power to diffuse through all of American manufacturing. The development of the microprocessor in 1971, 35 years after COLLOSUS [COLOSSI?], seems to be the analogous event of the "computer age." The nation is already more than ten years into the microprocessor era, and the diffusion of computers is far from complete. . . .

Computers' Effects on Jobs

A commonly expressed fear about computers is that they will either replace workers in their jobs or change job requirements so much that many workers and potential workers will not qualify for the jobs they hold or want. This fear is reflected in predictions about the "paperless office," the "factory of the future," "upskilling," and "mismatches" between what people know how to do and the skills that jobs require.

Determining the accuracy of such predictions can only occur through hindsight. Since . . . the United States is somewhere in the midst of the diffusion of computers, the best we can do at present is to analyze the changes that have occurred in offices and factories and add some of the lessons learned from experiences with previous innovations.

[We will now discuss] the crucial role that human resources decisions play in the introduction of computer-based equipment, [and] then [examine] the issues of job loss and work displacement associated with changes occurring in the economy. . . .

The Role of Human Resource Decisions

The installation of computer-based equipment sometimes compels, but more often permits, industries and individual firms to reorganize and to restructure jobs. The effects of these changes on workers vary, in part because of the nature of the equipment and in part because of personnel decisions.[9] For example, computerized equipment enables some firms to decrease their demand for workers in some low-skill jobs, such as spot painting in the automobile industry and taking inventory in supermarkets. Other types of computerized equipment can decrease the demand for higher-skill jobs, such as drafting.

Computers also permit workers in some occupations to reduce the amount of time spent on routine tasks and thus increase the amount of time available for problem diagnosis, problem solving, and in some cases, interacting with the public, as in some service representative jobs. This "upskilling" is in contrast to the "deskilling" possible in other occupations, where the effect of computers is to make the work less demanding, as with cash registers that calculate the change due a customer.

Many computer applications allow managers to monitor workers' performance more closely. The most publicized stories involve people who have been fired as a result of computer surveillance—telephone operators, word-processor operators, and people who clean hotel rooms. These cases have given rise to one of the fears of computers. However, there are also instances in which the capability for rapid feedback on performance has been incorporated into performance evaluation systems and has allowed workers to be rewarded for outstanding performance or to learn how they might improve their performance.

What happens to employees—whether their jobs are abolished, "deskilled," "upskilled" or increasingly monitored—is primarily determined by managerial decisions and worker/management relations. There are a number of ways that computer-based equipment can be installed within factories and offices. At one end of the continuum, management may unilaterally redesign an organization or individual jobs. The workers have no say in any aspect of implementation, including which workers are to be trained to use the equipment. Alternatively, management may involve workers at all stages of the planning and implementation process. This means that workers' concerns will be considered along with engineering and financial ones. . . .

Job Loss and Worker Displacement

The programs of workers who have been displaced due to computers have been matters of widespread concern over the past several years. Addressing these problems has been complicated by uncertainty over how many workers are involved and who they are.

Computers have been blamed for the decline of some occupations and simultaneously credited with the growth of others. However, it is impossible to separate the effects of technology on changes in employment from the effects of other factors operating simultaneously in the economy. Some of these other factors are structural, such as changes in patterns of international trade and changes in consumer preferences. There are also cyclical effects; for example, firms may be more financially able to keep and retrain workers in good times than in bad.

An important aspect of the simultaneity and interdependence of structural and cyclical changes is that a technology-caused decline in the number of jobs in an industry or an occupation does not necessarily result in layoffs. Further, not all laid off workers experience long-term unemployment—that is, become displaced workers. Some industries and areas of the U.S. may experience layoffs without workers becoming displaced because there are job openings in other parts of the local economy[10]; other industries and areas may experience a disproportionate amount of both layoffs and subsequent dislocation because there are few opportunities elsewhere in the local economy; and still other industries and localities may experience slow growth, but neither layoffs nor dislocation.

A lesson from history—that major innovations tend to be adopted over long periods of time—is important to remember. The length of the period of diffusion influences an innovation's effects on employment. . . . [T]he spread of computers is likely to continue for a number of years. This means that some computer-related job losses that might otherwise result in layoffs can be handled through attrition. It also means that the number of workers categorized as displaced may be lower than if computers were spreading more rapidly, simply because there are some job openings every year.

Nonetheless, policymakers need to realize that even if attrition relieves some problems, some worker displacement in the future is inevitable. It is also certain that a substantial number of these people will need some sort of assistance in finding work. To do so, many of them will need to acquire a good working knowledge of English, math, and problem solving—the basic skills.

A second policy issue is whether displacement, regardless of its cause, is manageable through existing mechanisms such as job search, mobility assistance, and income support programs or if new policies and programs will be necessary. At the root of this issue are the size and characteristics of the displaced worker population. Characteristics in this context refers to who they are, where they are, and the reasons for the long duration of their unemployment.

Current Estimates   Estimates of the number of displaced workers vary greatly.[11] [One study which] focused on workers who had the greatest difficulty finding jobs[12] . . . investigated dislocation problems in five States. [It] defined workers as displaced if they had lost jobs in industries experiencing decline in the workers' local areas, and had also exhausted their unemployment benefits. Under this definition, about 11 to 19 percent of all unemployment workers, depending upon the state, were considered displaced. Applying this range to the country as a whole means, for example, that of the more than 8 million people reported as unemployed in September 1985, between 900 thousand and 1.5 million . . . would be categorized as displaced.

The prototypical displaced worker is a middle-aged white male, living in the Great Lakes region, with at least ten years work experience in the auto or steel industries. However, that image is only part of the picture. There is, in fact, considerable diversity among displaced workers in terms of the industry in which they had worked, the region of the country in which they live, and their ages. This diversity becomes especially apparent when looking at local labor markets. National estimates mask particular regional or local dislocations because they combine dislocations resulting from plant closings in some areas of the country with openings or expansion in the same industry in other parts of the country.

Projections Only a few studies have predicted the effects of computers on the number of jobs available over the next several years. Two of the studies examined the employment implications of robotic systems in factories, but most focused on office and service work. The studies in robotics resulted in a broad range of estimates of the number of jobs potentially lost or not created because of their introduction: from 100,000 to 800,000 over the 1980s.[13] Similarly, the projections of clerical employment changes also varied widely: from an absolute decline of 250,000 to an increase of about 50,000 between 1982 and 1995.[14]

Despite the wide ranges in the numbers emerging from these studies, it is clear that there will be some decline in the number of jobs needed to produce the same level of output because of robotization and computer-based office equipment. To put these numbers in perspective, in 1980, about 3 million workers, or just over 3 percent of all employed people, were in those assembly-line occupations that are most susceptible to robotization: welding, assembling, painting, and machine loading and unloading. Most of these workers are men who live in the industrial Northeast or Midwest. About 19 million workers, or 20 percent of all employed people, are in those clerical occupations most susceptible to the introduction of computer-based office equipment. The occupations range from bank teller to office machine operator to postal clerk to secretary, reflecting the diversity of uses of the equipment. Most of these workers are women and the jobs are spread throughout the nation.

The preponderance of women or men in particular jobs brings up another aspect of the impact of computers on jobs, and one which has received relatively little attention. Computers are sex, age, race, and national origin "neutral." That is, personal characteristics are not necessarily related to the ability to use computer-based equipment. However, employment and training opportunities for different demographic groups differ when computers are introduced into the workplace. According to available research, women, minorities, and older workers are less likely than white males under the age of 45 to have access to computer-related training and to more-skilled jobs.[15] This reflects employers' (and sometimes employees') preconceptions about either what is "appropriate" for different groups of workers or what abilities and commitments to work different groups have. . . .

Preparing for a Change

Computers and computer-based equipment are one of the new technologies that are changing what we do and how we do it. Whether the process is termed "revolutionary" or "evolutionary" is largely a matter of personal and historical perspective rather than of fact. If *your* job is abolished or dramatically altered when new equipment is introduced into the workplace, it is a personal revolution, even if you are retrained to work with the new equipment or can move to another occupation. From a broader perspective, however, technological change is a continuous process, and the diffusion of computer-based equipment appears more evolutionary, as has been the case with other innovations.

Computerization is one of the forces causing permanent structural change in the economy, in contrast to cyclical changes, which tend to be reversed as economic conditions vary. A useful analogy is the distinction between changes in the weather (cyclical) and changes in the climate (structural). Most studies of the economy are like weather forecasts, whereas [our] goal . . . was to investigate the extent to which the climate is changing. Our conclusion is that while structural change is proceeding irreversibly, it is at a rate which is manageable given intelligent and well-informed preparation. . . .

The conclusion that structural adjustment is manageable means that individuals and firms should use this time to plan and adjust to new labor market realities. The political, educational, and social institutions which play their own roles in the labor market, such as government agencies, schools, and unions, must also use this time to adjust. Employment opportunities and job requirements will change regardless of whether this planning takes place: the hope is that preparation will reduce distress and delay. . . .

New technologies in general, and computers in particular, can only be implemented by human decisions. Human decisions are also central in the distribution of the benefits of change, including the economic growth made possible by greater productivity, and the improved quality of work life in safer and more pleasant surroundings. Some of these implementation decisions have imposed significant costs on the people least able to bear them, in the form of extended unemployment and reduced earnings prospects. Displaced workers are only the most obvious manifestations of these costs of change. Decisions which take human resource impacts into account would reduce these costs and redistribute them more in accordance with the distribution of the benefits of change, which could also be made more equitable.

Many other nations share with the United States the realization that change needs to be managed, not reversed. If an industry is to be viable in world competiton, it must adopt world-class technology. Aid to industries and individuals making the transition to the new technologies can be a legitimate policy instrument of governments at all levels. Those aided, however, also have the responsibility to make conscientious use of public and private resources to prepare themselves for the new environment into which we are moving. When the waters are uncharted, it is wise to know how to swim.

Notes

1. John A. James, *Perspectives on Technological Change: Historical Studies of Four Major Innovations,* Research Report 84-07 (Washington, DC: National Commission for Employment Policy, September 1984).
2. AFL-CIO Committee on the Evolution of Work, *The Future of Work* (Washington, DC: AFL-CIO, August 1983).
3. Henry M. Levin and Russell W. Rumberger, *The Educational Implications of High Technology,* Project Report 83-A4 (Stanford, CA: Institute for Research on Educational Finance and Governance, School of Education, Stanford University, February 1983).
4. Education Commission of the States, *The Information Society: Are High School Graduates Ready? (*Denver: Education Commission of the States, 1982).
5. H. Allan Hunt and Timothy L. Hunt, *Human Resource Implications of Robotics (*Kalamazoo, MI: W. E. Upjohn Institute for Employment Research, 1983).
6. Northwest Regional Educational Laboratory, "Technological Literacy Skills Everybody Should Learn," in *Ideas for Action in Education and Work (*Portland, OR: Northwest Regional Educational Laboratory, August 1984).
7. Georgia Tech Research Institute, *Impact of Office Automation on Office Workers (*Washington, DC: U.S. Department of Labor, Employment and Training Administration, Grant 21-13-82-13, 1984).
8. Robert Ayres and Steven M. Miller, *Robotics, Applications and Social Implications (*Cambridge, MA: Ballinger Publishing Co., 1983).
9. A firm's motivation for computerizing may have little to do with a desire to affect jobs. For example, computer control of office-building heating and cooling systems occurred in response to higher energy costs, and the

implications of these systems for building service jobs is unclear. More sophisticated equipment may require more frequent and/or more skilled maintenance, even as it may reduce skill requirements for equipment operators. However, most decisions to computerize reflect, at least in part, a desire to lower labor costs.

10. Of course, new jobs may offer lower wages than those the displaced workers received in their previous jobs. A recent survey found that displaced workers who found a new job experienced an average loss in earnings of about 9 percent. See *Displaced Workers, 1979–83*, BLS Bulletin 2240 (Washington, DC: U.S. Department of Labor, Bureau of Labor Statistics, July 1985). . . .

11. These variations exist largely because there are several definitions of the term displaced worker, each with different criteria for inclusion. The different criteria include layoff from a declining industry or from a declining occupation, or the duration of unemployment.

   The BLS report noted in the previous endnote estimated that about 5 million workers were displaced between 1979 and 1984. The criteria used in making this estimate were that the workers had been with their previous employer for at least three years; had lost their jobs due to plant closings, layoffs, or plant relocations; and were at least twenty years old at the time they lost their jobs. . . .

12. Robert Crosslin, James Hanna, and David Stevens, *Identification of Dislocated Workers Utilizing Unemployment Insurance Administrative Data: Results of a Five-State Analysis,* Research Report 84-03 (Washington DC: National Commission for Employment Policy, April 1984).

13. One study, which emphasized what tasks robots are technically capable of performing, suggested that up to 800,000 jobs could be replaced by the equipment by 1990. Another study projected 100,000 to 200,000 jobs lost during the 1980s. This estimate appears to be more realistic since the authors attempted to take into account the economic feasibility of robotic systems for firms. Robert Ayres and Steven Miller, *Robotics, Applications and Social Implications* (Cambridge, MA: Ballinger Publishing Co., 1983); and H. Allan Hunt and Timothy Hunt, *Human Resource Implications of Robotics.*

14. For example, one study estimated a decline of about 250,000 clerical workers between 1982 and 1995, while another study projected growth, albeit below average, for the period 1978 to 1990. A third study, which looked at clerical employment in the banking and insurance industries specifically, projected slow growth to 1990 and a decline thereafter. (A survey of these studies is given in H. Allan Hunt and Timothy Hunt, "Clerical Employment and Technological Change: A Review of Recent Trends and Projections.")

   In contrast to these studies, the Bureau of Labor Statistics (BLS) estimates that clerical employment will increase about 25 percent between 1982 and 1995, the same rate as employment generally. The BLS projections are the most useful. According to an examination of these projections, if technological change occurred but the economy did not grow, clerical employment might decline in some industries. Employment opportunities for clerical workers are expected to increase because overall economic growth is projected, as well as growth in those industries in which clerical workers are concentrated, such as finance and services.

15. Patricia M. Flynn, *The Impact of Technological Change on Jobs and Workers* (Washington, DC: U.S. Department of Labor, Employment and Training Administration, Grant 21-25-82-16, March 1985).

— 21 —

Artificial Intelligence, Employment, and Income

Nils J. Nilsson

Artificial intelligence (AI) and other developments in computer science are giving birth to a dramatically different class of machines—machines that can perform tasks requiring reasoning, judgment, and perception that previously could be done only by humans. Will these machines reduce the need for human toil and thus cause unemployment? There are two opposing views in response to this question. Some claim that AI is not really very different from other technologies that have supported automation and increased productivity—technologies such as mechanical engineering, electronics, control engineering, and operations research. Like them, AI may also lead ultimately to an expanding economy with a concomitant expansion of employment opportunities. At worst, according to this view, there will be some, perhaps even substantial shifts in the types of jobs, but certainly no overall reduction in the total number of jobs. In my opinion, however, such an outcome is based on an overly conservative appraisal of the real potential of artificial intelligence.

Others accept a rather strong hypothesis with regard to AI—one that sets AI far apart from previous labor-saving technologies. Quite simply, this hypothesis affirms that anything people can do, AI can do as well. Certainly AI has not yet achieved human-level performance in many important functions, but many AI scientists believe that artificial intelligence inevitably will equal and surpass human mental abilities—if not in twenty years, then surely in fifty. The main conclusion of this view of AI is that, even if AI does create more work, this work can also be performed by AI devices without necessarily implying more jobs for humans.

Of course, the mere fact that some work can be performed automatically does not make it inevitable that it will be. Automation depends on many factors—economic, political, and social. The major economic parameter would seem to be the relative cost of having either people or machines execute a given task (at a specified rate and level of quality). In this respect too, AI differs from many previous labor-saving technologies in that it is relatively very inexpensive and will undoubtedly become even more so in the future. Yet, even granting an economic rationale for replacing human labor with machines, we as a society may choose not to do so. That is, we may decide to continue to employ humans in jobs "next to the window" (as the Japanese say), simply as a way to distribute income and to give people something tangible to do.

In this paper I examine the potential economic effects of artificial intelligence. I conclude that AI does indeed offer the potential for achieving massive reductions in the amount of human labor needed to produce the world's goods and services. While acknowledging that there are understandable reasons people might feel threatened by this outcome, it seems to me that we should view it as a blessing rather than a curse. As John Maynard Keynes said over fifty years ago:

> All this means in the long run [is] that mankind is solving its economic problem. . . . The economic problem is not—if we look into the future—the permanent problem of the human race. (Keynes, 1933)

From this standpoint, I then review some suggestions for disassociating income from employment so that people will be able to benefit from the elimination of unnecessary toil.

The Diminishing Need For Human Labor

Commonsense Arguments

Before beginning a more technical discussion of the economic effects of AI, it is worth considering a few general statements that are being made about the consequences of automation.

First, let's look at some of the arguments supporting the view that automation (including AI) will not result in unemployment. In a [1983] interview, James Albus, a leading researcher in robotics, made several important points. He stated, for example:

> There is no historical evidence that rapid productivity growth leads to loss of jobs. In fact, quite the contrary. In general, industries that use the most efficient production techniques grow and prosper, and hire more workers. Markets for their products expand and they diversify into new product lines. (Albus, 1983)

A related argument is based on the observation that unemployment is worse in the developing countries than in the industrialized ones. Since automation is much less pervasive in the Third World and unemployment is still so acute there, automation obviously cannot be the principal cause of unemployment.

Even if automation makes it possible to perform every task with fewer workers, there are a great many needs to satisfy. Albus expands on this point by observing that

> there is not a fixed amount of work. More work can always be created . . . Work is easy to create . . . There is always more work to be done than people to do it. . . . The problem is not in finding plenty of work for both humans and robots. The problem is in finding mechanisms by which the wealth created by robot technology can be distributed as income to the people who need it. If this were done, markets would explode, demand would increase, and there would be plenty of work for all able-bodied humans, plus as many robots as we could build. (Albus, 1983)

There are several industries that have pursued automation aggressively without reducing overall employment. In U.S. banking, for example, because the increased productivity resulting from automation has been accompanied by a relatively even higher demand for bank services, employment grew by 50 percent between 1970 and 1980 (Ernest, 1982). (On the other hand, however, we note that most of the jobs in the banking industry involve "knowledge work" of one sort or another—*i.e.,* the very category that is succumbing most rapidly to automation by AI techniques. In fact, the Bank of America . . . announced [in 1984] that it is now seeking to reduce its employment levels significantly (Gartner, 1984).)

Even if automation proceeds rapidly, the task of converting to automatic factories and offices will itself require considerable labor. According to Albus, "building the automatic factories . . . is a Herculean task that will provide employment to millions of workers for several generations" (Albus, 1983).

Critics of the hypothesis that artificial intelligence will be able to do anything argue that there is a large number of tasks that simply can never be completely automated. For instance, some people believe that it will prove impossible or undesirable to automate such services as marriage counseling, child care, and primary-school teaching. They might also claim that machines will never be able to generate truly excellent music, literature, and other art forms. Some, such as Professor Thorne McCarty of Rutgers, have suggested that the economy of the future might be based on these specialized kinds of "human-oriented" and creative services, just as much of our present-day economy is based on a more general array of services.

On the other hand, those who argue that the more advanced forms of automation (like robotics and AI) will cause increasing unemployment have several reasonable arguments on their side. For example, they point to the fact that over the past decade or so unemployment in the technically advanced societies does appear to have grown. With each successive business cycle, the "troughs" in the unemployment graph move upward. Although many people lose their jobs at times of recession, there are many others among the jobless who can blame their plight on robots and other automatic devices. Some economists think that we are already in the initial stages of a critical period in which large-scale unemployment due to automation is inevitable. For example, Dr. Gail Garfield Schwartz, an economic consultant in Washington, D.C., was quoted as saying, "perhaps as much as 20 percent or more of the work force will be out of work in a generation" (Neikirk, 1982).

Nobel prize winning economist Wassily Leontief, director of the Institute for Economic Analysis at New York University, adds weight to this prediction. He says that

> we are beginning a gradual process whereby over the next 30–40 years many people will be displaced, creating massive problems of unemployment and dislocation. . . . In the last century, there was an analogous problem with horses. They became unnecessary with the advent of tractors, automobiles, and trucks. . . . So what happened to horses will happen to people, unless the government can redistribute the fruits of the new technology. (Leontief, 1983)

We should also realize that employment data, as collected and published by the Bureau of Labor Statistics of the U.S. Department of Labor, include all the people who are ordinarily considered to be working. We must be honest enough with ourselves to admit that probably not all of these people are really presently required to produce the goods and services that we need. Some might not actually be needed—but are being paid anyway because of labor contracts that have set excessive standards as to the number of persons it takes to perform certain jobs. Most of us are being paid, quite legitimately perhaps, for vacations—which is one way of spreading the available work around. Some are being paid because various governmental bodies have been persuaded that certain goods and services are "needed," despite the fact that they are quite controversial and might not even be desired by a majority of those of us who are *buying* them. Some are being paid for crops not produced. Some are being paid because of eliminable inefficiencies that we prefer to continue tolerating. Because statistical employment does not necessarily mean real employment, the magnitude of the unemployment problem may already be greater than we realize.

Another factor pointing toward future reductions (or at least shifts) in the labor force is the rapid progress in automating much "white collar" work. It has been estimated that more than half of all American workers are engaged in "information-processing activities." Included in this category are many managerial functions, such as decision making, reporting, communicating and coordinating, fact gathering, and the supervision of similar activities by subordinates. Also included are many paper-handling clerical functions. The "expert systems" and automatic planning programs currently being developed in AI research laboratories will be able to perform many of these tasks, with a consequent drastic reduction in the need for human involvement or intervention.

Others have argued that the majority of new jobs created by automation will require only low-skilled labor. In a recent Stanford study Levin and Rumberger conclude the following:

> Most new jobs will not be in high-technology occupations, nor will the application of high technology in existing jobs require a vast upgrading of the skills of the American labor force. To the contrary, the expansion of the lowest-skilled jobs in the American economy will vastly outstrip the growth of high-technology ones. And the proliferation of high-technology industries and their products is far more likely to reduce the skill requirements for jobs in the U.S. economy than to upgrade them. . . . About 150,000 new jobs for computer programmers are expected to emerge during [the next] 12-year period, a level of growth vastly outpaced by the 800,000 new jobs expected for fast-food workers and kitchen helpers. . . . Past applications of technology in the workplace as well as present evidence suggest that future technologies will further simplify and routinize work tasks and reduce opportunities for worker individuality and judgment. Moreover, the displacement in jobs and the downgrading of skill requirements for most of the new positions

will undermine employment generally, and especially the employment of skilled workers. (Levin and Rumberger, 1983)

So we see that there are many more or less reasonable arguments on both sides of this issue. It is likely that we will continue, almost daily, to hear conflicting opinions about the prospective impact of AI on employment.

. . . . .

What's So Bad About Unemployment?

Two Fears

Instead of welcoming the arrival of mechanical slaves to perform much of the world's toil, most people view the prospect of increasing unemployment with great alarm. Leontief puts this paradox in sharp relief:

> Adam and Eve enjoyed, before they were expelled from Paradise, a high standard of living without working. After their expulsion, they and their successors were condemned to eke out a miserable existence, working from dawn to dusk. The history of technological progress over the past 200 years is essentially the story of the human species working its way slowly and steadily back into Paradise. What would happen, however, if we suddenly found ourselves in it? With all goods and services provided without work, no one would be gainfully employed. Being unemployed means receiving no wages. As a result, until appropriate new income policies were formulated to fit the changed technological conditions, everyone would starve in Paradise. (Leontief, 1982)

Leontief's story highlights one of the fears that people have about unemployment, namely, that they will lose their incomes. Presumably this economic fear would evaporate if people could obtain an income in some other manner so that they could purchase goods and services produced by the machines. Many economists, as well as others, have proposed various schemes that separate income from employment; I shall examine some of them in this section.

Another cause for apprehension has to do with social and psychological needs of human beings rather than with their economic requirements. What will people do with their "free time"? What activities will be as fulfilling and rewarding as jobs? Some people are pessimistic about the ability of their fellows (but not of themselves) to adjust to "becoming rich." Others, like John McCarthy, an AI pioneer, opine facetiously that this adjustment "could take all of ten minutes" (McCarthy, 1983). I shall also have some comments about this problem.

Allaying the Economic Fear

There are several ways of dealing with the economically motivated fear of unemployment. They range from rather crude approaches, like attempting to slow down or halt technological change so as to delay or prevent unemployment, to more sophisticated and possibly impractical reorientations of our economic system.

Placing obstacles in the path of either using or abetting technology might be called a "Luddite approach" to the economic problem of unemployment. This approach is unfair to humanity because it condemns us to continue toiling when toil is technologically unnecessary. To use Leontief's metaphor, it is equivalent to disrupting our attempts to reenter Paradise. In any case, the approach would inevitably fail because, fortunately, no government or other group has sufficient repressive power to prevent technical progress. Even if technology were temporarily slowed in one country, so much the worse for that country; its foreign competitors would soon outrace it and it would have unemployment anyway—unemployment and poverty.

Another way to solve the economic problem posed by technological unemployment is to invent jobs that are either unnecessary (that is, they do not contribute to absorbable consumables) or could be performed by machines. This approach may be one way of distributing income, but it is unfair because it condemns some people to unnecessary toil. There is reason to fear that the conventional goal of full employment (espoused by both political parties in the United States) can be achieved only through such "makework" schemes.

Separating income from employment would seem a better way to solve the economic problem of unemployment. This solution actually suggests itself as a corollary of our earlier economic analysis; if income is not derived principally from labor, it must instead come from either capital investments, sale of materials, or transfer payments.

．　．　．　．　．

As regards transfer payments, the industrialized nations already have a great deal of experience with government techniques for distributing income independently of work. Social security, "welfare" payments of various kinds, farm subsidies, and the "negative income tax" have all been used in the United States. Expansion of these programs is one possible way of decoupling income from work. I shall not attempt here to give arguments for or against transfer payments, except to note that many people fear the pernicious effect of some types of transfer payments—i.e., that they might subvert the American ideal of a free and independent citizenry.

There have been many intriguing proposals for more of us to obtain more of our income from a return on capital investment. Louis Kelso and Mortimer Adler have written a book that proposes an imaginative, capitalistic "society in which machines do all or most of the mechanical work that must be done to provide the wealth necessary both for subsistence and for civilization." They recommend a diffuse, private ownership of the means of production so that:

> every man, or every family, has a sufficient share in the private ownership of machines to derive sufficient subsistence from their productivity. In this automated industrial society, each man, as an owner of machines, would be in the same position as an owner of slaves in a slave society. As a capitalist, he would be an economically-free man, free from exploitation by other men, free from destitution or want, free from the drudgery of mechanical work—and so free to live well if he has the virtue to do so. (Kelso and Adler, 1958)

Kelso and Adler envision that people would receive most of their income from dividends on common stock. To achieve this situation, they make proposals that would:

1. Broaden the ownership of existing enterprises.
2. Encourage the formation of new capital and the organization of new enterprises by new capitalists.
3. Discourage concentration of the ownership of capital by households in which such concentration has passed beyond the point determined to be the maximum consistent with the just organization of a completely capitalistic economy.
(Kelso and Adler, 1958)

They also list specific recommendations, including the use of tax and credit devices, whereby families may begin to accumulate stock ownership in corporations.

James Albus has suggested the formation of a National Mutual Fund [NMF], which would use credit from the Federal Reserve System to finance private investment in automated industries. Ultimately this fund would invest about $300 billion a year, which would double the then current (1980) rate of investment in plant and equipment. This extra investment in private companies would earn profits that would be distributed by the NMF to the general public as dividends to stockholders. "Everyone would receive a substantial income from invested capital. Everyone would be a capitalist, not just the wealthy" (Albus, 1981).

To offset the short-term inflationary effect caused by the investment of this newly created money, Albus suggests that short-term demand be restrained through a mandatory savings bond program. These bonds would bear interest and be redeemable after five years.

> The key idea in this plan, which might be called an Industrial Development Bond program, is to index the mandatory savings rate to the leading indicators for inflation on a monthly basis. If inflation is predicted, mandatory savings go up for the next month and reduce consumer demand. As soon as prices stabilize or decline, mandatory savings are reduced. This policy would effectively divert short-term demand from consumption into savings and compensate for increased investment. At the same time, it would assure that the purchasing power to distribute the fruits of investment in highly productive technology would be available once the new plants and modernized machinery began to produce increased output.

As Albus notes, separating income from employment explicitly acknowledges

> that the primary goal of an economic system is not to create work, but to create and distribute wealth, i.e., goods and services that people want and need.

He goes on to say:

> I believe we have it within our power to create an everyperson's aristocracy based on robot labor. (Albus, 1983)

The process of converting to an economic system that separates income from employment will face major, perhaps unsurmountable political, psychological, and social obstacles. "Earning a living" is a very deeply ingrained notion in our culture. Different levels of skill, luck, and hard work in earning a living allow a spread of incomes, from low to high, that many people regard as equitable and desirable. Even if most consumables were being produced automatically, material and energy limits might not allow everyone to consume at the rate he would like. Since some of us (perhaps many) will still need to work, the lure of higher incomes might provide the necessary incentive—even as it so often does now.

Allaying the Sociopsychological Fear

There are many people who are fortunate enough to gain many psychologically valuable benefits from their jobs in addition to those of a strictly economic nature. Job satisfaction, the joy of achievement, an enhanced personal identity, opportunities for growth and learning, and social interaction are among the things that many of us derive from our work. Clearly, humans need such beneficial activities, but must they be tied to the production of income? There is already a large number of people who gain fulfillment and psychological rewards from activities they pursue in retirement (at which time their income is derived from pensions, social security, investments, etc.), or from volunteer or public-service activities (with income perhaps provided by a spouse's job or from inherited wealth). Many people also forego a chance at higher incomes so that they can fulfill themselves in artistic and creative pursuits in which the potential for income might be very low or even nonexistent.

Although many of us fear the prospect of losing a job, do we really fear more the loss of psychological rewards than economic ones? One simple test is to ask, "Suppose you inherited one million dollars. Would you go back to your old job, or would you do something else with your time?" Probably not many people are fortunate enough to have a job they would want to continue if they were suddenly to become wealthy.

Margaret Boden argues quite convincingly that the new age of automation could be "rehumanizing" rather than "dehumanizing." She foresees a "Polynesian-type" culture based on artificial intelligence. In Polynesia (at least in precolonial times) no one worried very much about the fact that freely available mangos caused unemployment. Professor Boden states:

> AI could be the Westerner's mango tree. Its contribution to our food, shelter, and manufactured goods, and to the running of our administrative bureaucracies can free us not only from drudgery but for humanity. It will lead to an increased number of "service" jobs—in the caring professions, education, craft, sport, and entertainment. Such jobs are human rather than inhuman, giving satisfaction not only to those for whom the service is provided, but also to those who provide it. And because even these jobs will very likely not be full-time, people both in and out of work will have time to devote to each other which today they do not enjoy. Friendship could become a living art again. (Boden, 1983)

It should also be noted that, besides providing people with time for human-oriented activities, automatic devices can be utilized in support of these activities to make them richer and more enjoyable.

Thus, it seems that there is no real reason to believe that a paying job is essential for a rewarding life. There is abundant evidence that people can receive important life-fulfilling benefits from a wide variety of activities that do not generate income. Some, like Willis Harman, envision a new conception of work made possible by our growing ability to produce goods and services automatically (Harman, 1981).

Before leaving this topic, however, we might mention another possible function of employment. In addition to the positive benefits that accrue to a jobholder, some observers, citing the correlation between crime and unemployment rates, see compulsory employment as a way to keep people out of trouble. Such a view not only seems inordinately pessimistic with respect to human nature and the human potential, but is probably wrong about the underlying causes of criminal activity. In any case, there are probably more humane ways to maintain civil tranquility than chaining people to work they dislike. Also, as Herb Simon has pointed out:

> most people who are alarmed at [the prospect of too much leisure time] do not find that they themselves are endowed with too much leisure. But there are 'many people,' it is argued, who would not know what to do with leisure time, and who, presumably, would lend their hands to the Devil. (Simon, 1977)

The Transition

For those who are willing to grant that artificial intelligence and related technologies will eventually reduce the total need for human labor and that there are stable and desirable socioeconomic systems that separate employment from income, there still remains one very difficult question: how do we get there from here? Now some might say that we have plenty of time to worry about that problem and that now is too early to think about a transition. In the first place, it might be a long time before we develop the ultimate systems that will be able to perform the new jobs created by currently emerging AI systems. Secondly, a huge amount of human labor will be required to convert present-day industrial societies to fully automated ones (not to mention the labor needed to lift the living standards of the third world).

Nevertheless, I think there are good reasons people should now start concerning themselves with this problem.

- *First,* the pace of technical change is accelerating. While it is true that the technical problems involved in creating artificially intelligent systems are still immense, we may solve most of them within the next generation.
- *Second,* if we begin to welcome rather than fear the "unemployment" consequences of AI, we can avoid the technological lethargy that unwarranted anxiety might otherwise induce.
- *Third,* socioeconomic changes are extremely slow (compared with technical progress). We must allow time for the several stages needed for the transition to new systems of distributing income. There will be at least five to ten years of discussion and argument

among intellectuals and other social thinkers. Next, the voting public must have sufficient confidence in some of these ideas to approve any necessary legislation. At the same time, we must anticipate an inevitable reaction against these changes, stimulated by a general yearning to return to the "good old days" in which everyone did an honest day's work for an honest day's pay. People may blame these economic experiments for one or more of the expected future slumps in the business cycle. Taking all of these processes into account, it may well require one or two generations before the necessary changes can be made in our economic system, even if concerned people begin thinking earnestly about the problem right now.

• *Fourth,* starting to think about the problem and instituting some transitional measures now will minimize the discomfort of workers who are already being affected by automation. There are grounds for believing that the current high unemployment rates of the industrialized countries are not completely explainable by business cycles and will be cured by neither *supply-side* nor *demand-side* economic policies. This unemployment is rather a symptom of the "new automation," and it will continue to worsen even as business conditions improve. If the root causes of high unemployment are in actuality related to automation, policies that recognize this fact will have a better chance of alleviating the misery and poverty of the unemployed.

There are several things that we can begin to do now to prepare for these effects of AI. First, we must convince our leaders that they should give up the notion of "full employment" as a goal for the postindustrial economic system. It is unachievable, unnecessary, and undesirable—and it keeps us from focusing on the real problem.

Retraining is critically important, but we must not assume that everyone who now holds a blue-collar or middle-management job can become a computer scientist or programmer when his present job disappears. We must begin training for such automation-resistant "human service" jobs as teaching, family counseling, day care and health care. We must also educate people in arts, crafts, literature, writing, and sports so that they will benefit more from their increasing leisure time. People cannot become "Polynesians" without training. Many community colleges already give adult education courses with this orientation; these programs should be expanded.

We should also begin to work much more earnestly on the many transition projects required to move us into the computerized, postindustrial age. Probably our most important task is to improve the living standards of people in third world countries. I agree with James Albus that without rapid economic growth, a world of growing shortages will become an increasingly dangerous place. Nations competing over a shrinking stock of wealth and resources will inevitably come to military confrontation. The world's best hope is a great surge of industrial productivity that can outstrip the present population explosion and give us one more period of affluence in which we will have another chance at bringing the human population into stable equilibrium with the finite living space aboard the planet Earth. (Albus, 1981)

Another transition task is to design and build new automated equipment and factories. This work should be preceded by national projects, like those sponsored by the Japanese, that plan and develop the necessary technology. Additional projects could be initiated to improve education and health care in all parts of the world. Communities throughout the United States have been concerned about the problem of aging highways, bridges, and other transportation and communication facilities. Upgrading this "infrastructure" would absorb surplus labor during the transition stage. The postindustrial information age will need another infrastructure—one consisting of computer systems, data bases, and networks. Putting all of this in place and maintaining it will require human labor for several more years.

Much of the work I have just mentioned can be funded only by governments. Because such work accomplishes goals that need to be satisfied, it should not be thought of as "made-work." But it does have the desirable side effect of giving people employment during the transition from an economy in which most income is derived from employment to one in which most income is derived from other sources. In order to undertake these large public projects, we need to think differently about the matter of spending public funds. Instead of asking the rather outmoded question, "Can we afford such expenditures?" we need to learn to ask instead, "Should otherwise idle human labor be employed to accomplish socially desirable tasks?" The use of terms like "spending public money" and "affordability" focuses on arbitrary accounting conventions rather than on reality. What really counts is not an abstraction like money but whether or not people who could be working on these projects would otherwise be idle and whether or not the rest of society can produce enough goods and services to satisfy the demand of people working on public projects.

In seeking to analyze the financial aspect of these public works, one could begin by observing that laborers are idle because the type of work they would have been doing in producing consumables in the private sector is now being done by machines. During the transition—that is, before these idled laborers receive fully compensating income from sources other than employment—part of the automation-derived benefits realized by businesses and consumers should be used to help pay the salaries of the workers on public projects. Simply put, the public projects can be financed by taxes levied on automation and consumption. Salaries paid to workers on public projects will increase demand (beyond what it would have been if those workers had remained unemployed and unpaid), but this demand can be met by the increased productivity of the automated industries. Of course, the taxes levied on automation must not be so high as to destroy the incentive to automate. Furthermore, they should decline as the cost of labor for the public projects goes down because of automation.

As automation takes over more and more of the work heretofore performable only by humans, we need to take steps to ensure that people become unemployed in a gradual and nondisruptive fashion. New approaches to work, such as job-sharing, should be encouraged. Shrinking of the workweek and a compensating increase in income derived from nonemployment sources, such as stock ownership and transfer payments, should proceed in step.

References

Albus, J. 1981. *Brains, behavior and robotics.* Peterborough: BYTE Books, Subsidiary of McGraw-Hill.

———. 1983. The robot revolution: An interview with James Albus. *Communications of the ACM,* March.

Boden M. A. 1983. Artificial intelligence as a humanizing force. *IJCAI,* Los Altos: William Kaufmann, Inc. August.

Duchin, F. 1983. Private communication.

Ernest, M. L. 1982. The mechanization of commerce. *Scientific American,* September.

Gartner, T. 1984. B-of-A is offering to pay workers who resign early. *San Francisco Chronicle,* January 17: 1.

Harman, W. W. 1981. Work. In A. Villoldo and K. Dychtwald, eds. *Millenium.* Published by Jeremy Tarchure.

Kelso, L. O., and M. J. Adler. 1958. *The capitalist manifesto.* New York: Random House.

Keynes, J. M. 1933. Economic possibilities for our grandchildren (1930). *Essays in persuasion.* London: Macmillan and Co.

Leontief, W., and F. Duchin. 1983. The impacts of automation on employment, 1963–2000. Draft final report. New York University: Institute for Economic Analysis. September.

Leontief, W. W. 1966. *Input-output economics.* Oxford University Press.

————. 1982. The distribution of work and income. *Scientific American,* September.

————. 1983. The new new age that's coming is already here. *Bottom Line/ Personal,* 4 (April): 1.

Levin, H. M., and R. W. Rumberger. 1983. The educational implications of high technology. Institute for Research on Educational Finance and Government. Stanford University: School of Education, February.

McCarthy, J. 1983. Interview with John McCarthy. *Omni,* April.

Neikirk, W. 1982. Recovery could be a jobless one. *Chicago Tribune,* August 15.

Simon, H. A. 1977. *The new science of management decision.* Englewood Cliffs: Prentice-Hall.

—— **22** ———————————————————————

Information Technology and the Third World

Juan Rada

Introduction

The effects of information technology on North–South relations, the international division of labor and development, is a subject of growing interest to specialists as well as policymakers (Mitterand, 1982). I will attempt to review here the reasons for this interest by analyzing aspects of the effects of the technology in some of these areas. In a brief article all aspects cannot be covered, and the intention is to highlight some of the main issues.

Since the beginning of the debate in developed countries about the effects of information technology on their societies and economies, consideration has been given to potential impacts, both positive and negative, on developing countries. In the early stages of discussions and research, the main concern was related to the assessment of economic

From "Information Technology and the Third World," by Juan Rada. Paper read at IFAC seminar, Vienna, Austria, March 1983. Reprinted by permission.

effects, especially the potential erosion of one of the developing countries' perceived comparative advantages, namely low labor costs (Rada,
1980).

This concern remains central today, but it is qualified by a number of
considerations as the understanding of the effects of technology grows.
Such considerations include questions of quality, product cycle, change in
the function of products, and a better understanding of the behavior and
structure of the suppliers of the new type of equipment. In addition, the
early concern was especially centered around the erosion of advantages of
offshore assembly, especially in those countries with export-oriented
economies—which, of course, constitute a rather exceptional group of
developing countries.

From the almost exclusive focus on automation in manufacturing and
economic effects, new and important elements have been added. From an
economic point of view, two of these deserve special attention. The first
concerns attempts to understand the effects of automation within developing countries, particularly in those cases where development is geared
primarily toward the internal market. This is the case, for instance, with
most of the Latin American economies (UNIDO/ECLA, 1982).

The second relates to one aspect of the effects of information technology that, although implicit in early research, was somehow overlooked in
terms of its real importance: namely, that it allows the transportability of
services to an extent and depth not dreamed of some years ago. This has
opened up new questions, especially for developing countries.

Although I shall focus on these two points, the social, cultural, and
political effects of automation are of equal, if not of greater, importance.
Lately, substantial attention has been paid to these points, particularly in
the context of searching for alternative development strategies. This
search is not new, but it differs from earlier attempts in its desire to see
whether, given the nature of the technology, some entirely new concept
of development can be pursued, heavily based on human resources, information, and knowledge-intensive activities: in brief, to determine how
developing countries can appropriate information technology in a qualitatively different form, and aim at some sort of leapfrogging.

These views are based on a prospective assessment of the form human
activities could take as a result of the changing technological profile of
society. This view has been developed out of a mechanical extrapolation
of the potential of the technology and is most creatively expressed in
works such as Alvin Toffler's The Third Wave.

In other words, would it not be better for LDCs [less developed coun-
tries] to aim at societal models that will truly consider the prospective
"information society," rather than to evolve (if possible) through the traditional lines? This view might be tempting for many, but often fails to

acknowledge not only the unequal distribution of resources and knowledge, but also the fact that "informatization" is the *consequence* of development and not its cause, although the technology can be used for development purposes.

This is not to negate the fact that development needs to be conceived in a completely different form in order to account for the current technological mutation, which indeed questions the very core of currently pursued development strategies. Such questioning, however, involves developed as well as developing countries, since a global approach will sooner or later be necessary to readdress the direction that current changes seem to be taking. The starting point in this discussion is to list briefly the main areas of the economic impact of information technology (IT).

IT: The Main Areas of Impact

Effects in Production

1. Substitution of mechanical components (e.g., watches)
2. Substitution of electromechanical components (e.g., cash registers and calculators)
3. Substitution of electric and older electronics (e.g., computers)
4. Upgrading of traditional products, creating entirely new capabilities (e.g., word processors)
5. Upgrading of control systems and substitution by electronic ones (e.g., machine tools)
6. New products (e.g., games)

In the case of products, the use of electronics can alter the very function of the product. A good example is cash registers, which, from being only adding machines, have become data entry terminals, potentially part of a system of accounting, control, and ordering. Other products, such as machine tools or word processors, if supplied with the adequate hardware and software, can communicate to other machines, data bases, or computers.

Effects in Manufacturing Processes

1. Increase in the flexibility, adaptability, and economy of production (e.g., CAD/CAM)
2. Incorporation of skills and functions into equipment (e.g., CNC machines or robots)

In the case of processes, the important developments are essentially two. First there is the programmability, which leads to flexibility, since the same equipment can be reprogrammed to perform a different task rather than changing the machine. This in turn leads to the second important point, which is that there is great resistance to obsolescence and thus an extension of the life cycle of manufacturing processes. Robots, for example, are very resistant to obsolescence since in most cases it is sufficient to change the program and/or the "hand" to use the equipment for a different operation. This has an important capital-saving effect. In addition, there are savings on downtime for retooling and changing tasks.

Effects in the Office

1. Automation of routine clerical work (e.g., data and word processing)
2. Increase in the efficiency and effectiveness of communications, especially in those areas where work is less formalized, such as in professional and managerial areas.

Automation in the office or, more accurately, the use of electronic tools as aids or facilitators of the work is and will have a far-reaching effect on economic activities in general and especially in the production of services. In developed countries the percentage of the labor force working in offices or information activities is constantly growing, in some cases reaching 50 percent. The office sector has traditionally been undercapitalized and its productivity has been low as compared with manufacturing and agriculture.

It is not known how the current process of rapid diffusion of electronic technology in the office, in either clerical or managerial activities, might affect companies' competitiveness, productivity, and the international division of labor.

It is safe to say at this stage that, increasingly, most employees, whether in agriculture or in manufacturing, are concerned with information-processing activities rather than production. In fact, the absolute number of people employed in manufacturing has been decreasing in the U.S. since 1964 and in Europe since the early 1970s. This trend will continue, as it did with agriculture in the past. The decrease in the labor content of agriculture was accompanied by substantial reductions in hours worked and important increases in output.

Effects in Services

1. Transportability of services (e.g., remote access to data bases, banks, archives, etc.). This also leads to new services (e.g., Prestel)
2. Increase in self-service (e.g., gas stations, banking, etc.)

These in turn lead to the replacement of human-to-human services by machines. The impact of information technology on services is perhaps the most important in the long run because it is creating entirely new possibilities that are different from past activities. This point will be discussed later.

The Reinvention of Industry

The first question to ask in terms of the impact of information technology on developing countries is how technology affects developed ones. This question is pertinent owing to the fact that it is in these countries that the technology was first used to increase productivity in manufacturing and services. This problem has been extensively studied, and the main conclusion is that a reinvention of industry will be necessary, that is, a radical change in the technological profile of productive activities.

This change naturally affects countries that have geared their efforts toward industrialization either through import substitution for the internal market or through exports.

The basic hypothesis is that the industrial utilization of IT leads to an erosion of developing countries' comparative advantages and international competitiveness, especially in traditional industries. The main reasons for this are as follows.

Decrease in the Relative Importance of Labor-Intensive Manufacturing and Cost of Labor

This is essentially due to the *automation of production,* which tends thereby to erode the competitiveness of low labor costs. A good illustration is the comparison of Hong Kong and the U.S. in the manufacturing of electronic devices. Table 22.1 shows that when the process is manual the difference in cost is about 1 to 3, decreasing drastically with semiautomatic processes and automatic ones. While this example refers to components, a similar process is taking place for systems and consumer products. In the case of TV sets, the chairman of Electronics Industry of Korea stated that owing to the automation of assembly and technological change,

> the manufacturing costs of a TV set in Korea and that of the U.S. are practically comparable to each other. Rapid advancement of industrial technology is eliminating labor-intensive portions of the electronics industry; this tends to make it harder for Korea to earn enough foreign currency to import expensive new technology. (Kim, 1980)

Table 22.1 Manufacturing Cost Per Device (U.S. $)

Process	Hong Kong	U.S.
Manual	0.0248	0.0753
Semiautomatic	0.0183	0.0293
Automatic	0.0163	0.0178

Source: Global Electronics Information Newsletter, No. 25, October 1982.

It should be mentioned that TV sets as a product have changed substantially in the last few years and are bound to change more in the future when they become digital and the CRT (cathode-ray tube) is replaced by some other form of display. In other areas preliminary evidence shows similar trends. For instance, the expansion of automation in Japan has contributed to a recent reduction of investment in the Asia/Pacific region involving firms in electronics, assembly parts, and textiles (*Business Asia,* 1982).

In the case of garments, a trend to systems optimization and automation is clearly underway. Although not yet a "perfect fit," Hoffman and Rush (1982) conclude that: "Although it has not happened yet, to a great extent there is a feeling among the large producers that a large share of offshore production will be brought back [to the developed countries]."

Value-added is pushed out of assembly and into components as integration increases. This occurs at the product level, while in systems value-added, it is pushed upward toward servicing. This process is proper to the industries where electronics has substituted other components in products (as described above) and implies that functions previously obtained by assembling pieces are incorporated in the electronic component itself.

In this category fall, for instance, calculators, telexes, sewing machines, and precision engineering in general. In the case of electronic components the amount of value-added obtained in offshore assembly has been decreasing constantly. The dutiable value of components imported under the U.S. tariff arrangements, that is, the value-added in offshore plants to U.S. products diminished from 57 percent in 1974 to 39 percent in 1978 (USITC, 1979).

The main explanation for this is the increasing value of the parts produced in the U.S. as a result of the growing complexity of devices. This process has continued since: one only needs to see that in 1978 the level of technology was LSI (large-scale integration) rather than the current VSLI (very large-scale integration). As the level of integration of components increases, the value-added obtained in front-end operations also increases. Furthermore, the assembling of chips is being automated and moved, in the case of sophisticated devices, to "clean rooms."

Changes in Product Cycles

Product cycles in many areas have been considerably condensed while process cycles have increased, owing to the resistance to obsolescence of programmable machines and equipment. Typically, product cycles have been shortened in some industries (e.g., office equipment) from twelve to three or four years. This has led to a concentration of manufacturing investment in capital-intensive flexible manufacturing, and partly explains the erosion of the advantage of developing countries in so-called "mature" or "semimature" products.

In other words, the "product cycle" view of international trade needs review, since formerly mature industries or products are being completely revitalized. A case in point is a European company that closed a plant in South America because the short product cycle did not justify the investment. In the past, the payback time was far longer with a more stable technology. At the time of mechanical or electromechanical technology, local manufacturing was justified because of the large amount of value-added obtained in assembly; and, because of the longer cycle, there was a relatively small incidence of amortization and development cost in the final cost of products (Cohen, 1981).

Quality Considerations

These are growing in importance as markets become more segmented and competition increases under conditions of low growth. This in itself leads to what has been called the "hands-off" approach in manufacturing or automation, coupled with a change in the skill-mix at the shop floor. One of the main reasons why Japanese manufacturers in the field of semiconductors make very limited use of offshore facilities in developing countries is precisely the perception that the required level of quality cannot be obtained.

Quality has a cost, and requires an infrastructure and substantial managerial know-how. In some cases the old manufacturing system of assembly is inconsistent with quality requirements and new methods have to be used, notably modular or group-work schemes, with emphasis on a highly multiskilled labor force. This phenomenon highlights the dilemma for export as well as import substitution strategies.

The elements mentioned above are the tip of the iceberg of current change. Further down the inevitable question is: What about access to technology, its production and application? Three comments are necessary here.

First, almost by definition, advanced and rapidly changing technology is not properly documented and therefore its transfer tends to have peculiarities. In fact, transfer of technology in the area of concern of this paper takes place essentially through three main mechanisms:

1. mobility of personnel, which take with them their own knowledge (this is the so-called "Silicon Valley syndrome"). This accounts for a large part of transfers in the U.S.;

2. second sourcing, which is the agreement between two producers to manufacture fully compatible products. This might or might not entail full exchange of technology. An agreement of this nature implies a partner that can produce at similar technical, economic, and quality standards;

3. cross-licensing agreements, which assumes a mutual exchange of technology.

In brief, transfer tends to take place among established or important producers, and furthermore, the technology is tightly guarded as trade secrets. Many companies in the software area, for instance, do not patent or copyright their products because it entails disclosure of valuable information.

The second comment is that the issue is access not only to a given technology but to the *process of technological change,* because of the dynamism of it. This leads to a number of questions that I shall not discuss in this paper, notably about the innovation environment. The point that I wish to make is that access to the process of technological change in advanced areas (and not only IT) seems to take place essentially, as European companies have discovered, with participation in the equity of companies. The possibility of some developing countries doing this is relatively small, as some exploration has shown, essentially because of the high mobility of the personnel and also because of political considerations. In this respect one should simply mention that in many areas of electronics civilian applications have surpassed military ones, creating an additional obstacle to prospects for transfer of technology.

The third comment refers to production and applications. In terms of production, few LDCs are in a position even to raise the question. Some have implemented policies in this field (i.e., India and Brazil), but their performance cannot be evaluated at this stage, except to say that they at least provide the countries with the capacity to follow the technologies closely. Success will depend largely on the targeting of market segments and technologies (e.g., uncommitted logic arrays and custom circuits in general). At the systems (e.g., minis, micros, etc.) and software level, the situation is different. The assembly of equipment from components that are bought practically off-the-shelf is taking place in many countries, and this is likely to continue for some time to come. But as the level of integration of components grows, the amount of software incorporated into chips (firmware) will also grow, taking value-added away from the assembly of systems.

To illustrate this further, the trend in microcomputers is to incorporate into the hardware as many "utilities" as possible, such as word processing, Visicalc, and others, in a similar fashion to what happened with the pocket calculator. At the beginning the calculator featured four arithmetic operations; as integration increased, more were added, making the machine more useful and also less expensive. This trend implies that, in the not-too-distant future, the source of value-added will go to systems software, design, and service in a far more pronounced way than today.

The assembly of systems will continue, especially when protected by tariff barriers, incentives of industrial policies, or both (i.e., Brazil). These types of equipment will be used largely in internal markets, and are unlikely to make a dent internationally. In addition, systems in developing countries tend to be far more expensive than in the international marketplace, making their application less economical, especially when labor costs are lower. In one Latin American country the cost of word processors and microcomputers was, respectively, double and triple that of the U.S. This is explained by a number of reasons, the most important being that suppliers have to cover maintenance, software development, and overheads, selling a rather limited amount of equipment as they operate as "profit centers." Installation and use of the equipment is also more costly in some cases because of expensive auxiliary installations (i.e., electrical generators), subutilization, or lack of adequate skills, especially managerial ones.

Optimization of Systems and Office/Service Automation

One of the most important effects of IT is that it leads to the optimization of business activities as a result of rapid and timely processing of information and the relative ease of communications. I prefer to use the term "service automation" rather than "office automation" because it truly accounts for the nature of the change. The impact of office automation is measurable not at the workstation level (as is done with word processors), but rather at the level of total systems performance, mainly because of the effects on management information systems and managerial effectiveness. To illustrate the point, consider that secretaries account for only 7 percent of the total clerical costs in the U.S. and spend only 20 percent of their time typing.

It is not, then, in the workstation that the effects of office automation are to be found, but rather on the synergies, greater numbers of options, faster response, and more informed decisions that are derived from it. This is not to negate important productivity increases at the workstation but to treat them as one, and perhaps not the most important, component of current improvement. Research conducted at IMI-Geneva shows precisely this, and furthermore confirms the optimization of systems that takes place within companies utilizing IT.

The effects of these processes on the international division of labor and developing countries is yet unknown, but it is possible to make some tentative hypotheses. First, given the composition of the labor force in the advanced countries and also within manufacturing companies, an improvement in systems performance will further reinforce the advantages derived from automation and product change. For instance, in pharmaceuticals today, typically only about 30 percent of the labor force is employed in production, and the proportion is expected to decrease from current levels by as much as 40 percent by 1990. This implies that manufacturing is decreasing in importance (as measured in total cost) while performances at the systems level and innovation are becoming the key to profit, growth, and survival.

Second, and most important, is the increase in productivity of services, which for the most part are information-processing activities. The transportability of services is the most important long-term effect of the technology; thus, more efficient production of them reinforces the great advantage that developed countries have in this area. I shall elaborate on this point later.

Changes in Skill-Mix and Conditions for Absorption of Technology

I mentioned earlier that important changes in skills accompany the product/process changes. In some cases there are significant skill-saving effects (e.g., CNC machine tools), which can be beneficial to developing countries. In general, the trend seems to be toward higher skills, especially at the systems and design level and not the least in software. Most of the developing countries' labor forces have low skills, or skills of a mechanical nature that in many cases are being substantially altered (for instance, for interface types of work). This again calls for a more active policy on the part of the developing countries in terms of training and education. The absorption of technology is also changing, not only for the reasons stated above about transfer, but also because knowledge tends to be of a more abstract nature.

It is not by chance that much innovation in electronics has taken place around universities (Stanford, Berkeley, MIT, and now Cambridge in the U.K.). This means that the links between scientific and technological knowledge are becoming tighter, and the neat categories of the past that distinguished invention from innovation are not always tenable.

The need for scientific policies is obvious, especially in areas where these types of knowledge are closely related to technological development (e.g., physics of materials in electronics or genetic engineering), but these policies require a clear focus. It is true that current and future technological progress is based on science, but at the same time innovation does not necessarily require a sophisticated scientific base, as Japan has proven so convincingly during the 1960s and the early 1970s. This is particularly true in relation to process and systems innovation.

In the first case changes tend to be incremental (unless the product changes), and in the second case (systems) changes tend to be of a conceptual and organizational nature. A classical example of this latter type of innovation is self-service, which has boosted productivity in many sectors with little "hardware" investment of R & D work. Credit cards, marketing systems, financial services, leasing and rental operations are other examples. In this respect it would be interesting to compare innovation in banking with the familiar innovation curves in electronics; we would probably find, to our surprise, that they would not differ much.

The skill-mix is changing while new skill requirements are emerging, particularly in software, systems design, and (an almost forgotten one) management and organization. Two different companies or countries with similar skills and other endowments may perform quite differently simply because of differences in their management and organization which lead to varying degrees of technology absorption. Often, skills of different natures can be obtained through training and retraining, but a precondition is the action that creates the *need* for them: that is the answer to the questions what and why, and know-what and know-why. If these two questions are not answered, obtaining know-how will make no difference to performance—on the contrary, it will not even be possible to obtain the proper know-how.

These changes are to an extent already occurring in LDCs, but in some fields the die is cast while in others selectional decisions can still be made. Developing countries can obtain immense benefits from technology if it is applied in the context of a development strategy. For instance, IT is capital-saving in manufacturing and services (low entry barriers) per unit of output. This leads to the traditional dilemma of technologically induced labor-saving effects, since capital intensity tends to increase.

The main issue for developing countries in terms of employment/technology lies in the field of agriculture rather than computerization in manufacturing or services. The reasons are rather simple. The number of computers currently in use has a practically insignificant effect on the overall volume of employment when they do cause displacement. The two largest users of computers in the developing world are India (about 1,000) and Brazil (about 10,000), which in the context of their economies is minimal. The applications to which they are put remain traditional, and they tend to optimize administrative systems that in turn create beneficial effects in the rest of the country's economy.

An illustrative case is the informatization of the postal check system in Algeria, where 176 people were made redundant through voluntary retirement. The general effects can be seen in Table 22.2. Examples like this illustrate the trade-off between employment and the use of technology and show how beneficial IT can be when applied in critical bottlenecks, especially in relation to infrastructure and services. It needs to be done with the normal criteria of appropriateness of technology, that is, selectivity. It is in the fields of administrative services and infrastructure that short-term benefits can be realized rather than in totally new types of applications, which will take a long time to mature and are heavily dependent on equipment performance and characteristics of the human–machine interface.

Capital-saving effects also take place in manufacturing and agricultural applications, together with skill-saving effects. The real challenge in this field is to combine traditional and low technologies with advanced ones, and much needs to be done here.

Developing countries are far more heterogeneous than developed ones, and sometimes this shorthand concept masks tremendous differences. In some countries (e.g., Southeast Asia) companies are combining advanced technologies with lower labor costs, and future developments will depend on the competitive reaction of developing countries'

Table 22.2 Computerization of the Postal Check System in Algeria, 1974–77

	Manual 1974	Computerized 1977
No. of operations	24,360,000	33,620,000
Volume (millions DA)	109.55	210.8
No. of accounts	452,000	709,000
Waiting time at centers before processing of document	15 days	2 days
Payment at cash desk	3–6 hours	2 min.
Saturation ratio	95%	50%
Employment	856	680

Source: Secretariat d'Etat au Plan, Commissariat National a l'Informatique, "L'Informatique en Algerie," Algiers, 1978.

producers. Others are following a policy of technological upgrading in the exporting industries, and even authorization for acquiring foreign computerized equipment depends in some cases on potential export performance. Furthermore, many developing countries have large pools of educated labor in areas most appropriate to current change, such as software, and great potential exists in this field. I insist on the word "potential" since production or export of software is not as easy as it might sound; in some cases import substitution of software might be far more economical than attempting exports.

Possible policies and measures to maximize the benefits of IT for developing countries are not only necessary but urgent. But this should not mask the equally urgent need to search for different development models, South–South cooperation, regionalism, and, most importantly, some sort of social command of technology. *Command* of technology differs substantially from *control* in the sense that it maintains the relative autonomy of action and creativity at the technological and scientific level, but provides a direction for the application of that creativity based on fundamental human and social needs. A policy of this nature will emerge only if, at country, regional, and global levels, the priorities of IT are identified as being based on a normative concept of development.

Notwithstanding short-term policies and strategies, the fact remains that the gap between developed and developing countries will increase, as will the gaps *within* developing countries, which is one of the important structural causes of their present state. The gap between countries has been shown in many ways. For our purpose what matters is the situation in IT, which can be seen from Tables 22.3 and 22.4.

Three important conclusions can be derived from this purely quantitative check. First, the participation of developing countries in the process of "informatization" is indeed small. Second, . . . telecommunications investments are far more predictable than computer investments, owing to longer planning cycles. If anything, the figures in tables 22.3 and 22.4 are optimistic, and would probably need to be adjusted downwards in the

Table 22.3 Value of Data Processing Equipment* (U.S. $1,000)

	1978	%	1983	%	1988	%
Developed countries (U.S., W. Europe, and Japan)	110	83	180	82	250	80
Other countries (incl. centrally planned economies)	22.5	17	40	18	61	20
Total	132.5	100	220	100	311	100

*Micros are not included.

Table 22.4 Worldwide Telecommunications Equipment Market*
(U.S. $1,000)

	1980	%	1985	%	1990	%
Developed countries	36	90	53.5	89	75.4	86
Developing countries	4	10	6.7	11	12.1	14
Total	40	100	60.2	100	87.5	100

*Includes: telephone, telegraph, telex, data communications, satellite communications, mobile radio and radio telephone, radio paging, and cable TV.

Source: Arthur D. Little, Inc.

Table 22.5 Transmission and Satellite Markets (U.S. $1 million)

	Telegraph, telex & data transmission		Satellite communications	
	1980	1990	1980	1990
Africa	48.4	97.3	3.0	10.5
Latin America	106.5	189.0	14.3	34.9
North America	2481.4	6000.5	122.8	463.7
Europe	733.9	1984.1	59.0	189.2

Source: Arthur D. Little, Inc.

light of the severe financial problems of developing countries that had large telecom and informatic projects (e.g., Nigeria and Mexico).

More precise "informatic indicators," particulary in terms of data satellite communications, are revealed in the data of Table 22.5. The gap in transmission and satellites is even larger when one considers the entire telecom market (Table 22.4), and it points to a qualitative difference in priorities for investment and the type of emerging infrastructure. The telecommunications infrastructure, especially in data transmission, is the one that will largely determine the "multiplier effects" of information technology, particularly in terms of knowledge and information-intensive activities.

The third conclusion that can be drawn from tables 22.3–22.5 is that investment in the field of communications and transmission equipment increases almost mechanically the possibility of optimizing systems, increasing office and service productivity, raising the efficiency of production, and furthering conditions for capital-saving effects (e.g., optimization of stocks). This implies that the relative position of developing countries in terms of leapfrogging into the "information age" is even lower than thought if taking traditional indicators of the "industrial age."

This is why I stated earlier that IT is the consequence rather than the cause of development, and leapfrogging can be possible only within a global rather than a purely national or regional strategy.

It should be understood that the process of "informatization" of society is one in which greater amounts of knowledge and information are incorporated into goods and services. This also means that knowledge and information activities acquire a dynamism of their own right and become sources of wealth creation and value-added (e.g., design, programming and R & D). As the amount of information and knowledge incorporated into products, processes, and services increases, the relative amount of energy, materials, labor, and capital decreases. Technologies diffuse through society precisely because they are factor-saving, and IT saves simultaneously in all directions while increasing the capacity to create and process information, and therefore contributing to the accumulation of knowledge.

The empirical evidence shows precisely that the current process is one where greater amounts of information and knowledge are going into production, and not, as some might suggest, that we shall live off information exchanges (Gershuny and Miles, 1983; Jonscher, 1983). Indeed, the greater consumers of robots or even computers are those industries such as automobiles or telecommunications that a few years ago were considered "traditional" sectors.

The nature of the technology calls for a more detailed understanding of "knowledge and information-intensive activities." I have chosen here to examine the services (there are other aspects, such as knowledge and information transfers), because perhaps the most far-reaching effect of technology is that it allows the transportability of services: instead of going to a bank or library, we can transport the bank or library to our own terminals in our offices or homes.

. . . . . .

Conclusions

IT is an expanding reality with far-reaching consequences for the relationships within and between countries. In this context it should be clear that we are at the beginning of the development of IT, not in the middle or at the end.

From the point of view of developing countries, four general conclusions seem to be valid.

1. An erosion of their advantages in low labor costs and "mature industries" is taking place because of changes in products and manufactur-

ing processes. The final outcome of this trend will depend larg
LDCs' response at the macro and micro level.

2. Developments of the service sector and in particular of value-add
services depend on the industrial base, as it is not an autonomous sector.
This means that a deterioration of the industrial base will inevitably affect
the creation and development of services. In turn, as the service content
of industry increases, the lack of development in services could further
affect the possibilities of industrial development.

3. The use of IT in developing countries is limited, and the gap in this
field is increasing. With few exceptions, most LDCs have not developed
policies to confront the challenge at the different levels required (e.g.,
skills).

4. A new reality and opportunity is emerging with the internationaliza-
tion of services. Redressing trends here seem to be a priority, essentially
because the situation is still in a state of flux. The lower barrier to entry in
services offers some developing countries opportunities that were impossible to imagine with tangible products, among them the possibility of
reaching consumers directly.

The final outcome of the effects of IT in developing countries and the
global realities will depend largely on the willingness of the actors to
approach the problems and opportunities in a global context. The command of technology based on human needs seems to be a priority for
what remains of the century.

References

Business Asia. 1982. March 12.

Cohen, E. 1981. Modificaciones provocadas por la microelectronica en el rol
de las empresas transnacionales electronicas en los nacionales electronicas en los paises en vias de desarrollo. Analisis de dos casos en el
area de maquinas de oficina, Primer Seminario Latinoamericano
sobre Microelectronica y Desàrrollo, Buenos Aires (mimeo).

Gershuny, J. and I. Miles. 1983. *The new service economy.* London.

Hoffman, K. and H. Rush. 1982. Microelectronics and the garment industry:
Not yet a perfect fit. In IDS Bulletin, *Comparative Advantage in an
Automating World*, 2.

Jonscher, C. 1983. Information resources and economic productivity. *Information Economics and Policy* 1, 1.

Kim, W.H. 1980. Challenge to U.S. domination: The promise of technology
for newly industrialized countries. In *Financial times conferences:
World electronics strategies for success*. London. 99.

Mitterrand, F. (1982), *Report to the summit of industrialized countries: Technology, employment and growth*. Paris.

...es Trade Representative. 1983. *A U.S. national study*
...ces. Washington, DC.

...ociates. 1981. Review of economic implications of Ca-
...order data flows. Reported by *Transnational Data Re-*

...*impact of microelectronics,* chap. 7. Geneva.

...1982. Expert group meeting on implications of microelec-
...r the ECLA region. Mexico City, June 7–11, 1982. *Confer-*
...*ceedings,* Vienna.

USITC. ...*Competitive factors influencing world trade in integrated circuits.* Washington, DC: 14.

--- **23** ---

Technology and Human Freedom

Philip Bereano

Most of us have been brought up to believe that the term "technology"
refers to physical artifacts, like a typewriter or a heating system. But that
view is not sufficiently helpful in analyzing technologies in terms of their
social, political, cultural, and economic ramifications. I prefer to define
"technologies" as the things *and* the institutional (the social, political,
cultural, and economic) mechanisms which produce them and are affected by them.

Human beings have been involved in producing technologies and using
and exploiting them for a long time. But now many of the effects and
ramifications are much more massive than they were in the past and, in
certain ways, not readily reversible. New terms such as "postindustrial
society" or "technotronic society" are attempts to indicate that there is
something qualitatively different about what is currently going on.

Emmanuel Mesthane of Harvard's former technology and society program wrote:

> New technology creates new opportunities for men and societies and it
> also generates new problems for them. It has both positive and negative

From Philip Bereano, "Technology and Human Freedom," pp. 132–43, *Science for the
People,* November/December 1984. Reprinted by permission.

effects and it usually has the two at the same time and in virtue of each other.

In certain aspects I think this observation is pretty shrewd, but I fundamentally disagree with his position that technology is neutral. David Dickson has called this the "use/abuse" model of technology. For example, I have a pen in my pocket which I can use to sign someone's death warrant or to write the Declaration of Independence. The uses and abuses of the pen are many, but the pen itself is neutral. Although this might be true about some very simple technologies such as ballpoint pens, I maintain that it is not true about most of the substantial and important technological phenomena which we find in our civilization.

The notion that technology is neutral is very important to the corporate ideology in American. This free enterprise model says that the problems associated with technology are what the economists call "externalities"—the unexpected, unintended side effects of things. The factory which is manufacturing something that we all want may be polluting the air or the water, but pollution is a side effect and is not intentional. Until society creates air pollution laws which internalize these external factors, such side effects will continue.

Because technologies are the result of human interventions into the otherwise natural progression of activities, they themselves are imbued with intentions or purposes. Current technologies, however, are not intended to equally benefit all segments of society. We are not all equally involved. Our society is a class society in which different people have different access to wealth, to power, to decision making, to responsibility, to education, et cetera. We live in a society in which such access is differentiated on the basis of gender, of color, and so on. Because technologies are intentional or purposeful interventions into the environment, those people with more power can determine the kinds of technological interventions which occur. Because of their size, their scale, their requirements for capital investments and for knowledge, modern technologies are powerful interventions into the natural order. They tend to be the mechanisms by which previously powerful groups extend, manifest, and further exacerbate their powers. These technologies are not neutral; they are social and political phenomena.

The Appearance of Choice

These social and political aspects of technologies are often hidden behind the appearance of decentralized "choice." On the surface, modern technology offers society many choices, many sources of information.

Television, for example, appears to be a great decentralized resource, with 60 to 70 percent of Americans using TV as their primary source of news. Yet as a technological system, television is one of the most highly centralized phenomena that we have. It is literally true that a very small number of people are able to determine what *is* and what *is not* news; how material classified as news shall be presented and how not; whether it will get thirty seconds or fifteen seconds or no time at all.

Census data are also available in a decentralized way to many people. Any person can walk into the library and get access to the computer printout. But the census itself is not really decentralized. The actual forming of the data pool, the decisions as to what questions will be asked and how they will be formulated are very centralized. These centralized decisions reflect the power differentials which exist in our society. Census takers ask how many bathtubs there are in a household (of interest to the American Porcelain Institute) but they don't ask questions which are of particular interest to me or to you. This appearance of access to information and of choice also occurs in the transportation system.

As David Dickson has said about the automobile, they give you tremendous numbers of choices: color, white or black wall tires, digital or sweephand clock. But the important decisions, like what kind of propulsion system it's going to have, you don't have any choice about. The fact is there have been propulsion systems, such as electric or steam, that have been technologically feasible for over half a century. Yet they do not in any real sense exist for people. In fact, it is not practical to have electric cars today because technologies are not individual components but systems. The automotive system is designed for gas combustion cars. We would need to have a totally different kind of support network—completely different service stations—if a hundred million electric cars were on the road. This happened to a small degree with an increase in diesel cars. One's ability to get fuel, top service, and knowledgeable mechanics changed dramatically. Without the whole technological infrastructure, which is as much a part of the technology as the artifact of the car, you cannot have an electric car. It is not a real choice. But I *can* have a car with whitewalls if I want. Dickson claims that this is a very common manifestation of modern technology. One's choices only appear to be decentralized.

Control and Understanding of Technology

We live in a society which styles itself to be democratic. How are we to reconcile the fact that the technological values of efficiency, expediency, and high-powered knowledge and science, tend to involve a relatively small number of people? Academics, government, and corporate officials routinely make important decisions that have impacts upon all of us, but over which most of us have relatively little control. And it is not only control. I think that our society is historically unique because for the first time the overwhelming majority of people do not even pretend to understand how their life-support systems operate. What actually happens when you flip the light switch on the wall? In many earlier societies, whether we may now ridicule their beliefs or not, people thought they understood how things important to them and to their culture worked and why. The reason this is important is that what technology has really produced—and I think this also has relevance for human freedom—is a very profound sense of alienation. I mean it in the Marxian sense, not in the pop-psychology or pop-sociology sense of alienation. Alienation is the sense that something is going on which is "other," apart from what I am. Most people have a very pervasive, inchoate, unrealized alienation in their day-to-day life.

The workplace is a good example of a situation where most of the technology that people use they are powerless to make choices about. Each week thousands and thousands of women are told that they are going to become word processors and that their typewriter is going to be replaced by a word processor. They have absolutely no control over the phenomenon. And that phenomenon is more than just getting a new high-powered machine to do what they used to do. Technology, in this case, is not just a machine. It is a whole social milieu and involves a very important redefinition of roles and functions. A woman who did typing and filing, answered the phone and interacted with people, also had a certain measure of control over the arrangement, flow and pacing of the various activities. In this example, she is now being transformed into a person who will sit eight hours a day in front of a cathode ray tube and "word process." She will do so whether or not it hurts her eyes or her overall health. This person's job is being substantially degraded; the whole notion of control, the sense of autonomy, no matter how limited it might have been under the earlier situation, is being taken away, all under the guise of a new technology.

Most of us learned that, in the Industrial Revolution, people invented productive machines and then gathered workers together to use them in factories. But actually the factory was a social system which *preceded* many of the new technological mechanisms. It was designed for the social goal of controlling the workers, regulating and rationalizing production (at the very least because the entrepreneur did not know how to make cloth and wanted to control the operation of those who did).

There are two objectives a capitalist has: productivity, and control of the workers. Only one of them has been generally presented as being the reason for all these changes. We can see that today in the arguments being made for things like word processing are these neutral-sounding "increased productivity" arguments. When corporations advertise in the general press—the *New York Times, Atlantic* or *Harper's*—they talk about productivity in such a way that the readers will not conclude that these people are actually scheming to further control workers.

For example, high tech industries offer a limited range of jobs in which average pay levels are low. Most of these industries, largely un-union-ized, have lots of low-paying, boring, repetitive, unskilled jobs and a very few flashy engineering positions. Yet, when the promoters of high tech talk about the need to increase productivity in this society, they want people to view that position as neutral, good, and progressive. So they say things like, "progress is our most important product." But they do not talk about how the industry will affect the workers and their workplace. We have all been subjected to a tremendous barrage of attempts to sell us computers. Such efforts inevitably engender in us a fear that our children will be technologically inadequate if they are not "computer literate." But most people do not need computers. They are not writing books, analyzing large masses of data with correlation and regression statistics. What are the companies telling these people? They are telling these people that a computer will help manage their finances, which, for most people, means balancing a checkbook. This is a third-grade skill: the addition and subtraction of whole numbers. The mistakes made are mostly entry mistakes which computers will not avoid. The computer is a two thousand-dollar abacus.

I believe that most computer users of the future will be word proces-sors and not highly educated high tech people. There will be some of the latter but there will be ten unskilled laborers plugged into a computer for every one creative person who is working on a novel and wants to be able to justify the margins as the work progresses.

Another area in which I have done research is household technology— or kitchen technology, for instance. Without painting any kind of conspir-acy theory, the overwhelming decisions about household technology, their development, their deployment, have been made by men who do not use, have never used, and do not want to use these technologies. Here again, there is a tremendous dichotomy between the people who are making those kinds of choices and, at least demographically speak-ing, a totally different group of users.

Utopian Visions Versus Decreased Possibilities

There are writers such as Cullenbach, LeGuin, and Bookchin who offer a political, utopian vision of a different kind of society and a different way to organize the "good life" socially. They would use technological systems very differently from those which are currently manifest around us. They would be much more conducive to the fulfillment of human values by a large number of people, increase human autonomy and decrease alienation, put more of a premium on altruism and less on selfishness and privatism. I think they are structured on a set of values preferable to those I see imbedded in the dominant technology around us.

But utopian means "nowhere." You cannot wake up one morning and find that liberation has occurred. It is a very long and intricate kind of process to raise the consciousness of people so that they can develop that kind of autonomy. When people criticize Marcuse, for example, they say he is elitist because he claims he knows better what people want than they themselves. The point these critics miss, however, is that Marcuse is quite firm about the fact that humans have the potential for autonomous decision making. But he also realizes that in this highly industrialized society, most people have had that sense of their power and their ability systematically stripped from them, not only through their socialization (so that the ideologies they believe tend to disempower them), but through the realities in which they find themselves, which give them relatively little freedom of movement.

I will conclude with a quotation by Lewis Mumford. Mumford was very romantic about technology and values, with the result that he is not terribly helpful to us. But in this quotation I think he shows tremendous insight. He is talking about automation, but it is really about technology in the larger sense. He states:

> It has a colossal qualitative defect that springs directly from its quantitative virtues. It increases probability and it decreases possibility.

In other words, there is something wrong about the qualitative aspects of technological phenomena, something, he says, which springs directly from "their qualitative virtues." That is to say, the power that technology has in the quantitative sense reduces quality. One of the things that modern technology claims to do, for example, is to make available to masses of people experiences which were once reserved for the few, such as the opportunity to have tomatoes in January. In the early part of this century you had to be someone like Andrew Carnegie to have a tomato in January. Now anyone can have a tomato in January just by going to the supermarket. But the quantitative virtue—the ability to produce week after week millions of tomatoes—has altered, must alter, the quality of the tomatoes you can buy. The tomatoes we get at Safeway are intentionally not the same as the tomatoes that Carnegie ate, because the tomatoes he ate were grown in Cuba or Mexico and specially transported, or grown in special hothouses. But you cannot do a million of those a week.

In order to have the mass phenomenon of tomatoes in January, the technological adventure had to change the essence of what the tomato is. And the mass phenomenon means that certain technological events become very probable, and alternative possibilities are decreased (e.g., the internal combustion engine overwhelms the electric car).

Since technologies are systems of hardware *and* social institutions, the phenomenon is linked increasingly to concentrations of power—a threat to our existence as a truly free people.

References

Munford, Lewis. 1964. Automation of knowledge. *Vital Speeches of the Day,* May 1: 442.

Mesthane, Emmanuel. 1976. Social change. In *Technology as a social and political phenomenon,* Philip Bereano, New York: John Wiley & Sons, p. 69.

Dickson, David. 1974. *The Politics of Alternative Technology.* New York: Universe Books.

— III

Issues Facing Computer Professionals

Justice: The Distribution of Rewards

Two essays (readings 24 and 25) in this section deal with issues that are currently the subject of intense debate within the computer community: how and whether to protect legally the creations of computer professionals. That is, who is to own the copyrights to and earn income from computer software or programs? In reading 24 Rosch presents the evolving legal status of this issue. The law appears to be moving away from the position that only the code of a program is protected by copyright; more recent cases give hope that the "look and feel" of a program is also protected. In reading 25 Machrone argues that such protection would impede the development of more powerful software by preventing programmers from incorporating the useful ideas of their predecessors.

Pamela Samuelson (reading 26) takes on the special case of ownership rights when the work in question has been generated by a computer program. She reviews legal issues in this new category of inventiveness, and considers whether the computer, the user, or the programmer should be considered the author of a computer-generated work.

Richard Stallman (reading 27) takes an unconventional position equating "freedom of programming" with freedom or speech or thought. His 'GNU Manifesto" is full of ethical contentions, mostly with Kantian overtones. For instance, if everybody were to use destructive means to become wealthier, the resulting chaos would make everybody poorer. Putting a price on computer technology, Stallman says, produces the same outcome.

The Look-and-Feel Issue:
The Copyright Law on Trial

Winn L. Rosch

It looks like *1-2-3*. It works like *1-2-3*. All the commands are the same, all your old Lotus worksheets load into it without a hitch, and all your hard-earned macros run just fine. It's a near-perfect clone. At $50, it sounds too good to be true. Or at least too good to be legal.

That's what Lotus Development Corporation claimed when it filed a lawsuit in the Boston Federal District Court on January 12, 1987. In two separate civil suits, Lotus claimed that both *VP-Planner,* published by Paperback Software, and *The Twin,* from Mosaic Software, infringe on its *1-2-3* copyrights. (The lawsuits also allege unfair competition under federal and state laws.)

Glance at today's PC and software markets and you're apt to think Lotus is just trying to frighten the industry, hoping to keep it in line and prevent anyone else from thinking about such a heresy as cloning *1-2-3.* Although that conclusion might have been justified 10 years ago, the rules have since changed. Not only does Lotus have a solid case, but other program clones are susceptible to similar copyright infringement actions. Even PC-compatible computers are in danger—the current re-thinking of the copyright laws and vigorous prosecution of clone makers could change the face of the entire industry.

The driving force behind this change in traditional copyright concepts is a new appreciation for the creativity and effort involved in making a successful computer program.

Under the old rules, a huge industry sprang up to make and sell clone computers and software. Everyone seemed willing to offer a cheaper clone than the next guy. Long before Lotus filed its lawsuit, other companies had brought out programs that so closely resembled successful forebears that you needed a price list to tell them apart.

One classic example was the word processing program *NewWord.* Not just a *WordStar* clone, *NewWord* was actually written by former employees of *WordStar*'s publisher, MicroPro. There was no legal action against *NewWord.* In fact, the program code was recently sold to MicroPro for a

From "Taking the Stand: The Look-and-Feel Issue Examined," pp. 157–58, 160, 164. *PC Magazine,* May 26, 1987. Reprinted by permission.

multimillion-dollar sum. Or look at PC clones in general. Every one of them requires a BIOS [Basic Input/Output System] to emulate an IBM PC, and every one of them has it. Yet, until now, IBM seems to have accepted the existence of the multitude of clone makers even though they have severely cut into its PC business.

This freedom-to-clone policy was a product of the contemporary interpretation of American copyright law. Until some recent court decisions were announced, the law was read by both product makers and their attorneys to restrict the copying of program code but not of program concepts. If you could duplicate the operation of a program without copying the underlying code, you could clone anything you wanted.

The current legal viewpoint is quite different. Copyrights are no longer considered code specific. Under the emerging interpretation of copyright law, one computer program can infringe on another even if it is written in a different language and designed for an entirely different operating system or computer. The "total concept and feel" or, alternately, the "look and feel" of two programs rather than their underlying code, determines whether one is similar enough to be a copy.

Although the words "look and feel" are absent from the complaint in the lawsuit Lotus filed, its arguments follow this new doctrine. The company contends that the way *1-2-3* appears on the screen and the way it works is an intrinsic part of the company's copyright. By copying everything from the exact command names and the screen arrangement to the way the program works, Paperback and Mosaic are alleged to have infringed on the Lotus copyright.

If the subtle issues confuse you—such as exactly what constitutes the "look and feel" of a program—you're not alone. Various courts have struggled to sort out all the issues involved with software and copyrights. Although most have come up with answers, none is truly definitive. At least one leading case (*Whelan Associates* v. *Jaslow Dental Laboratory,* which will be discussed later) has been appealed to the United States Supreme Court, and its outcome may determine the fate of software copyrights—and the PC industry.

Ideal Situation

The problems begin with the purposes underlying copyrights. Copyrights, as well as patents, are legal rights granted by the Constitution that protect your creative work. But, contrary to what most people believe, neither copyrights nor patents offer protection of your ideas. One of the foundations of our country and society is that ideas are to be freely shared, that the exchange of ideas (and, one hopes, their critical assessment) leads to enlightenment. All ideas by their nature belong in the public forum. An exclusive right to any idea that removes that idea from public access and use is contrary to the entire philosophy of enlightenment which the Constitution promotes.

But automatically dumping all the fruit of one's creativity into the public domain removes much of the incentive for creative thought— the profits. So provisions for patents and copyrights were written into the Constitution.

Patents and copyrights deal with the manifestations of ideas rather than the ideas themselves. Patents prevent the appropriation of your implementation of an idea, putting your brainstorm to work either as a specific product (say a servo-controlled knurling machine) or as a process in accomplishing some end (for instance, a method of making fertilizer from Congressional debates). Copyrights deal solely with the expression of ideas—how you go about communicating your idea to the rest of the world.

In exchange for granting a limited monopoly on the products of your creativity, patents and copyrights ensure that eventually the embodiments of your ideas will be in the public domain. Laws put definite limits on what can be patented and copyrighted, how long those respective rights survive, and the protections they provide. Patent protection is comprehensive but short. It prohibits unauthorized people or organizations from using your product or process—even if they should independently stumble upon it—for 18 years. Copyright protection is longer—up to your life-time plus 50 years—but prohibits only copying your expression. You have no recourse against anyone who independently develops the same expression without copying yours.

The copyright law also limits its own subject matter and does not protect every expression of every idea. Although your idea can be retread and reused, the expression must be original.

Discrimination

Functionality is another distinction between patent and copyright. Under traditional theories, patents covered functional applications. The subject matter of copyrights had to be nonfunctional.

This dichotomy presented immense problems in protecting programs. Established patent law has long held that mathematical algorithms cannot be patented because they are merely ideas. Computer programs are merely algorithms so they, too, were unpatentable. On the other hand, copyright law held that computer programs were functional and therefore were not proper subject matter for copyright. Furthermore, doubts existed as to whether software was an expression that could be copyrighted, notwithstanding its functionality, or whether it was an algorithm, purely an unprotectable idea.

These roadblocks to the protection of computer programs have fallen away in the last 20 years. As automation swept through industries, the law recognized that software could be an essential part of a machine or a process and could be protected by a patent on the overall application. More recently, the patentability of software itself has been recognized.

Copyright protection for computer programs was assured in 1980 when Congress officially amended the law to include them. Under the new provisions, software is an expression, and copyrights on computer programs are just like any other, save a few additional rules. Unauthorized copying is forbidden unless it is necessary for the program's normal use. The law also provides that a single archival copy of a program can be made as a backup.

As with other copyrights, the new law puts no limits on the use of copyrighted material once it is sold. The purchaser of a copyrighted program—absent other agreements and legal entanglements—can dispose of it as he pleases, using, selling, or renting it. He is forbidden only from copying it.

Thus, when you buy an IBM PC with copyrighted material in its built-in ROM, you can later sell the computer with no worry that IBM will elbow in and try to collect an additional royalty. However, you cannot copy the protected BIOS code on which IBM's copyright subsists.

Is it Real?

To establish copyright infringements, courts require that you prove two things—that you are the owner of the copyright and that the work you allege infringes on your copyright is indeed a copy of it. While the former is relatively straightforward, the latter can be troubling. In most cases you won't have a videotape of a scribe furtively glancing at your original, then scratching out his copy.

Consequently, the law has evolved a twofold test that substitutes for direct evidence of copying to relieve you of need for such espionage. Instead of showing the copying itself, the copyright owner can prove to the court that the alleged infringer had access to the original work and that a "substantial similarity" exists between the original and the alleged copy.

The recent change in copyright law for computer programs and the

Lotus lawsuits hinges on the issue of exactly what constitutes this substantial similarity.

The old school of thought held that the expression of a computer program was the source code and, by extension, its machine-readable object code (the functioning program). The written code obviously expresses the underlying idea of the program. For two programs to be substantially similar under this philosophy, the actual program code would have to be the same.

For instance, under this interpretation of expression IBM could not copyright the idea of a Basic Input/Output System (BIOS) for its PC, and other people would be free to dream up their own input/output systems so long as they didn't peek into IBM's ROM and copy the exact instructions IBM used. (So-called reverse engineering is allowed, but only to glean the ideas, not the actual expressions.) Each developer of an IBM-compatible BIOS would need merely to use different code to accomplish the same ends as the IBM original.

As a result, BIOS developers have gone to great lengths to be certain their engineers have not been contaminated with knowledge of the IBM BIOS and do not duplicate any IBM code. (When access and substantial similarity can be proved, even unintentional copying is forbidden.)

BIOS developers who have attempted to take the easy way out have suffered dire consequences. The ROMs in certain Apple-compatible computers made by Franklin Computer Corporation were found to have been copied from the Apple original, primarily because the Apple ROM contained hidden and otherwise nonfunctional code that, to the chagrin of the copycats, unambiguously identified its origins.

Touch Test

One way around the need to prove the literal transcription of program code is the audiovisual copyright. The images that make up a slide show, motion picture, or television program are copyrightable expressions. The same protection has been extended to the video displays generated by computer programs.

More important to the protections provided for computer programs is the emerging philosophy of "total concept and feel." Under this doctrine the expression of the idea inherent in computer software is more widely spread. It's more than just the underlying program code or the video screens.

According to one court, the expression of an idea in software is the manner in which the program operates, controls, and regulates the computer in conceiving, assembling, calculating, retaining, correlating, and producing useful information either on a screen, in a printout, or by audio communication.

Under this doctrine, less reliance is placed on the exact verbal or visual similarity of the works. The nonverbal expression becomes paramount. The order in which the various screens are presented and the way the user interacts with the software become an acknowledged part of the way the software expresses its underlying idea.

The "concept and feel" legal copyright philosophy appeared long before the personal computer. As early as 1970 it was used in the prosecution of a copyright infringement action, *Roth Greeting Cards* v. *United Card Co.,* which involved commercial greeting cards. The court found infringement even though the copies were not identical. The copies merely conveyed the same mood and character—in some instances, nearly identical typestyles—as the originals.

The first major computer case to use the phrase "total concept and feel" was filed by Atari against North American Philips Consumer Electronics Corporation in 1982. In this action, Atari alleged that the eat-the-dots game K. C. Munchkin infringed upon its copyrighted game PAC-MAN. Although the layout of the mazes, the number of gobbled-up dots on the screen, their arrangement, and the sounds and colors used by the two games were different, the court found that Atari's copyright had been infringed upon.

Among other conclusions, the court noted that slight differences between a copyright-protected work and an accused work will not preclude a finding of infringement. The overall similarities are more important than minute differences.

Probably the most important computer case to deal with these issues thus far is *Whelan Associates* v. *Jaslow Dental Laboratory,* which is being appealed to the Supreme Court. The *Whelan* case concerns a dental office management program that was originally written for use on IBM Series One computers. The infringing copy was designed for the IBM PC and was written in an entirely different source language than the original. The trial court and an appellate court have held that the detailed structure of a program is part of its expression, not part of its idea, and that copyright protection is thus not limited to the literal elements of the program, that is, the source code itself.

Inherent in these decisions is the growing recognition that the work and creative effort involved in writing computer software requires more than merely arranging program code. Developing specifications for file structures, the user interface, and other intangible aspects of the program may require more effort than the mechanical work of coding. Under the new software copyright interpretations, this substantial development effort is protected.

No More Clones?

The implications may seem more ominous than they really are. For instance, in the *Whelan* case, the court based a portion of its finding of substantial similarity on the common file structure used by the original and infringing copy. However, this conclusion does not mean that the file structure itself is protected by copyright. Rather, the duplication of file structure was merely one element of the evidence used to show that copying had occurred. Other programs might use the same file structure if they are otherwise not substantially similar to the original. For instance, a data base should be able to use the same file structure as a spreadsheet without fear of infringement.

When two programs are designed to accomplish the same task in essentially the same manner, however, any similarity between them can be used as evidence pointing toward infringement. For instance, in the . . . case of *Broderbund Software and Pixellite Software* v. *Unison World,* such design choices as the duplication of command names in the programs *Print Show* and *Printmaster* were evidence of substantial similarity between the two packages (the latter being the copy).

Even advertised claims can be used as evidence of substantial similarity and, hence, copying, as they were were used in the *Whelan* case, in which the maker of the infringing program advertised that it worked just like the original.

The *1-2-3* clones subject to the Lotus lawsuit demonstrate many of these same similarities. They use the same commands, the same macros, the same file structures, and have even been advertised as being much the same as *1-2-3*. Whether these similarities are substantial and the "total concept and feel" of the programs is, in fact, the same will have to be decided at trial.

For software developers, the emerging interpretation of the copyright law means that duplicating the operation of another program is ill-advised. Rewording or writing in a different programming language may be insufficient to dispel the substantial similarity between the new program and the original. And such acts could lead to a finding of copyright infringement.

So far the "total concept and feel" doctrine has not been applied to the computer BIOS. Whether it will be seems to depend on court interpretation of the copyright statute and case law. Certainly the elements of substantial similarity are present in the ROMs of IBM-compatible computers. For instance, they use exactly the same interrupt entry vectors (if they didn't, they wouldn't be compatible). Certainly the clones embody the same total concept as the original IBM BIOS. Although other copyright issues cloud the picture, copyright infringement lawsuits by IBM against the makers of compatibile ROM are indeed possible.

You may feel predisposed against Lotus in its lawsuits. The company seems to be running against the hallowed traditions of the PC software community. Lotus is big and fat and has made the millions we wish we had. Moreover, many people see Lotus as the bad guy because the company clings to copy protection and seems to be trying to chase competition off the market to keep its price high. But the law is blind to such prejudices. Besides, if you were Lotus, you would want to pursue your legal rights, too.

And though you might think that Lotus's suits will be bad for competition and for the future of bargain-priced programs, it can be good, too. The "look and feel" copyright doctrine should provide incentive for program developers to be more creative. Not only will writing more creative programs help programmers avoid infringement, but the increased copyright protection afforded them should give them more monetary incentive to pursue creative work.

[In another important look-and-feel case, Apple is suing Microsoft and Hewlett-Packard, claiming that the "desktop" on the computer screen produced by Windows and New Wave resemble too closely Apple's copyrighted "desktop" on the Macintosh. The March 1989 ruling on a preliminary motion in this case was a significant victory for Apple.—Eds.]

—— 25 ——————————————————————

The Look-and-Feel Issue:
The Evolution of Innovation
Bill Machrone

Those most violently opposed to Lotus's lawsuits say that Lotus is playing a one-sided game—that it is attempting to stop others from copying aspects of its products, even as it has freely copied others.

From "Taking the Stand. The Look-and-Feel Issue Examined," pp. 166, 168, 174. *PC Magazine,* May 26, 1987. Reprinted by permission.

This, they say, is counterproductive, since incremental progress is still progress. Users would be far worse off if they had to wait for conceptual breakthroughs or new paradigms instead of refinement. Dan Bricklin summed up this argument best in a wry twist of Isaac Newton's immortal words: "If you are going to see farther than others, make sure you're not standing on the shoulders of any giants."

Here is a brief genealogy of some of the most important products in our industry.

First and foremost, there is DOS. Back in the late 1970s, the 8086 was a little-used oddity in Intel's arsenal. The 8-bit 8080 and its derivative, Zilog's Z80, owned the market, and the operating system of choice was Digital Research's CP/M. Seattle Computer Products, a small manufacturer of personal computer boards, built a CPU around the 8086 instead of the 8080 and Z80 boards that dominated the market at the time. It needed an operating system, and Seattle's Tim Patterson built a CP/M workalike, which he called 86-DOS. Every function call of CP/M was faithfully duplicated, taking into account the differences in register design of the 16-bit 8086 and the 8-bit 8080.

Programmers familiar with CP/M could move their software over to 86-DOS with little or no trouble. Seattle even offered a translation of Microsoft BASIC 5.21 for the 8086 (I still have a copy on an 8-inch floppy in my basement). This precursor to BASICA opened the 8086 to casual programmers in addition to serious developers.

The story of how Microsoft acquired 86-DOS and used it to convince IBM to build a 16-bit personal computer is an oft-told tale, and not germane to our discussion here. MS-DOS has changed completely since then, as it goes through endless enhancement and rewrite. But the story didn't begin with CP/M.

Gary Kildall, Digital Research's founder, found much to admire and emulate in Digital Equipment Corporation's operating systems. Commands in RT-11 and RSTS are familiar to an MS-DOS user, and even more so to a CP/M user. The copy command, for instance, is a streamlined version of PIP (Peripheral Interchange Program) in CP/M. Not surprisingly, many DEC operating systems sport a PIP command, too. DOS's DEBUG is command compatible with CP/M's DDT (Dynamic Debugging Tool). And so on.

Coming full circle, Digital Research took its multitasking operating system, Concurrent CP/M-86, and made it DOS compatible in the version called Concurrent PC-DOS. It's still available. In fact, IBM uses it as the operating system in its PC AT-based point of sale and intelligent cash register system. While I don't know where DEC's ideas might have come from, I'm reasonably sure they weren't all original.

War of Words

Word processors have tended to borrow heavily from the past. Some of them based their whole reason for being on their similarity to the ubiquitous Wang word processor. Wangs, as you will recall, were among the first office automation tools, and the secretarial schools turned out thousands of Wang-trained secretaries each year. As PCs replaced hardwired, dedicated word processors in the workplace, it made perfect sense for software to capitalize on this tremendous pool of trained, talented workers. *MultiMate* was the first Wang imitator, and *OfficeWriter* followed soon thereafter. Of course, each made the necessary concessions to the PC's lack of dedicated keys, and each picked up some benefits from the PC's flexible design.

Both products have moved beyond the original Wang paradigm, especially *OfficeWriter*. While secretaries of sound mind would never move back to the old beast, they remember their roots.

WordStar Wizardry

Back in 1979, Seymour Rubenstein created *WordStar.* Through Rob Barnaby's programming wizardry, it became marvelously suited to the unruly world of dissimilar terminals, often without cursor controls, on CP/M systems. Rubenstein invented the cursor control diamond, using the E, S, D, and X keys in conjunction with the Ctrl key to move the cursor up, left, right, and down. It made good ergonomic sense. Prior to this, the best idea anyone had had for using the keyboard for cursor control was the Ctrl-H-J-K-L scheme pioneered in UNIX and adopted by *Magic Writer,* which later became *PeachText.*

Suddenly it was fashionable to be keystroke compatible with *WordStar.* The control sequences, while in no danger of being mnemonic, were efficient. *dBASE II*'s built-in editor was *WordStar* compatible, as are the editors in all the Borland languages. You might have thought that the dominance of the PC, with its well-defined set of cursor control keys on the numeric keypad, would eliminate the need for the control diamond. But it keeps surfacing in new products, even the sophisticated *NewViews* accounting system.

Data-Base Descendants

Two of today's most popular data bases are descended from larger systems. Wayne Erickson is the chairman of Microrim, the *R:base-System V* company. Not coincidentally, he was also the chief architect of *RIM*, a mainframe data-base program that Boeing Computer Services offered on its time-sharing system. *R:base Series 4000* begat *R:base Series 5000*, which begat *R:base System V.* And they kept it all in the family.

Wayne Ratliff played a similar role in the genesis of *dBASE III.* While he was at Jet Propulsion Labs, Ratliff built *Vulcan*, a mainframe data base. JPL used it for tracking things on the Voyager flights, and more. Ratliff got interested in microcomputers and rewrote *Vulcan* for CP/M. Then advertising genius Hal Pawluk renamed it *dBASE II,* ran a famous ad asserting that all other data-base products were "bilge pumps," and the rest is history.

dBASE has spawned a sizable aftermarket. First came the enhance-ment products that either generated applications or made *dBASE* more flexible. Then came the compilers. While they stole some unit sales from Ashton-Tate, they also catapulted *dBASE* into new respectability. Com-petition dignifies a market and helped make *dBASE III* a real program-ming language. The existence of the compilers is still a problem for Ashton-Tate, but a nice one to have.

And then there was *VisiCalc.* First conceived for the Apple II, it was the original reason for businesspeople to buy PCs. Before *VisiCalc* there were financial-planning languages and there were manual spreadsheets. The stroke of genius was the idea of an electronic spreadsheet; every-thing since has just been refinement.

Lotus's founder, Mitch Kapor, designed the original spreadsheet add-on product, a graphics program for *VisiCalc.* From then on, he worked very closely with Dan Bricklin and Bob Frankston, *VisiCalc*'s designers. It's only natural that *1-2-3* owed *VisiCalc* a large debt of gratitude. More-over, it went to great lengths to be a superset of *VisiCalc*. . . . Originally, *1-2-3* used the F10 key to get you to the menu, but it was changed to the Slash key in order to be *VisiCalc* compatible and to overcome the possi-ble objections of *VisiCalc* users. As an interesting side note, the F10 key has reemerged as Lotus's command invoker. Both *Metro* and *Manuscript* use it.

Dozens more stories could be told about products, innovations, flank-ing maneuvers, and head-on assaults by competitors. But through them all, you would find one dominant theme: people. Developers are influ-enced by powerful ideas and by one another. Marketers are influenced by powerful products and by market opportunities. Is it illegal when they join forces? We'll see.

Can a Computer Be an Author?

Pamela Samuelson

As early as 1965 the Register of Copyrights expressed concern about whether a computer could own rights in computer-generated works. In a report to Congress, the Register of Copyrights raised several difficult questions: Would computer-generated works have a human "author"? Was the computer merely an assisting instrument of its human user or was what copyright law had traditionally regarded as "authorship" actually conceived and executed by a machine and not by a human? The questions were apparently raised in the register's mind because of contemporaneous attempts to register works created by computers. The register did not report on how those applications for registration were handled, or how he thought they should be handled. Rather, the register simply posed the question to Congress.

At the time the register made this report, Congress was in the midst of a major revision of the copyright laws. Congress apparently found the register's questions to be sufficiently disturbing and perplexing as to require more thorough investigation, for in 1974 Congress created the National Commission on New Technological Uses of Copyrighted Works (CONTU) to study a variety of new technology issues, among them, the issue of authorship of computer-generated works.

In 1978, CONTU made its final report to Congress. Most of the commission's attention had been devoted to the photocopying and software copyrightability problems, but one short section of the CONTU final report addresses the issue of authorship of computer-generated works. CONTU seems to have found the issue to be a simple one for it opines that the answer it gives, which is that the user of the program is the author of the computer-generated work, is the "obvious" one.

Concerning the question whether a computer could be said to be "the author" (or "an author") of works created through its use, CONTU expressed certainty that it could not: "On the basis of its investigations and society's experience with the computer, the Commission believes that there is no reasonable basis for considering that a computer *in any way*

From Pamela Samuelson, "Ownership Rights to Computer-Generated Work," *University of Pittsburgh Law Review,* Vol. 47, No. 4, pp. 1192–1209, Summer 1986. Reprinted by permission. [Original footnotes have been deleted—Eds.]

contributes authorship to a work produced through its use." CONTU likened the computer to a camera or a typewriter, all three being instrumentalities that in themselves were completely lacking in creative capabilities while requiring human direction to bring about a creative result. CONTU reviewed a number of instances of works generated through the use of a computer. "In every case," said CONTU, "the work produced will result from the contents of the data base, the instructions indirectly provided in the program, and the direct discretionary intervention of a human involved in the process." CONTU regarded as too speculative to require serious consideration the proposition that computers could or would soon be able to exhibit creative authorship. In CONTU's view, the computer was not and could not be "the author" of anything.

Some authors have taken issue with CONTU's assumption that computers are incapable of exhibiting sufficient originality or creativity to support a copyright. To demonstrate this point, one author began an article with a somewhat wacky but coherent paragraph that had been written by computer. The . . . Office of Technology Assessment Report on "Intellectual Property Rights in an Age of Electronics and Information" responds to this aspect of the CONTU final report by saying:

> It is misleading . . . to think of programs as inert tools of creation, in the sense that cameras, typewriters, or any other tools of creation are inert. Moreover, CONTU's comparison of a computer to other instruments of creation begs the question of whether interactive computing employs the computer as a co-creator, rather than as an instrument of creation.

[In 1950] the eminent British mathematician and computer scientist Alan Turing stated the challenge that has led to a vast scientific inquiry into artificial intelligence. As paraphrased by another author, "In the last analysis, . . . the question of whether a computer can 'think' or not can be answered in the affirmative if a human being, by asking it questions, could not tell from the answer whether he were interrogating a man or a machine." No deep study of the literature on artificial intelligence is necessary to observe that a great many brilliant scientists take the idea of machine intelligence very seriously. While there may be some debate about how advanced the state of the art currently is, there is no question but that many machine-generated works are already available, and that in the future they can be expected to become ever more complex, sophisticated, and valuable.

If a machine can think, it would seem logical that it could also compose or design a work. If a machine does compose something, such as a piece of music, and it is impossible to tell by hearing the music whether it was composed by a computer or by a human, one might wonder whether the notion of machine authorship ought to be accepted.

What then is the "case" for allocating authorship rights to a computer under the copyright statute? While Congress may never have anticipated machine authorship, the statute itself says nothing about what kind of being one has to be in order to qualify as an author. To qualify, there must only be a category of work that is eligible, some tangible expression of it, and some minimal quantum of originality (and perhaps creativity). Let us grant that a category of eligible work (say, music) and a tangible expression of it (a printed score) can be generated by a computer program. In that case, copyright protection should be available if the last of the requirements, originality, can be shown.

"Originality" is required by statute to qualify for copyright protection. Only "original works of authorship" are statutory subject matter of copyright, and then only if they are "fixed" in some "tangible medium of expression." The legislative history of the 1976 act indicates that originality and fixation are the two "fundamental criteria of copyright protection."

Despite its fundamental importance, the statute does not define what is meant by originality. The legislative history indicates that it was "purposely left undefined" and that Congress "intended to incorporate without change the standard of originality established by the courts under the present statute." The legislative history goes on to say that the standard "does not include requirements of novelty, ingenuity, or esthetic merit, and there is no intention to enlarge the standard of copyright protection to require them."

The case law reflects a very minimal standard of copyright originality. In one of the most famous originality cases, *Alfred Bell & Co.* v. *Catalda Fine Arts, Inc.*, a lithographer who made and distributed copies of the plaintiff's mezzotint engravings of "great master" paintings attacked the copyright of the mezzotintist by arguing there was no originality in the expression of the mezzotints. The lithographer argued that the aim of the mezzotintist had been to copy as exactly as was humanly possible the expression of the great masters—which expression was in the public domain. The great masters had expressed originality, argued the lithographer, not the mezzotintist. The court disagreed, observing that "far less exacting standards" of originality had been required for copyrights as compared with patents. Although it was true that in the common sense of the term, "original" could mean "startling, novel, or unusual, a marked departure from the past," that was not the meaning of the term in copyright. In copyright, "original" means only "that the particular work 'owes its origin' to the author.' " Because of this, the court recognized that it was possible for identical or very similar works created independently by different authors to be separately copyrighted without either infringing the other.

Specifically as to the mezzotints, the court observed that so long as there were discernible differences between the old masters' and the mezzotintists' works,

> even if their substantial departures from the paintings were inadvertent, the copyrights would be valid. A copyist's bad eyesight or defective musculature, or a shock caused by a clap of thunder, may yield sufficiently distinguishable variations. Having hit upon such a variation unintentionally, the "author" may adopt it as his and copyright it.

The copyright standard of originality is sufficiently low that computer-generated works, even if found to be created solely by a machine, might seem able to qualify for protection. Although both the programmer and user might contribute to the framework within which the computer makes its selections or arrangements of data, the computer actually makes the selection. Trying various combinations of data is one of the things that computers do best. Consequently, unless the Constitution were construed to bar machine authorship, perhaps the copyright statute should be construed to permit it.

Machines may be capable of exhibiting sufficient originality to qualify for copyright, and may be able to express that originality in a tangible form. What basis, then, would there be for denying a copyright to a computer? Despite the fact that the statute does not *require* that one be human to qualify as an author, it is still fair to say that it was not within Congress' contemplation to grant intellectual property rights to machines. In the long history of the copyright system, rights have been allocated only to humans.

The system has allocated rights only to humans for a very good reason: it simply does not make any sense to allocate intellectual property rights to machines because they do not need to be given incentives to generate output. All it takes is electricity (or some other motive force) to get the machines into production. The whole purpose of the intellectual property system is to grant rights to creators to induce them to innovate. The system has assumed that if such incentives are not necessary, rights should not be granted. Only those stuck in the doctrinal mud could even think that computers could be "authors."

In future sections, we will attempt to consider the potential solutions not only in terms of whether they make sense from a doctrinal standpoint, but also whether they make sense in terms of the realities of the world in which the problem exists.

Is the User of a Generator Program the Author of Computer-Generated Output?

If a computer is not a meaningful candidate to be considered for the authorship rewards of copyright, who is? CONTU found this to be a simple question: "The obvious answer is that the author is one who employs the computer." CONTU favored allocating authorship rights to the user because of the commission's perception that the user would always have made a very substantial contribution to shaping the output.

Of course, there are many instances in which computer-generated works will have been "written" entirely (or virtually entirely) by the human user. In other instances the user's directions for the computer manipulations will have been so extensive and detailed that stating that the user is the author does not present a conceptual problem. But when the user's instructions become increasingly brief or general and the role of the computer in the design or arrangement process becomes correspondingly greater, the authorship of the user becomes increasingly difficult to defend. It is difficult to justify user authorship when the role of the user of a generator program has been reduced to merely causing the output to be generated (for example, typing the word "compose" in a music generator program). There can, of course, be no user authorship if the output of a generator program is, say, a valuable piece of music that was encoded in the program in this precise arrangement by the programmer. The programmer must be considered the author in that case, or the composer whose work the programmer borrowed. But if the generator program instead produces raw output dissimilar in text from the generator program, can anyone be said to be the "author" of the raw output? Whatever the user does to the text thereafter to edit it or change it will, of course, create a basis for saying that the user may be an author of those portions of the text that he modified. But can he be the author of the unmodified portions? Or if the raw output is "perfect," can the user claim any intellectual property rights in it at all?

Under the traditional paradigm of copyright, the answer to these questions would seem to be "no." A human claiming authorship rights has traditionally had to have "tinkered" with the subject matter, even if by accident, to make the work indisputably "his own." The Copyright Office has previously rejected machine-generated works as copyrightable by the firm that owned the machine that did the generating.

Under these circumstances, it may seem indefensible to allocate ownership rights to the user who has merely typed the word "compose." There are some statutory and some policy reasons why it may nonetheless be sensible to allocate rights in the output to the user. For one thing, the user will have been the instrument of fixation for the work, that is, the person who most immediately caused the work to be brought into being. Copyright law has traditionally considered the person who has "fixed" a work in a tangible medium as the "author" of it. Since it is the user of a generator program who most immediately and directly causes music or a story to be generated, the user would seem to have the strongest claim to owning what is produced by his instruction.

The copyright standard of originality, which is low, also supports the position that the user is entitled to the rights in the output of a generator program. One who tape-records a live performance of improvised jazz, for example, is considered the "author" of the sound recording produced thereby under copyright law, even though the creative input by the user of the recorder might be limited to pressing the "record" button. If this is sufficient originality to support a copyright, there would seem no insuperable difficulty with designating the user of a generator program as the author of a work generated by the program even if all the user did was type the word "compose." That the user will often also select, arrange, edit, and polish the raw output reinforces the existence of sufficient originality necessary to support a user copyright in computer-generated works.

Allocating the copyright in computer-generated works to the user would not be the first time the law allocated rights to those who were responsibile for causing a creative work to be brought into the world, although they might not have been directly involved in the creative effort. The "work made for hire" rule, for example, gives a direct copyright interest to employers for all works prepared by their employees; the employer need not have had any direct role in the creative process. Since one who buys or licenses a generator program has in some sense "employed" the computer and its programs for his creative endeavors, similar considerations to those that underlie the work made for hire rule support allocation of rights in computer-generated works to users.

From a policy standpoint, there are several reasons it would make sense to designate the user of a generator program as the "author" of its output, even when the user's contribution is minimal. For one thing, the user will generally have already tithed to the owner of the program for rights to use it, either by purchase, lease, or license. This provides the programmer with some reward for the value of what he has created (that is, the program). It is not unfair in these circumstances to give some rights to a person who uses the work for its intended purpose of creating additional works.

Furthermore, the user will often play a much greater role in shaping the output into a commercially valuable form than merely pressing a button or typing a simple instruction that triggers the generative process.

Often, the user will have to provide relatively elaborate instructions to the machine and the raw output will need substantial modification to be made valuable. Moreover, even the process of recognizing the quality of the output and/or selecting and identifying those parts of the output that are valuable and should be retained, or must be changed to fit the pattern the user envisioned for the work, may be a substantial creative act.

In addition, the user may use a program for functions that are beyond the programmer's expertise. For example, a programmer may have worked with an experienced architect (or group of architects) to develop a program capable of generating architectural plans. The programmer himself may not be an architect, and may not be able to utilize his own program to create a comparable architectural design that an experienced architect using the program could develop with its aid. Similarly, though a programmer may have studied musical theory and written a program that generates very fine musical compositions, the programmer himself may not, in fact, be able to assess accurately which of the pieces generated by the program are musically superior to the others, or which parts of the raw output are better than other parts, let alone what to do to fix the parts that are not very good. It may be that an experienced composer must use the program in order to create the quality of music that the programmer had hoped for.

In summary, from a doctrinal standpoint, the fact that the user of a generator program will have been the human instrument of fixation of computer-generated output and will have often contributed substantially to the originality of expression in such output supports recognizing authorship rights in the user. There are also both doctrinal and policy reasons to allocate ownership rights to the user even in the "hard case" of minimal input by the user.

.   .   .   .   .

Should the Programmer be Considered the Author of a Computer-Generated Work?

.   .   .   .   .

The primary reason the programmer deserves serious consideration as a claimant for ownership rights in the output of his copyrighted program is that the programmer will have been a substantial contributor to the production of any output generated through use of the program. But for the programmer's creativity, the output might never have been brought into existence. If the output produced by a particular generator program is of excellent quality, it will be fair to attribute at least part of the excellence to the programmer. Creating an excellent generator program is intellectually demanding, as well as time-consuming and expensive for the programmer. Furthermore, it is fair to reward the programmer for the value attributable to this fruit of his intellectual labor, even though it may be a fruit he had not envisioned.

Also, by comparison with the user's input, which may be limited to uncreative acts such as typing "compose" into a music generator program, the programmer's contribution to the generated work will often seem to be the more substantial and significant. The computer, after all, simply follows the instructions of the programmer. It was the programmer who wrote into the program the capability that allowed the user to produce the output by entering general instructions into the computer. The more intricate and precise guidance to the computer, however, will have come from the programmer. From the programmer's point of view, the user of a generator program might seem no more entitled to be considered the "author" of the output than the user of a video-game machine would be considered the author of the audiovisual work created by his control of the play when he hits a button to "shoot" at images of attacking spaceships. In many cases, the program would have generated the same output no matter which human user caused the output to be generated.

Moreover, while the programmer might be willing to tolerate private noncommercial use of the output by the user, he might object if the user attempts to cash in on the value of the output by selling it in the public market. This may lead the programmer to insist on recognition of some intellectual property rights in him.

As strong as these pro-programmer arguments may seem, there are some strong reasons why the computer program author should not be given rights in all works generated by use of his program. For one thing, the programmer can protect his interests legitimately by not distributing the program. By keeping the program to himself and copyrighting every piece of music or patenting every nonobvious chemical formula that the program generates, the programmer would be able to prevent others from obtaining interests in the program's output. If he does this, of course, the programmer will not make any money directly from the program, although he may profit from selling the output that the program generates. Thus, the programmer has a choice, and should not complain about the consequences of his choice to market the program.

If the programmer chooses to exploit the value of the program by charging a significant fee for its acquisition, it seems only fair that he agree to yield some of his rights to those who have paid for that right. Generating output is the purpose of such programs. Indeed, the programmer must cede some rights to the output produced through use of the program to create incentives for users to use the technology.

Furthermore, the purchaser of a generator program can reasonably assume that acquisition of the program brings with it the right to use it to create output. A purchaser is likely to feel defrauded by the programmer if the programmer demanded rights to computer-generated works after the user made an arrangement to sell rights in the output to someone else. At the very least, the programmer should be required to give ample notice of an intent to assert ownership in works generated from his program.

Granting all rights to the programmer would mean that the programmer would automatically own everything the program was *capable* of generating. This solution overrewards the programmer, particularly in light of the fact that the programmer is no more able to anticipate the output than anyone else.

Additionally, allocating ownership rights exclusively to the programmer would cause some serious enforceability problems. First, the output will be in the hands of or under the control of the user and the user would have a strong interest in *not* reporting back to the programmer that a new piece of property has been created in which the programmer has rights. Second, it will often be difficult if not impossible to discern whether a particular work was generated by a program at all—let alone by this particular program. As a result, enforceability problems would be particularly acute if rights are allocated to the programmer.

From a doctrinal standpoint, regarding the programmer as the legal "author" of whatever output might be generated through use of his program leads to serious problems. The first doctrinal problem arises because the programmer may not be the instrument of fixation of the work. Copyright law tends to treat the person who causes a work to be fixed in a tangible medium as its "author," and in the case of computer-generated works, the person who uses the generator program will cause the output to be "fixed." The programmer creates the *potentiality* for the creation of the output, but not its *actuality*. It would be a substantial break from copyright tradition to award rights to a person who merely creates a potentiality for, but not the actuality of, creation of a work.

Another doctrinal problem with designating the programmer as the author of all the output from his program is connected to the problem of unpredictability of output. If the programmer has not conceived—indeed, cannot conceive—what output will be created, it does not seem appropriate to designate him as the author of the output. Conceiving a work is part of what traditional copyright doctrine has meant by authorship and creativity, without which rights should not inure in the programmer.

— 27 —————————————————

The GNU Manifesto

Richard M. Stallman

What's GNU? Gnu's Not Unix!

GNU, which stands for Gnu's Not Unix, is the name for the complete Unix-compatible software system which I am writing so that I can give it away free to everyone who can use it. Several other volunteers are helping me. Contributions of time, money, programs, and equipment are greatly needed.

So far we have a portable C and Pascal compiler which compiles for Vax and 68000 (though needing much rewriting), an Emacs-like text editor with Lisp for writing editor commands, a yacc-compatible parser generator, a linker, and around 35 utilities. A shell (command interpreter) is nearly completed. When the kernel and a debugger are written . . . it will be possible to distribute a GNU system suitable for program development. After this we will add a text formatter, an Empire game, a spreadsheet, and hundreds of other things, plus on-line documentation. We hope to supply, eventually, everything useful that normally comes with a Unix system, and more.

GNU will be able to run Unix programs, but will not be identical to Unix. We will make all improvements that are convenient, based on our experience with other operating systems. In particular, we plan to have longer filenames, file version numbers, a crashproof file system, filename completion perhaps, terminal-independent display support, and eventually a Lisp-based window system through which several Lisp programs and ordinary Unix programs can share a screen. Both C and Lisp will be available as system programming languages. We will try to support UUCP, MIT Chaosnet, and Internet protocols for communication.

GNU is aimed initially at machines in the 68000/16000 class, with virtual memory, because they are the easiest machines to make it run on. The extra effort to make it run on smaller machines will be left to someone who wants to use it on them.

From Richard M. Stallman, "The GNU Manifesto," *GNU Emacs Manual*, pp. 175–84.
Copyright © 1987 Richard Stallman. Reprinted by permission.

Why I Must Write GNU

I consider that the golden rule requires that if I like a program I must share it with other people who like it. Software sellers want to divide the users and conquer them, making each user agree not to share with others. I refuse to break solidarity with other users in this way. I cannot in good conscience sign a nondisclosure agreement or a software license agreement. For years I worked within the artificial intelligence lab [at the Massachusetts Institute of Technology] to resist such tendencies and other inhospitalities, but now they have gone too far: I cannot remain in an institution where such things are done for me against my will.

So that I can continue to use computers without dishonor, I have decided to put together a sufficient body of free software so that I will be able to get along without any software that is not free. I have resigned from the AI lab to deny MIT any legal excuse to prevent me from giving GNU away.

Why GNU Will Be Compatible with Unix

Unix is not my ideal system, but it is not too bad. The essential features of Unix seem to be good ones, and I think I can fill in what Unix lacks without spoiling them. And a system compatible with Unix would be convenient for many other people to adopt.

How GNU Will Be Available

GNU is not in the public domain. Everyone will be permitted to modify and redistribute GNU, but no distributor will be allowed to restrict its further redistribution. That is to say, proprietary modifications will not be allowed. I want to make sure that all versions of GNU remain free.

Why Many Other Programmers Want to Help

I have found many other programmers who are excited about GNU and want to help.

Many programmers are unhappy about the commercialization of system software. It may enable them to make more money, but it requires them to feel in conflict with other programmers in general rather than feel as comrades. The fundamental act of friendship among programmers is the sharing of programs; marketing arrangements now typically used essentially forbid programmers to treat others as friends. The purchaser of software must choose between friendship and obeying the law. Naturally, many decide that friendship is more important. But those who believe in law often do not feel at ease with either choice. They become cynical and think that programming is just a way of making money.

By working on and using GNU rather than proprietary programs, we can be hospitable to everyone and obey the law. In addition, GNU serves as an example to inspire and a banner to rally others to join us in sharing. This can give us a feeling of harmony which is impossible if we use software that is not free. For about half the programmers I talk to, this is an important happiness that money cannot replace.

How You Can Contribute

I am asking computer manufacturers for donations of machines and money. I'm asking individuals for donations of programs and work.

One consequence you can expect if you donate machines is that GNU will run on them at an early date. The machines should be complete, ready-to-use systems, approved for use in a residential area, and not in need of sophisticated cooling or power.

I have found very many programmers eager to contribute part-time work for GNU. For most projects, such part-time distributed work would be very hard to coordinate; the independently written parts would not work together. But for the particular task of replacing Unix, this problem is absent. A complete Unix system contains hundreds of utility programs, each of which is documented separately. Most interface specifications are fixed by Unix compatibility. If each contributor can write a compatible replacement for a single Unix utility, and make it work properly in place of the original on a Unix system, then these utilities will work right when put together. Even allowing for Murphy to create a few unexpected problems, assembling these components will be a feasible task. (The kernel will require closer communication and will be worked on by a small, tight group.)

If I get donations of money, I may be able to hire a few people full- or part-time. The salary won't be high by programmers' standards, but I'm looking for people for whom building community spirit is as important as making money. I view this as a way of enabling dedicated people to devote their full energies to working on GNU by sparing them the need to make a living in another way.

Why All Computer Users Will Benefit

Once GNU is written, everyone will be able to obtain good system software free, just like air.

This means much more than just saving everyone the price of a Unix license. It means that much wasteful duplication of system programming effort will be avoided. This effort can go instead into advancing the state of the art.

Complete system sources will be available to everyone. As a result, a user who needs changes in the system will always be free to make them himself, or hire any available programmer or company to make them for him. Users will no longer be at the mercy of one programmer or company which owns the sources and is in sole position to make changes.

Schools will be able to provide a much more educational environment by encouraging all students to study and improve the system code. Harvard's computer lab used to have the policy that no program could be installed on the system if its sources were not on public display, and upheld it by actually refusing to install certain programs. I was very much inspired by this.

Finally, the overhead of considering who owns the system software and what one is or is not entitled to do with it will be lifted.

Arrangements to make people pay for using a program, including licensing of copies, always incur a tremendous cost to society through the cumbersome mechanisms necessary to figure out how much (that is, which programs) a person must pay for. And only a police state can force everyone to obey them. Consider a space station where air must be manufactured at great cost: charging each breather per liter of air may be fair, but wearing the metered gas mask all day and all night is intolerable even if everyone can afford to pay the air bill. And the TV cameras everywhere to see if you ever take the mask off are outrageous. It's better to support the air plant with a head tax and chuck the masks.

Copying all or parts of a program is as natural to a programmer as breathing, and as productive. It ought to be as free.

Some Easily Rebutted Objections to GNU's Goals

"Nobody will use it if it is free, because that means they can't rely on any support."

"You have to charge for the program to pay for providing the support."

If people would rather pay for GNU plus service than get GNU free without service, a company to provide just service to people who have obtained GNU free ought to be profitable.

We must distinguish between support in the form of real programming work and mere hand-holding. The former is something one cannot rely on from a software vendor. If your problem is not shared by enough people, the vendor will tell you to get lost.

If your business needs to be able to rely on support, the only way is to have all the necessary sources and tools. Then you can hire any available person to fix your problem; you are not at the mercy of any individual. With Unix, the price of sources puts this out of consideration for most businesses. With GNU this will be easy. It is still possible for there to be no available competent person, but this problem cannot be blamed on distribution arrangements. GNU does not eliminate all the world's problems, only some of them.

Meanwhile, the users who know nothing about computers need hand-holding: doing things for them which they could easily do themselves but don't know how.

Such services could be provided by companies that sell just hand-holding and repair service. If it is true that users would rather spend money and get a product with service, they will also be willing to buy the service having got the product free. The service companies will compete in quality and price; users will not be tied to any particular one. Meanwhile, those of us who don't need the service should be able to use the program without paying for the service.

"You cannot reach many people without advertising, and you must charge for the program to support that."

"It's no use advertising a program people can get free."

There are various forms of free or very cheap publicity that can be used to inform numbers of computer users about something like GNU. But it may be true that one can reach more microcomputer users with advertising. If this is really so, a business which advertises the service of copying and mailing GNU for a fee ought to be successful enough to pay for its advertising and more. This way, only the users who benefit from the advertising pay for it.

On the other hand, if many people get GNU from their friends, and such companies don't succeed, this will show that advertising was not really necessary to spread GNU. Why is it that free market advocates don't want to let the free market decide this?

"My company needs a proprietary operating system to get a competitive edge."

GNU will remove operating system software from the realm of competition. You will not be able to get an edge in this area, but neither will your competitors be able to get an edge over you. You and they will compete in other areas, while benefiting mutually in this one. If your business is selling an operating system, you will not like GNU, but that's tough on you. If your business is something else, GNU can save you from being pushed into the expensive business of selling operating systems.

I would like to see GNU development supported by gifts from many manufacturers and users, reducing the cost to each.

"Don't programmers deserve a reward for their creativity?"

If anything deserves a reward, it is social contribution. Creativity can be a social contribution, but only in so far as society is free to use the results. If programmers deserve to be rewarded for creating innovative programs, by the same token they deserve to be punished if they restrict the use of these programs.

"Shouldn't a programmer be able to ask for a reward for his creativity?"

There is nothing wrong with wanting pay for work, or seeking to maximize one's income, as long as one does not use means that are destructive. But the means customary in the field of software today are based on destruction.

Extracting money from users of a program by restricting their use of it is destructive because the restrictions reduce the amount and the ways that the program can be used. This reduces the amount of wealth that humanity derives from the program. When there is a deliberate choice to restrict, the harmful consequences are deliberate destruction.

The reason a good citizen does not use such destructive means to become wealthier is that, if everyone did so, we would all become poorer from the mutual destructiveness. This is Kantian ethics; or, the Golden Rule. Since I do not like the consequences that result if everyone hoards information, I am required to consider it wrong for one to do so. Specifically, the desire to be rewarded for one's creativity does not justify depriving the world in general of all or part of that creativity.

"Won't programmers starve?"

I could answer that nobody is forced to be a programmer. Most of us cannot manage to get any money for standing on the street and making faces. But we are not, as a result, condemned to spend our lives standing on the street making faces, and starving. We do something else.

But that is the wrong answer because it accepts the questioner's implicit assumption: that without ownership of software, programmers cannot possibly be paid a cent. Supposedly it is all or nothing.

The real reason programmers will not starve is that it will still be possible for them to get paid for programming; just not paid as much as now.

Restricting copying is not the only basis for business in software. It is the most common basis because it brings in the most money. If it were prohibited, or rejected by the customer, software business would move to other bases of organization which are now used less often. There are always numerous ways to organize any kind of business.

Probably programming will not be as lucrative on the new basis as it is now. But that is not an argument against the change. It is not considered an injustice that salesclerks make the salaries that they now do. If programmers made the same, that would not be an injustice either. (In practice they would still make considerably more than that.)

"Don't people have a right to control how their creativity is used?"

"Control over the use of one's ideas" really constitutes control over other people's lives; and it is usually used to make their lives more difficult.

People who have studied the issue of intellectual property rights carefully (such as lawyers) say that there is no intrinsic right to intellectual property. The kinds of supposed intellectual property rights that the government recognizes were created by specific acts of legislation for specific purposes.

For example, the patent system was established to encourage inventors to disclose the details of their inventions. Its purpose was to help society rather than to help inventors. At the time, the life span of 17 years for a patent was short compared with the rate of advance of the state of the art. Since patents are an issue only among manufacturers, for whom the cost and effort of a license agreement are small compared with setting up production, the patents often do not do much harm. They do not obstruct most individuals who use patented products.

The idea of copyright did not exist in ancient times, when authors frequently copied other authors at length in works of nonfiction. This practice was useful, and is the only way many authors' works have survived even in part. The copyright system was created expressly for the purpose of encouraging authorship. In the domain for which it was invented—books, which could be copied economically only on a printing press—it did little harm, and did not obstruct most of the individuals who read the books.

All intellectual property rights are just licenses granted by society because it was thought, rightly or wrongly, that society as a whole would benefit by granting them. But in any particular situation, we have to ask: are we really better off granting such license? What kind of act are we licensing a person to do?

The case of programs today is very different from that of books a hundred years ago. The fact that the easiest way to copy a program is from one neighbor to another, the fact that a program has both source code and object code which are distinct, and the fact that a program is used rather than read and enjoyed, combine to create a situation in which a person who enforces a copyright is harming society as a whole both materially and spiritually; in which a person should not do so regardless of whether the law enables him to.

"Competition makes things get done better."

The paradigm of competition is a race: by rewarding the winner, we encourage everyone to run faster. When capitalism really works this way, it does a good job; but its defenders are wrong in assuming it always works this way. If the runners forget why the reward is offered and become intent on winning, no matter how, they may find other strategies—such as attacking other runners. If the runners get into a fist fight, they will all finish late.

Proprietary and secret software is the moral equivalent of runners in a fist fight. Sad to say, the only referee we've got does not seem to object to fights; he just regulates them ("For every ten yards you run, you can fire one shot"). He really ought to break them up, and penalize runners for even trying to fight.

"Won't everyone stop programming without a monetary incentive?"

Actually, many people will program with absolutely no monetary incentive. Programming has an irresistible fascination for some people, usually the people who are best at it. There is no shortage of professional musicians who keep at it even though they have no hope of making a living that way.

But really this question, though commonly asked, is not appropriate to the situation. Pay for programmers will not disappear, only become less. So the right question is, will anyone program with a reduced monetary incentive? My experience shows that they will.

For more than ten years, many of the world's best programmers worked at the artificial intelligence lab for far less money than they could have had anywhere else. They got many kinds of nonmonetary rewards: fame and appreciation, for example. And creativity is also fun, a reward in itself.

Then most of them left when offered a chance to do the same interesting work for a lot of money.

What the facts show is that people will program for reasons other than riches; but if given a chance to make a lot of money as well, they will come to expect and demand it. Low-paying organizations do poorly in competition with high-paying ones, but they do not have to do badly if the high-paying ones are banned.

> *"We need the programmers desperately. If they demand that we stop helping our neighbors, we have to obey."*

You're never so desperate that you have to obey this sort of demand. Remember: millions for defense, but not a cent for tribute!

> *"Programmers need to make a living somehow."*

In the short run, this is true. However, there are plenty of ways that programmers could make a living without selling the right to use a program. This way is customary now because it brings programmers and businessmen the most money, not because it is the only way to make a living. It is easy to find other ways if you want to find them. Here are a number of examples.

A manufacturer introducing a new computer will pay for the porting of operating systems onto the new hardware.

The sale of teaching, hand-holding, and maintenance services could also employ programmers.

People with new ideas could distribute programs as freeware, asking for donations from satisfied users, or selling hand-holding services. I have met people who are already working this way successfully.

Users with related needs can form users' groups and pay dues. A group would contract with programming companies to write programs that the group's members would like to use.

All sorts of development can be funded with a Software Tax: Suppose everyone who buys a computer has to pay x percent of the price as a software tax. The government gives this to an agency like the NSF [National Science Foundation] to spend on software development. But if the computer buyer makes a donation to software development himself, he can take a credit against the tax. He can donate to the project of his own choosing—often chosen because he hopes to use the results when it is done. He can take a credit for any amount of donation up to the total tax he had to pay. The total tax rate could be decided by a vote of the payers of the tax, weighted according to the amount they will be taxed on.

The consequences:

- The computer-using community supports software development.
- This community decides what level of support is needed.
- Users who care which projects their share is spent on can choose this for themselves.

In the long run, making programs free is a step toward the post-scarcity world, where nobody will have to work very hard just to make a living. People will be free to devote themselves to activities that are fun, such as programming, after spending the necessary ten hours a week on required tasks such as legislation, family counseling, robot repair, and asteroid prospecting. There will be no need to be able to make a living from programming.

We have already greatly reduced the amount of work that the whole society must do for its actual productivity, but only a little of this has translated itself into leisure for workers because much nonproductive activity is required to accompany productive activity. The main causes of this are bureaucracy and isometric struggles against competition. Free software will greatly reduce these drains in the area of software production. We must do this, in order for technical gains in productivity to translate into less work for us.

[The GNU project has continued to prosper, with the support of "good citizens" who are agreeable to sharing programs (source and all) with others. It is growing under the auspices of the Free Software Foundation, which is "dedicated to eliminating restriction on copying, redistribution, understanding and modification of computer programs." The word "free" in this context refers to (a) freedom to copy a program and give it away to friends and coworkers and (b) freedom to change a program by having full access to source code. Programs parallel to Unix utilities have continued to multiply (but, remember, "GNU's Not Unix"), and include such advanced programs as C++ compilers. When this book went to press anyone with access to Internet could get the latest GNU software from the host "prep.ai.mit.edu." For more information, read the file "/u2/emacs/GETTING.GNU.SOFTWARE'—Eds.]

Other Legal Issues

Deborah Johnson (reading 28) discusses the problem of deciding under what circumstances a software developer ought to be legally liable if the software malfunctions. She shows how the legal concepts of contract, misrepresentation, implied warranty, and disclaimers in an express warranty apply in a situation involving computer software.

Steven Frank (reading 29) deals with issues of legal liability as they pertain to artificial intelligence. He reviews the current application of tort law, explaining problems courts have had in applying laws not specifically tailored to computers. He also examines the question of whether computer programs are products or services.

David Tinnin (reading 30) tells the story of how IBM "stung" Hitachi when Hitachi tried to buy secret IBM notebooks for a planned computer. This absorbing report sets the computer within the larger world of commercial competition and espionage.

Steven Mandell (reading 31) writes about crimes in which the use of computers compounds the harm done to the victims. He analyzes various types of computer crimes (such as sabotage and theft) and illustrates each type with actual cases. In addition, he explains why computers may be particularly vulnerable to criminal misuse.

28

Liability for Malfunctions in Computer Programs

Deborah G. Johnson

One of the relationships . . . most central to professional conduct is that between the computer professional and a client or consumer. I say client *or* consumer because I want to leave open, for the moment, an issue that will arise later, as to whether those who develop software are providing a product or a service. Though there are no hard and fast rules, if software is a product, then the relationship is appropriately characterized as that between a professional and a consumer. If computer professionals are viewed as providing a service, then the relationship is probably better characterized as that between a professional and a client. In either case, the relationship is one of exchange, of buying and selling. The vendor offers something for sale and the buyer purchases it.

Exchanges of this kind are, of course, nothing new, and there are many laws that apply to such activities. Nevertheless, because computer products are new and in some ways unique, and because they are continuously being used in new contexts or sectors, questions arise about which laws apply and what constitutes fair practices. What can the seller be expected to guarantee? What risks must the buyer accept? What should be legally required, and what should be left negotiable?

Ambiguities that result from the newness of computer products are intensified by the fact that often the buyers of computer products are ignorant of computers and, thus, do not know what problems to anticipate, or what claims to be cautious about, when they are making a purchase. Yet, the purchase of computer products can have a tremendous impact on the buyer. If a company purchases a good budget and management system, it can increase the productivity of the company enormously by streamlining its operations. On the other hand, if the system is poorly designed, it can cause a company no end of headaches, making for inefficiency, interfering with customer services, and so on. Indeed, poorly designed or malfunctioning systems can lead to serious harm. Consider the implications of an error in software used in an airplane, in a nuclear power plant, or in a dangerous industrial process.

From Deborah G. Johnson, *Computer Ethics,* © 1985, pp. 39–47. Reprinted by permission of Prentice-Hall, Inc., Englewood Cliffs, N.J.

Legal Liability

The primary concern of this [essay] is with the legal liability of software creators, and of vendors of that software, particularly for errors and malfunctions. This is a moral issue both because it involves the rights and duties of professionals and their clients or consumers and because it raises complex questions of fairness. For example, it might be unfair to hold computer professionals liable for errors that they could not foresee. Thus we must ask: what is it fair to expect a software vendor to do; and when is it right to hold the vendor liable to return someone's money, or pay damages, or repair a system?

Two points may be helpful before we get into this matter. Who are the parties usually involved in buying and selling transactions? The buyer may be an individual, company, or agency of any size. The vendor may be an individual, a large company, or a small consulting firm that has designed a system for a special purpose. When the case involves a company being sued for damages that result from an error, secondary issues may be raised about whether the company can turn around and sue the employee who made the error. We will not be concerned with this issue. Furthermore, while the transfer of computer software from vendor to buyer most often simply involves buyer and seller, sometimes it is third parties who are harmed. In such cases, a difficult question may arise as to who should be liable for the harm. For example, suppose a software firm develops a medical diagnostic system and sells it to a hospital. A doctor working at the hospital uses the system, and it misdiagnoses a patient. The patient is harmed as a result of the treatment recommended by the computer diagnosis. Who should be liable for the patient's injury? The doctor? The hospital? The vendor? This is an interesting question but, again, not one that we will take up here.

We will be concerned with the core of these situations, the relationship between software vendors and buyers. Though the creators of software may be different people or companies than the vendors, we will assume here that they are one and the same.

As a final preliminary, it may be helpful to clarify uses of terms like "responsible" and "liable." We can distinguish at least four different senses. First, persons may be said to be responsible in the sense that they have a duty or an obligation to behave in a certain way. We might say that a person has several highly varied responsibilities in a job. Another example of this use of "responsibility" is found in the Disciplinary Rules of the ACM [Association for Computing Machinery] Code which impose duties on members to behave or refrain from behaving in certain ways. Take Disciplinary Rule 4.2.1, which states that an ACM member has a responsibility (duty) *not* to "maliciously injure the professional reputation of any other person." In this use, saying that someone has a responsibility is equivalant to saying that he or she has a duty.

A second use of "responsibility" involves causality or causal production of some thing or event. Causal responsibility may be attributed to nonpersons, that is, to natural events or conditions, as well as to persons. An example of the nonpersonal attribution of responsibility is the following: "The wind caused the damage to my roof." Here we pick out one circumstance or event (the wind) and claim that it is responsible for another situation or event (the damage to my roof). Causal responsibility is attributed to a person when the person is singled out as having caused or produced a situation or event. Examples are: "John Doe lit the match that started the fire, so, he is responsible for the fire"; or "Her alarming speech is responsible for the panic." Attributions of causal responsibility are rarely simple. In almost any situation that we can imagine, an event or effect will be the result of a multitude of factors, but attributions of "responsibility," in this use of the term, single out persons (their action or inaction) as *the* cause. In attributing this kind of responsibility, we do not say that persons have a duty, but rather that their action or inaction produced a certain result.

A third use of "responsible" or "responsibility" can be associated with blame or blameworthiness. When we use "responsible" in this way, we claim or deny that a person is rightfully blameworthy for an event or circumstance. For example, we might say of someone who was tortured to reveal state secrets: "He is not blameworthy; he showed great courage." It should be fairly obvious that blameworthiness is often (but not always) associated with one of the other senses of responsibility just mentioned. A person may be considered responsible (blameworthy) if he fails to fulfill one of his responsibilities (duties) or a person may be considered responsible (blameworthy) if he is causally responsible for (his action produced) an event.

In this chapter, we will be concerned primarily with a fourth but related sense of responsibility. We will be concerned with liability for an event or action. One is responsible in this sense when one can be held to account for an action or event. In the law (and we will be concerned essentially with *legal* liability), one is liable when one can be required to go to jail, pay a fine or penalty, pay damages, and so on, for one's action. Liability is often related to the other senses of responsibility just mentioned. Often one can be held legally liable when one fails to fulfill a legal duty. For example, we will see in a moment that vendors have a duty to be honest about the products they sell. If they fail, they can be required to compensate the buyer. Also, to prove that someone is liable, that is, should be compelled to pay damages, often one must show that the person was causally responsible for an event. For example, to win a court case in which, say, I sue for damages done to my crops, I would have to show that the defendant did in fact spray my crops with toxic chemicals, and that it was those chemicals that caused the damage to my crops.

. . . . .

Contract

At base, buying and selling relationships are contractual in nature. One party (Party A) agrees to do certain things, usually to provide a product or service; and the other party (Party B) agrees to give money in return for the product or service. Often this relationship between vendor and buyer is established by means of a formal, written contract. When the contract is formal, the vendor can be sued for breach of contract—if the software fails to meet what was specified in the contract. One might expect this to be a very simple matter, but it is not always. For one thing, the matter over which the buyer has a complaint may not be something that was specified in the contract. There are also elements of representation at the time of the contract negotiation which can, in effect, void the contract.

Misrepresentation

A buyer of software has legal recourse in the courts on grounds of misrepresentation. That is, even if a contract was signed and the vendor has not violated the terms stated in the contract, the buyer may have a case if he signed the contract in reliance on a misrepresentation by the vendor. If, for example, the vendor suggested that the computer system could do things that are beyond its capability, then the buyer may win in court. To be sure, some puffery seems to be allowed by the courts, but not misrepresentation. For example, in *Clements Auto Company* v. *Service Bureau Corporation* (SBC), Clements sued on a number of grounds including fraudulent misrepresentation.[1] SBC had provided Clements with an accounting system and an inventory control system. Clements Auto claimed that the system was unsatisfactory. The system was slow to handle the volume Clements generated, and it was full of errors and inadequate controls. The trial court found that SBC had made misrepresentations to Clements because salesmen had told Clements that it could only obtain an inventory control system (which was what it originally wanted) if it automated its entire accounting system. The court found that this was not true. In fact, the difficulties encountered with the system arose from SBC's attempts to automate the accounting area. So, even though Clements Auto had agreed to purchase an accounting system, the contract was not considered binding since it was made on the basis of a misrepresentation.

The laws regarding misrepresentation are, of course, based on the duty of honesty mentioned above. Misrepresentation does not involve an error or malfunction in a program, but it is important here because, oftentimes when a buyer thinks a program or system is faulty, it turns out that it is not faulty at all but simply incapable of doing what the buyer was told it would do.

The legal rulings on misrepresentation impose a duty on software vendors to accurately represent what they sell. They may be liable to take back what they sold and return the buyer's payment, to supply what they said they would, and sometimes to pay damages when the buyer has lost business or suffered in some other way from the inadequacies of the system. Imposing this duty on software vendors—to accurately represent what they sell—and holding them liable when they fail, does not, in any case, seem an unfair practice. If anything, this pracitce seems necessary to insure that the customer is treated fairly. Thus, this part of the legal liability of a software vendor is not morally controversial.

Implied Warranty

Misrepresentation aside, you might think that people should be allowed to make whatever contract they like. That is, two parties ought to be allowed to agree to whatever conditions they find acceptable, assuming that each is honest with the other and neither is coerced. To a certain extent, this is allowed by law, but there are limits. For one thing, there are laws which protect the customer by insuring that he or she does not have to negotiate certain conditions of a purchase. They are always part of the purchase agreement, even when the agreement is not in writing. Implied warranty is an example of this, though it is provided for in the Uniform Commercial Code and this code only covers *products*. In a moment we will see that some software can or should be considered a product, so this law applies, at least, to some software.

Under the category of implied warranty, two principles may be used as grounds for a legal suit. A product sold must have fitness for its "intended purpose" and it must be "merchantable." Cases employing the principle of intended purpose often rest on evidence that the vendor knew what the buyer planned to use the system for, and yet sold the buyer a system that was not suited for that purpose. In *Public Utilities Commission for*

City of Waterloo v. *Burroughs Machines Ltd.*, the court found that Burroughs had breached its implied warranty when it supplied hardware that was useless without software that it did not supply.[2] The court held that the computer system was, therefore, unfit for the intended purpose. Burroughs was required to supply the software as part of the sale.

Merchantability involves fitness for the ordinary purposes for which the product is used. On this principle, something sold as a program must be capable of doing what programs ordinarily do. Breach of this warranty may be found when a system repeatedly fails or does not withstand normal shocks. The success of the merchantability claim often depends on additional factors such as a time period during which the purchaser is deemd to have accepted the system. But the courts have been fairly good in recognizing that buyers may spend some time and effort in trying to get a computer system working before they can conclude that the system is not merchantable.

Implied warranty, thus, imposes on software vendors a duty to sell software that is fit for its intended purpose and merchantable. Both of these legal principles seem to be connected with honesty in the sense that they require the vendor to provide the kind of product that the buyer expects he is getting. One might argue that the buyer is making assumptions that are never explicitly stated in the negotiation process. But the assumptions are quite reasonable, given the context; that is, if you go to a software vendor in search of an accounting system that can be used with your IBM machine, you explain this to the vendor, and the vendor suggests a product, it is reasonable to expect that the product will, in fact, do accounting and will operate, without repeated failure, on your machine. Thus, holding software vendors liable for selling software that is fit for its intended purpose and merchantable does not seem an unjust practice. Indeed, this again seems to be required for fairness to the buyer.

The courts' recognition of implied warranty, it should be added, seems to be in the long-term best interests of software vendors. In this sense, it can also be defended on utilitarian grounds. Implied warranties lessen the risk that a buyer has to take in buying computer software. You would be much more reluctant to buy software, or anything else for that matter, if you had no guarantee that what you were buying would come close to doing what you expected it to do. Thus, in the long run buyers *and* vendors benefit because computer software is used.

Many errors or malfunctions in software can be handled with the legal mechanism of implied warranty, but implied warranty applies only to software which is considered to be a product, and it only covers defects in software that make the software unfit for its intended purpose or unable to do what programs generally do. When it comes to the quality of a program, how efficiently it does what it is supposed to do, how long it lasts, how error-free it is, what machines it will work on, and so on, the law is much less specific. These matters, as well as agreements about provision of a service, are often handled in a formal agreement.

Express Warranty and Disclaimers

Express warranties are elements of an agreement that are specified in writing. Generally these have to do with the quality of the program and what the seller will do, if anything, to correct errors. An interesting and controversial element here is the use by software vendors of disclaimer clauses in the contract. Software vendors may put into the agreement statements of the following sort: "There are no other warranties, express or implied, including, but not limited to, the implied warranties of merchantability or fitness of a particular purpose." This means that the vendor makes no promises about the quality of the software and warns the buyer that he will not be liable for any faults beyond the implied warranties. The vendor may also try to limit his liability by means of statements to the effect that: "In no event will the vendor be liable for any lost profits or savings, or for other indirect, special, consequential, or similar damages arising out of any breach of this agreement or obligations under this agreement."[3]

The courts recognize the validity of such clauses except when they find them to be "unconscionable." This is not a precise notion, and consequently it is not easy for a buyer to void an exclusion clause. The courts' recognition of these clauses seems to be consistent with the idea that individuals—as rational, autonomous beings—should be allowed to take risks if they want, as long as they are not deceived or forced. The appearance of the disclaimer statement in the contract serves as proof that the buyer was aware of the risk.

As already mentioned, the use of such clauses is somewhat controversial. On the one hand, they seem unfair because they allow vendors to avert liability for what they sell. That is, with an exclusion clause a vendor is allowed to sell what he wants without incurring any liability-responsibility for the product or service. The risk is transferred to the buyer. Allowing this may actually discourage vendors from bringing their products or services up to a certain standard before they sell them for they are allowed to do much less testing. Thus, the effect of use of these clauses is not in general good for society.

On the other hand, the use of these clauses is not blatantly wrong since the buyer is not being deceived or manipulated. As long as the buyer understands and agrees to the clause, he is making a choice, and it would be wrong for the courts to interfere in this choice (except perhaps in extreme cases, as when the clauses are unconscionable). Also, tolerating disclaimer clauses allows software vendors to bring new software to the marketplace sooner. If the vendor could not avert liability, he might refrain from selling new, possibly useful or beneficial software because his risk would be too great. With disclaimers, buyers who are willing to take risks may get software sooner than otherwise.

Notes

1. See Larry W. Smith, "A Survey of Current Legal Issues Arising from Contracts for Computer Goods and Services," *Computer/Law Journal* 1 (1979):479.
2. Ibid., p. 488.
3. See Marvin N. Benn and Wayne H. Michaels, " 'Multi-Programming' Computer Litigation," *Chicago Bar Record 64,* 1 (July-August 1982):32–47.

—— 29 ——

What AI Practitioners Should Know About the Law

Steven J. Frank

Although recent computer-related disputes have involved a variety of legal issues—breach of contract specifications, misrepresentation of performance capability, and outright fraud, to name a few—it is the threat of tort liability that continues to strike the greatest fear in the heart of mighty industrialist and struggling start-up alike.

· · · · ·

Tort Liability

The system of tort compensation exists to provide victims of injury—physical, economic, or emotional—with the means to seek redress in the form of monetary damages. The purpose of the damage award is to "make the plaintiff whole" by compensating for all losses flowing from the defendant's wrongful act. Although this concept seems straightforward, damage awards have been known to vary enormously, even for identical injuries. The host of subjective factors that appear in the calculus of compensation precludes accurate forecasts in most cases, leading some defendants to financial ruin and others to surprised relief. However, injury alone does not guarantee recovery. Rules of tort liability mediate between the victim's need for recompense and the defendant's right to remain free from arbitrarily imposed obligation.

Although sensational cases involving large recoveries tend to generate the greatest alarm, the magnitude of damages in a particular case is actually far less important than the availability of any damages in similar cases. Potential tort defendants are primarily interested in their overall liability exposure, as determined by the evolving structure of case precedent. For software developers, expenditures for debugging, design safety analysis, and quality control assurance are necessarily affected by the extent of perceived vulnerability to tort actions.

More so than in any other legal field, the boundaries separating human from machine will be forced into focus by questions of tort liability. In addition to compensation, another function of tort law is to project proper standards of care in the conduct of activities that might cause harm. When the agent of injury is a tangible device or product, attention is currently directed toward three sets of possible culprits: manufacturers (Was the product designed or manufactured defectively?), sellers (Was the product sold in a defective condition?), and purchasers (Was the product used improperly?). Absent from the lineup is the injury-producing item itself, whose existence is relevant only insofar as it pertains to the conduct of human beings. As "devices" come to include electronic systems capable of judgment and behavior, they too will become objects of direct inquiry. Naturally, financial responsibility will ultimately rest with a human being or a corporation possessing a bank account; but the standards by which humans and machines are judged will begin to merge as the tasks they perform grow similar.

The question of whether liability accrues in a given instance invariably reduces to the application of two legal variables: the standard of care expected of the defendant and the requirement that a causal link exist between the defendant's substandard conduct and the plaintiff's injury.

The existence of a valid tort cause of action means that someone has suffered loss. This much is beyond change. The function of the liability standard is to allocate loss among all parties involved based on considerations of fairness, efficiency, and accepted behavioral norms. The terms of the standard prescribe the level of vigilance expected of individuals who conduct a particular type of activity. Although variations exist, most formulations derive from the fault-based concept of negligence: If the defendant failed to act in the manner of a reasonably prudent person under all circumstances, the loss falls on him or her. A special relationship between plaintiff and defendant or the performance of an unusual activity can prompt a court to adjust the standard. Innkeepers, for example, have been held liable to guests for the "slightest negligence" because of the high degree of trust placed in their hands. Certain activities have been identified as so inherently dangerous or unfamiliar that no degree of care can adequately prevent mishap. For these "strict liability" activities, the loss falls on the defendant, regardless of fault, as a cost of doing business.

This choice of liability standard calls for a decision based on public policy. Although through less obvious means, the parameters of causation are ultimately shaped by similar considerations. A scientific view of deterministic causality provides only a starting point for the legal notion of causation. Judges have recognized that too many events are logically interrelated for liability to rest solely on logic. Fairness to defendants requires that a line be drawn at some level of remoteness, and the location of this line reflects a policy-oriented value judgment. Liberal tests of causation shift a greater amount of loss toward tort defendants by including more distant effects of liability-producing conduct within the range of potential recovery. These tests generally focus on the existence of a chain of events, such that causation is based on proof of an unbroken progression of occurrences; the outer limits of connectedness are typically bounded only by the reluctance to impose liability for the extremely unpredictable. Restrictive tests of causation focus directly on the defendant's ability to foresee the possible harm arising from particular actions and reverse the shift in favor of the plaintiffs.

Returning to liability issues, the rule applied to parties engaged in commercial trade depends heavily on what is being sold. Providers of services have traditionally been held to a negligence standard based on reasonably prudent practice within a given area of specialization. In contrast, sellers and manufacturers of tangible commodities face strict liability for injuries to individual consumers caused by defects in their wares. The reason for the distinction does not lie in any perceived disparity of associated danger, but rather in the fact that product sellers are viewed as economically better able to spread a loss over a large number of users through price adjustments.

Courts have had difficulty fitting computer programs into this bifurcated world of products and services. Software can be supplied in a variety of forms, some more tangible than others. Programmers might write software for mass distribution, tailor an established package to the needs of a particular user, or design from scratch a custom system for an individual client. Although no court has yet faced this issue in the precise context of tort liability, most commentators (including this writer) believe that characterization as product or service is most properly determined by the supplier's degree of involvement with the customer. Greater availability of technical assistance and support make the overall transaction appear more like a service.

Because implementation of the high-level tasks performed by commerical AI software requires extensive immersion in the field of application, initial development contracts call most clearly for treatment as service arrangements. If the finished program proves suitable for an entire class of users, however, subsequent sales might appear to involve a product. Ambiguity is inevitable where the uncertain intrinsic character of software diverts attention to its mode of supply for purposes of tort liability.

The question of causation raises a different set of issues—those chiefly related to the manner in which the computer program actually makes contact with the end user. The factors that evoke a clear liability rule for mass-marketed software likewise provide the strongest link between defect and injury. Simple sales transactions force the consumer to rely solely on the purchased item for effective performance, and thus, any harm suffered can be traced directly to improper program operation. It seems doubtful, however, that AI programs will reach consumers through such direct market channels any time soon. The simplest current reason is cost: AI software is enormously expensive to produce. A longer-range consideration is the likely reluctance of human experts to relinquish control over the provision of their expertise. Professions shielded by licensure requirements, for example, have shown themselves to be well equipped to defend against unauthorized practice. For the foreseeable future, then, the most likely role for many applied knowledge programs is as an aid to the human expert.

Although perhaps depriving AI developers from access to the consumer market, such restrictions also relieve developers of a great deal of potential liability. The law will not treat an appurtenant factor as a causal agent of injury unless it materially contributes to this injury; yet material contribution is precisely what is prohibited by restrictions on unauthorized practice. For example, if a physician were to attempt to lay blame on a diagnostic expert system for improper treatment, the physician would thereby admit to allowing the computer to perform as a doctor.

The price of limiting the practice of a profession to a select group of peers is accepting complete responsibility for professional misjudgment.

Of course, if the source of the physician's error were indeed traceable to the expert system, the physician might sue the software developer to recover the money that must be payed to the injured patient. This possibility leads to the question: how should the law of product liability be applied to the creator of a device whose domain-specific capabilities might match or exceed those of human experts? The first step in any such lawsuit (whether based on strict liability or negligence) is to demonstrate that the product contains a defect. Should defectiveness be inferred from the mere fact of incorrect diagnosis? Human physicians are certainly not judged this harshly; their diagnoses must only be "reasonable." Separating programming errors from legitimate mistakes of judgment falling within professional discretion will indeed prove formidable.

A second, more practical obstacle facing this hypothetical physician is that a strict liability standard would probably not apply, regardless of whether the program is viewed as a product. Courts generally permit strict liability recovery only in actions for physical injury and property damage; economic loss is insufficient to trigger the doctrine. Unlike the hapless patient, the physician has personally suffered only financial impairment. The physician's tort suit, therefore, must be based on negligence. If the software developer has exercised reasonable care in debugging and packaging the product, including some statement warning of the system's limitations, the negligence burden might prove a difficult one to carry. The appropriate level of resources devoted to debugging efforts and the scope of the necessary warning depends on actual reliability and system design. For example, deep expert systems can generally be expected to deliver acceptable results over the entire useful range of the underlying causal model. Courts will undoubtedly expect greater vigilance from developers of shallow systems (or deep systems based on models that are not robust) simply as a consequence of the diminished reliability implied by program design.

To be sure, not all applications of AI techniques involve roles currently occupied by organized professions, nor must computer output actually touch consumers in order to affect their lives. The degree of contact necessary to trigger liability is once again determined by the nature of the commercial relationship. Personal interaction comprises an inherent feature of consultative service transactions, and a causal gap is probably inevitable unless the computer somehow communicates directly with the injured consumer; otherwise, the intervention of human judgment is likely to prove a sufficient superseding event to interrupt the nexus. Sales transactions, in contrast, are characteristically impersonal. The path of causation is likely to be far more direct if the computer's role involves assisting in the fabrication of a product. Any modicum of assistance not filtered through independent human oversight can furnish a link between computer operation and injury caused by defective manufacture. Hence, developers of computer-aided manufacturing (CAM) systems can expect increasing exposure to liability as their software assumes control over a greater portion of the manufacturing process.

Product design occupies a status somewhere between service and manufacture. Although the ultimate goal might be the production of a usable product, the design process involves early-stage development decisions of a far more basic and creative nature than those involved in automating production. The collaboration of computer-aided design programs with human engineers seems likely to persist for a much longer time than might be anticipated for CAM systems, which reduce design to actual practice, maintaining a greater opportunity for events that sever the chain of causation.

Increased trustworthiness inevitably results in heightened liability. As the role played by computers expands from passive assistant to independent practitioner, and as tasks delegated to computers begin to encompass a greater dimension of injury-producing activity, their owners will find themselves increasingly responsible for mishap traceable to improper operation.

30

How IBM Stung Hitachi

David B. Tinnin

When the FBI arrested two Hitachi employees in the act of buying IBM trade secrets in California [in June 1982], the curtain went up on an extraordinary spectacle of corporate warfare. For months, two of the world's mightiest, most respected, and most technologically advanced corporations had been stalking one another—Hitachi seeking to obtain secrets of its dominant competitor, IBM seeking to teach Hitachi a stinging, humiliating lesson. [Early in 1983], as Hitachi and two employees pleaded guilty in a federal court, the case ended on a hushed and anticlimactic note that gave no hint of the intricate saga that had preceded it—in which IBM helped the FBI catch Hitachi in a superbly executed sting.

During the sting operation the FBI used hidden cameras and listening devices to obtain 35 hours of videotape and 65 hours of audiotape. The tapes recorded numerous episodes in which Hitachi employees conspired to purchase IBM equipment and documents. After the arrests, one of Hitachi's foremost objectives in its legal maneuvers was to avoid a trial in which this embarrassing material would be displayed for the world to see and hear. For a while Hitachi's lawyers sought to persuade the court to quash the indictment on the ground that IBM, not the FBI, had controlled the sting. "IBM's goal was not a law enforcement goal," contended Hitachi's lawyers. "It was instead anticompetitive economic benefit for itself." When that argument failed to fly, Hitachi offered to plead *nolo contendere*—in effect, acquiescing to the charge—on condition that Hitachi employees escape trial and punishment. Even though the governments of both countries wanted a quick resolution of the case, the Justice Department balked at giving Hitachi quite that easy an exit.

Washington then instructed Special Prosecutor Herbert B. Hoffman to offer a plea-bargaining arrangement. The terms: plead guilty and nobody goes to jail. In late January [1983], Peter Fleming, Jr., the big-time Manhattan lawyer who was Hitachi's chief counsel in the case, flew to Tokyo for consultations. Hitachi's board of directors authorized him to accept the Justice Department's offer. An open admission of wrongdoing was no easy act for Hitachi, but ah, those tapes.

In the courtroom in San Francisco the scenario was so well arranged that suspense and drama were absent. Hitachi pleaded guilty to the one-count indictment of conspiring to transport stolen IBM property from the U.S. to Japan. That same day, Judge Spencer Williams imposed the maximum corporate penalty under the statute: a $10,000 fine. Hitachi senior planner Kenji Hayashi, who had been a major actor in the espionage drama, was fined $10,000 and placed on five years' probation. Isao Ohnishi, a Hitachi software expert, drew a $4,000 fine and two years' probation. "Confinement," declared the judge, "would not seem to serve any purpose."

Still pending is IBM's civil damage suit against Hitachi. IBM agreed to a 60-day delay while it attempted to reach a settlement. If the issues are not resolved to IBM's satisfaction, the company can be expected to revive the civil suit with a vengeance. In fact, moments after the criminal proceedings ended in the San Francisco courtroom, IBM lawyers hurried up to defendant Ohnishi to serve him a subpoena to appear as a witness in the civil action.

Even though the case did not come to trial, a substantial record of evidence and arguments was built up during the pretrial hearings. Prosecutor Hoffman put on the public record four affidavits by key prosecution witnesses, who chronicled the sting's operation in great detail. He also submitted to the court more than 250 pages of evidence collected by the government, largely through those clandestine cameras and microphones. The lawyers for Hitachi and individual defendants presented to the court a series of thick briefs in which they laid out their motions to dismiss the case and indicated the line of argument they would take before a jury. After examining that material and doing some investigating of its own, *Fortune* knit together the story of how IBM stung Hitachi.

A minor player touched off the chain of events from which all the rest of the drama unfolded. He was Raymond Cadet, 45, a Haitian-born computer scientist, who on November 20, 1980, resigned from IBM's computer labs in Poughkeepsie, New York. The parting was amicable. During the routine exit interviews, according to IBM, Cadet signed a pledge that he was taking no confidential material with him. But in reality, says the Justice Department, Cadet took with him ten of the 27 volumes that made up the so-called Adirondack workbooks. Adirondack was IBM's code name for its top-secret program to build a new generation of computers, the 308X. The first model of that series, the 3081, was shipped in October 1981. The workbooks, which were three-ring binders, contained 40 to 200 pages. The first page of each volume carried a warning that the contents were proprietary material, not to be divulged except to fellow IBM employees on a need-to-know basis. Printed in red diagonally across each page were the words DO NOT REPRODUCE.

After he left IBM, Cadet went to work for a computer firm near Washington. Then, on June 1, 1981, an Iranian named Barry Saffaie recruited him for a job in Silicon Valley. Saffaie was a manager at a California company called National Advanced Systems, or NAS for short, which is a subsidiary of National Semiconductor. NAS marketed Hitachi products in the U.S. and manufactured computer products of its own as well.

In light of what happened afterward, it is easy to leap to the conclusion that Saffaie recruited Cadet to get those workbooks for Hitachi, but the evidence does not support the leap. Saffaie may not even have known that Cadet had the documents. The reason he wanted Cadet at NAS is clear enough: that outfit closely tracks IBM, and Cadet's relatively current knowledge made him valuable. Once Cadet joined NAS, though, Saffaie soon got hold of all ten volumes, and many photocopies were run off. After the FBI closed in on Hitachi, the Justice Department brought charges against Cadet and Saffaie, but a federal judge threw out the indictments because Justice refused to supply all the documents the defense demanded. The Justice Department has appealed.

During the summer of 1981, Barry Saffaie shuttled across the Pacific to brief Hitachi experts on computer developments in the U.S. In August, according to the Justice Department, he delivered copies of the ten workbooks to Hitachi computer specialists. At first, it seems, they did not realize what they were getting.

Meanwhile, in San Jose, Hitachi was being offered a study of the 3081 by another source: Palyn Associates, a small consulting firm. Like NAS, Palyn keeps an eye on IBM. Palyn's president, Maxwell O. Paley . . . spend 21 years at IBM and rose to chief of the Advanced Computing Systems laboratory before leaving the company in 1970. Palyn's brochure boasts that the top executives possess "80 years' cumulative IBM experience."

Paley founded Palyn Associates in 1972, and almost from the start Hitachi was a major client. Hitachi was always on the lookout for information about IBM. The Japanese firm is one of the so-called IBM-compatible manufacturers, which build computers so they can operate with the same software and peripheral equipment as IBM computers. The other major outfits now doing that are Amdahl Corporation and Fujitsu; the rest of the competition, such as Control Data, Burroughs, and NCR, make systems that operate mainly on their own software.

A company in the business of making IBM-compatible computers has to keep pace with IBM or perish. The earlier a competitor can discover the design directions of a new IBM product, the earlier it can rush into production with a similar machine, usually selling for less than the IBM original.

One way competitors copy IBM is by "reverse engineering." They purchase an early model, take it apart, and design something similar, with comparable capabilities. But that can be a slow process, allowing IBM months of market dominance. It's much more advantageous, obviously, to acquire IBM designs far in advance of the new computer's shipment date.

And that, of course, was what made Paley and his fellow IBM veterans valuable to Hitachi. Using their own knowledge of IBM's techniques and design directions, Palyn Associates had compiled a study of the 3081, and the firm offered it to its regular contact man at Hitachi during one of his routine visits. He was Kenji Hayashi, a senior engineer in the computer project planning department. Hayashi took an index of the Palyn study back to Tokyo.

At that time, Hayashi knew nothing of Saffaie's delivery of the workbooks; he had never seen the man. But after he learned about the new material in Hitachi's possession, Hayashi sent Palyn Associates a telex, which the prosecution subsequently placed in the pretrial record. Hayashi wrote that his company was not interested in the Palyn study because, in his words, "we have already got Adirondack workbook that is similar to your covering [evidently meaning index]. But we have only Vol. 1, 3, 4, 8, 9, 10, 11, 12, 15, 22. If you have another Vol., let me know. We consider again . . . Please keep confidential. Regards." As an IBM veterran, Max Paley immediately recognized the message for what it was: dynamite waiting to explode.

In Silicon Valley, where trade secrets are easily spirited from firm to firm in the heads or attaché cases of job-hopping engineers, confidential information often ends up in the wrong place. Such problems are frequently settled by a phone call from one company president to another and the return of the filched material. Some Silicon Valley veterans fault Paley for not having turned first to Hitachi, telling his client that the workbooks had been improperly appropriated and should be returned to IBM. Instead, after intense and agonized discussions with his top associates, Paley decided to tell IBM.

That must have been a wrenching decision. As a consultant, Paley owed something to his client. Moreover, Hitachi had been an important customer in dollar terms. Hitachi and other Japanese companies accounted for around 20 percent of annual billings.

Immediately after the Hitachi arrests, before he went silent on lawyer's orders, Paley attempted to justify his behavior on patriotic and moral grounds. The Hitachi people, he said, "weren't fighting fair." Other possible motives come to mind. In his years at IBM, Paley developed loyalties and close personal friendships, some of which he kept up after he moved to California. Also, his business depends to some extent on retaining links to IBM. And he may also have thought that, if he had to choose, IBM was potentially a more important client than Hitachi.

From whatever mixture of motives, Paley put in a call to a close friend at IBM, Bob O. Evans, vice president for engineering, programming, and technology. "Bob," said Paley, "I think one of my Japanese clients has gotten your crown jewels."

Paley's message set off a high-level alarm at IBM. For a company that spends substantially more than $50 million per year on security and prides itself on employee loyalty, the warning conjured up nightmarish possibilities. Was a traitor or perhaps even an inside ring selling IBM secrets?

Only a few top executives—no more than eight—were informed of the threat. The man in immediate command was assistant general counsel

Donato Evangelista, a tall and robust cigar-chomper who oversees the highly sensitive areas of trade secrets and security. First off, Evangelista wanted to verify whether the Adirondack volumes in Hitachi's possession were genuine—they could have been a hoax contrived to extract money from Hitachi. For that mission he relied upon an agent whose performances in past emergencies inspired utmost confidence—Richard A. Callahan. A tall, white-haired, distinguished-looking man, . . . Callahan is IBM's top troubleshooter for security matters. "If you see Callahan in your building," says one former IBM executive, "you know something big is going on."

A Marine captain during the Korean War, Callahan flies American and Corps flags at his home in Pebble Beach, California. He joined the FBI in his early twenties and had an outstanding career in federal law enforcement. He served in counterintelligence and later held supervisory posts in the Treasury's investigative branch and the Bureau of Narcotics and Dangerous Drugs before joining IBM as a full-time employee in 1973.

Callahan met with Paley in San Jose and offered him a retainer if he would cooperate with IBM in determining whether Hitachi really possessed those "crown jewels." Paley agreed to take on the assignment. The amount of the payment remains disputed. Fleming, the lawyer for Hitachi, says it was $200,000. IBM says that's wrong, but declines to reveal the correct figure.

Acting on Callahan's instructions, Max Paley telexed a reply to Hitachi's query on the Adirondack workbooks. Wrote Paley: "I made a contact and was told information you requested is under rather strict security control but can be obtained." He proposed a meeting in Japan in early October "for positive exchange of information, terms, conditions, et cetera" On the flight to Tokyo, Paley and a colleague, Robert Domenico, were accompanied by Callahan. The three checked into Tokyo's Imperial Hotel, but Callahan stayed out of sight and did his coaching from the sidelines.

On October 2, Paley and Domenico met with Hayashi in a room at the Imperial Hotel. Well prepared, Hayashi handed over a five-page handwritten set of questions about the operating systems of the new 3081. Then Paley produced attractive bait, prepared by Callahan: a handwritten index of the entire 27-volume set of workbooks. Paley told Hayashi that Palyn Associates did not engage in acquiring confidential material, but that he might be able to find someone who did. According to Paley's affidavit, he asked to see the workbooks in Hitachi's possession so he could identify the genuine article after he returned to the U.S.

At a second meeting four day s later, Hayashi handed over copies of three workbooks, volumes 8, 11, and 22, each still bearing the words "IBM CONFIDENTIAL." Hayashi, who said Hitachi wanted four of the volumes "very badly," returned the index Paley had given him. Alongside the listing for each volume, he had placed a letter: A for the highest priority, B for the second, and C for a new copy of the books already in Hitachi's possession if an updated version had been issued. Hayashi volunteered that Hitachi had a man in the U.S. with "IBM friends." He emphasized, however, that the company had a constant need for new "pipelines" to provide more information on IBM designs. Hayashi had one special request: an early peek at IBM's most advanced disk-drive memory mechanism before volume shipments to customers began. Paley said he would see what he could do.

Paley immediately turned over the workbooks and other material to Callahan, who had been briefed about the appearance and contents of the workbooks. He recognized at once that they were genuine.

Basically, IBM had two choices. It could start a civil suit that would seek to prohibit Hitachi from making use of the stolen data and ask for punitive damages. Or it could turn its findings over to the Justice Department and, in effect, start a criminal case. For IBM, the selection of the second, tougher course came naturally.

As a matter of corporate conviction, IBM has no mercy on anyone, inside or outside, who steals its trade secrets—or tries to. Once the company has solid grounds for suspecting a theft or an intentional patent violation, pursuit is relentless. It has pulled off some undercover sting operations of its own within the company. Last September, IBM security men trapped three former employees peddling proprietary information about the company's new personal computer. What's more, IBM is not reticent about making such incidents known—it wants the message to be heard and heeded.

IBM was aware that the FBI had an undercover investigation under way in Silicon Valley. It was called Pengem, an acronym standing for Penetration of the Gray Electronics Market. One of the operation's objectives was to check the rapidly expanding flow of finished chips and sophisticated chip-production equipment to the Soviet Union and its allies. As a front for Pengem, the bureau rented an office in a two-story modern building in Santa Clara and established a realistic imitation of a consulting firm, called Glenmar Associates, staffed by FBI agents.

IBM had been involved with Pengem as far back as March 1981. The company's representative was none other than Richard Callahan. The IBM executive who authorized the company's participation was Evangelista. On August 27, 1981, Callahan had signed a three-page agreement in which IBM pledged to train no fewer than two and no more than seven FBI agents in "the area of purchasing processes, terminology, and industry testing procedures . . . for both civilian and military usage." IBM also agreed to establish cover for two agents by providing credentials and identification badges and to verify their employment records just in case anyone tried to check them out.

When Callahan told the FBI of IBM's findings about Hitachi's misdeeds, Pengem's focus shifted from the Russians to the Japanese. The FBI agents in the San Jose office, however, thought the Hitachi case would require no more than two weeks or so to clear up, and they could then return to their original mission. Hence, they wanted to avoid letting Paley, a well-known figure in Silicon Valley, discover the true nature of Glenmar Associates and thus blow their cover.

To sidetrack Paley, Callahan arranged a hand-over operation. Under instructions from Callahan, Paley telephoned Hayashi, then staying at the New York Hilton, with the message that a meeting directly with the source of IBM information could be arranged in early November in Las Vegas. Hayashi agreed. The FBI took care of the rest, installing listening devices in a room at the Las Vegas Hilton.

With electronic ears eavesdropping, Paley introduced Callahan to Hayashi, identifying Callahan as a retired lawyer by the name of Richard Kerrigan who, as Paley put it, had "done work for me in the past." Paley then walked out of the room. Later that day, Alan Garretson, the FBI agent in charge of Glenmar Associates, entered the scene. Callahan presented him to Hayashi as Al Harrison, a source who might be able to acquire the desired IBM information.

That day and the next, Hayashi spelled out Hitachi's wish list of IBM equipment and documents. In accordance with the Justice Department's guidelines on undercover operations, Garretson told Hayashi that the material would have to be illegally taken and that the person involved could be "put in jail for stealing." Hayashi, however, failed to recognize the warning signal.

Originally the FBI had intended for Callahan to leave the case as soon as he introduced Garretson to Hayashi. But things did not work out that way. One reason was that the FBI wanted the assurance of continuing IBM involvement. For another, Callahan—so viewers of the tapes testify—is a superb actor. Older and more experienced than the FBI agents, he dominated the scenes.

After the Las Vegas meetings the first follow-up conversation with the Japanese was conducted by Callahan, not the FBI's Garretson. On the line was a Hitachi memory-systems expert named Jun Naruse, who had been assigned to view IBM's new memory device. Naruse wanted assurances that all would go smoothly. According to affidavits, Naruse said that if there was any kind of trouble, "it's real trouble for Hitachi."

Three days later, Garretson met Naruse at 5 A.M. in the lobby of a hotel in Hartford, Connecticut. They drove to a parking lot near a Pratt & Whitney Aircraft plant, the jet engine division of United Technologies. In the parking lot, a Pratt & Whitney employee, whom the FBI had recruited for the mission, gave Garretson and Naruse identification badges that enabled them to gain admission to a high-security area in which one of the new memory devices was installed. In return for the badges, Garretson handed over an envelope apparently stuffed with money. "How much did you have to pay?" asked Naruse. "Plenty," whispered Garretson.

The three men passed uneventfully through the checkpoint at the plant's entrance. As they reached the door of the room containing the memory device, they found it closed by a combination lock. While Garretson and Naruse hid in a closet, the Pratt & Whitney employee fetched a guard, who opened the door. Once inside, Naruse was ecstatic. Both men carried cameras and began taking pictures. Garretson repeatedly reminded Naruse not to include any background that could identify the location. After each had shot many pictures, Naruse asked to be photographed hugging the device.

Back in the Hartford hotel, Naruse gave Garretson $3,000 in $100 bills. Then, on November 18 [1981] in Santa Clara, he delivered an additional $7,000 in cash in return for maintenance manuals for the memory device.

As a good hug should, the Hartford embrace produced heightened enthusiasm within Hitachi. Through various channels, more and more requests for secret IBM data and equipment began reaching Glenmar Associates. Max Paley also got a letter, with the memory-jogging date December 7, 1981, in which Hayashi stressed the need for confidentiality and reminded him: "I have no idea to pay your travel fee if you don't have the suitable information for us." In a separate letter to Garretson, Hayashi set out an expanded shopping list. He placed a code and price tag alongside each item Hitachi desired. D-14, for example, indicated the magnetic head and platters used to read and write data on the disk drive; for that Hitachi was willing to pay $10,000.

That letter crossed one from Callahan, in his role as lawyer Kerrigan. Callahan wrote to Hayashi that some IBM people were getting nervous about continuing to supply information: "As you well know, they risk disgrace and perhaps imprisonment if they are caught taking the IBM information you have been asking for. They are only willing to risk the consequences if the money rewards are great enough."

Hayashi quickly replied that "from the point of us, cost should depend on how we can use it. Except the rare case (D-14 head and media), our requesting information [he meant the information we are requesting] will be published in the future. Then timing is the best or most important as to decide the value." To make the point more graphic, Hayashi drew a chart that showed a sharply slanting line declining over a four-year period from a presumably high value to zero at the time of publication. He concluded by holding out the lure of a contract if Glenmar could obtain the complex microcode used to enhance the performance of one of IBM's older computers. "Our top management will understand your potential if you locate it by the end of January," Hayashi wrote.

With Callahan's help, Glenmar did better than that. On January 7, 1982, Garretson phoned Hayashi in Tokyo to report that he could deliver the microcode for $12,000. On January 18, at a meeting with Hayashi and Hitachi software expert Isao Ohnishi, Garretson was told that a secure money channel had now been established. Hitachi would send funds to Nissei Electronics Ltd., a Hitachi affiliate, which in turn would transfer the money to NCL Data Inc., where the president, Tom Yoshida, knows "how things are gotten" and would make the payments in bank transfers to Glenmar's account.

From then on, transactions occurred with almost dizzying rapidity. At most deliveries, a Hitachi expert was present to check the goods and often to make on-the-spot requests for an additional manual or part. A camaraderie was also developing between the Japanese and their American suppliers. Viewers of the tapes say that much of the time was taken up by friendly banter about baseball and where the men were going to have dinner together that night.

All the while, the FBI was seeking to lure higher-ranking Hitachi executives within range of its clandestine cameras. The opening came in March when Hayashi said that his company was very interested in hiring IBM executives who were about to retire as consultants. Garretson insisted that the IBM executives in question were at such high levels they required personal assurances about the security of the arrangement from a Hitachi executive of equal rank. The ploy worked. On April 23, Callahan and Garretson met in San Francisco with Kisaburo Nakazawa, general manager of Hitachi's Kanagawa Works, which produces the company's mainframe computers. According to an FBI affidavit, Nakazawa said he was aware of everything Garretson's company had provided and that it had been helpful. He also said he was aware of the risks involved.

As the Japanese shopping list grew ever longer, the sting operation became a great pain for IBM. Callahan had to have people scrambling between IBM plants and research labs collecting the desired items. The executives privy to the plot began to worry that they had been giving away too much. IBM told the FBI it could not allow another batch of material to leave the U.S.

From that point the sting rolled toward the climax. Hitachi was offered a package deal that would give the company just about everything it wanted. The Hitachi people requested a long list of items that included design documents and components for every major part of the 3081 computer. Garretson set the price at $700,000. After some haggling, the two sides settled on $525,000.

At precisely 9 A.M. on June 22, a brown Volkswagen van belonging to paymaster Tom Yoshida braked to a stop in front of the Glenmar offices. Hayashi and Ohnishi climbed out. Yoshida remained at the wheel.

As he entered the room in which the IBM booty was stacked on a table, Hayashi could hardly contain his delight. Triumphantly he seized one of the cartons and ripped off the sticky IBM label. With a flourish, he pasted the famed IBM logo on his notebook, souvenir of a job well done.

At that moment, two other men stepped into the room. "It is all over," said one of them. "We are FBI agents."

The two Japanese reacted with stunned disbelief. So too did their country—at first. Then Japan's mood quickly changed as Tokyo's sensationalist newspapers pounced on the story and sent reporters scurrying through Silicon Valley. Under a variety of titles and datelines, the theme was always the same: the entire episode was a dirty Yankee trick, aimed at bashing the Japanese. Against the backdrop of trade tensions between the U.S. and Japan, the public readily believed these stories—a readiness enhanced by a sense of outrage over the sting tactics. In Japan undercover operations are limited by law to the investigation of drug-related crimes.

Since Hitachi prides itself on being an innovative company, it was especially vulnerable to embarrassment. In the immediate aftermath of the arrests, it behaved as if it had indeed lost face. It suspended advertising, rejected press requests for interviews, and in general hid behind the shoji screen. Within a short time, however, Hitachi began to realize that the public was not crying out for corporate hara-kiri.

Quite the contrary—Hitachi started to benefit from its role as a victim. After sampling his countrymen's reactions, Yasou Naito, the editor in chief of Tokyo's respected bimonthly *Nikkei Computer,* found that "many users of IBM machines have told me they are thinking of switching to Hitachi." One Hitachi marketing executive claim[ed] sales of computers [rose] on a wave of sympathy. Hitachi even scored a heartening triumph in the U.S. when the Social Security Administration bought two Hitachi-manufactured computers for $7 million, instead of a more expensive IBM system.

As its confidence returned, Hitachi began to regard the sting as only a delay and not a derailment of its onward drive in the computer field. Its newest supercomputer—the S-810—was announced only one week behind schedule [in] August [1982]. The new Hitachi computer closest to the IBM 3081 also made its debut, despite the company's failure to get hold of all the IBM secrets it was after. But that does not mean the sought-after information was of little value after all. Far from it. What Hitachi was most interested in getting, it appears, was information on IBM's design directions. If Hitachi had obtained all it wanted, it might now have better prospects of narrowing IBM's lead in the development race.

Heartened by public support, Hitachi came out from behind the screen. It resumed advertising, with stress on innovation. One of its [1983] publicity releases point[ed] out that the company employs more Ph.D.s than the University of Tokyo.

Because of the public's perception that there was no guilt on the Japanese side, Hitachi finally felt able to do what otherwise would have been almost impossible for a Japanese company—plead guilty. At a press conference in Tokyo, a Hitachi spokesman said the guilty plea would have no effect on the future operations of the company.

For any company with valuable secrets, eternal vigilance is the price of keeping them. In starting a criminal case against Hitachi and participating in the sting, IBM was sending a message, not just to Hitachi but to the entire computer world, that it means to protect its secrets and will go to great lengths to do so. When Judge Williams imposed those fines in San Francisco . . . he said he hoped "that this example will be felt in other corporations throughout the world." A total of $24,000 in fines is not all that impressive a deterrent in a big-money game. The sting, however, is something IBM's competitors won't soon forget.

— 31 —————————————

Computer Crime

Steven L. Mandell

Introduction and Overview

.   .   .   .   .

Criminal activity is as old as civilization, and the advent of the computer age has created new opportunities for crime. Just how prevalent computer crime has become and how much goes undetected is not known,

Reprinted by permission from *Computers, Data Processing, and the Law* by Steven L. Mandell, pp. 154–62. Copyright © 1984 by West Publishing Company. All rights reserved.

but it is widely believed that only a fraction of such crimes are ever discovered. Furthermore, there is some evidence to suggest that the average computer crime involves greater sums of money than other, traditional crimes.

This [essay] examines the most common forms of computer crime, including those that occur both inside and outside an organization that uses data processing. . . .

Definitions

What is mean by the term "computer crime"? There is no consensus on this question although the legal community has been focusing more attention on it through legislation and court opinions. Some authors prefer the term "computer abuse" to computer crime because it encompasses a broader range of illicit activity and because existing laws are not equipped to provide adequate guidance in this emerging area of criminal activity.[1] Others may take the view that computer crime should be defined very narrowly to exclude crimes in which the criminal conduct is the same as that used in noncomputer crimes.[2] According to this view, for example, obtaining money by impersonating a bank officer over the phone and giving a secret code number would not be a pure computer crime, since the real essence of the wrongful conduct by which the crime was perpetrated was an impersonation, not a computer manipulation. According to this view, true computer crimes are so rare that they are almost mythical.

This [essay] takes a broad but pragmatic view toward defining computer crime: a computer crime is simply a criminal act that poses a greater threat to a computer user than to an otherwise similarly situated nonuser. Computer crime, as defined here, consists of two kinds of activity: (a) the use of a computer to perpetrate acts of deceit, theft, or concealment that are intended to provide financial, business-related, property, or service advantages; and (b) threats to the computer itself, such as theft of hardware or software, sabotage, and demands for ransom. Because computer crimes seldom involve acts of physical violence, they are generally classified as white-collar crimes.

Profile of the Computer Criminal

The popular view of the successful computer criminal is interesting and somewhat unsettling. Most companies would be eager to hire personnel who fit this description. Often such people are young and ambitious with impressive educational credentials. They tend to be technically competent and come from all levels of employees, including technicians, programmers, managers, and high-ranking executives. These people are often viewed as heroes challenging an impersonal computer as an opponent in a game. In contrast, the corporate victim of computer crime is not a sympathetic figure. The victim is often seen as one who is caught in a trap of the victim's own creation. Perhaps most unnerving, a commonly held belief is that many computer criminals have been discovered by chance, not by established detection techniques.

Prevalence of Computer Crime

Because so many computer crimes are discovered accidentally, there has been much speculation, and little consensus, about the actual extent of computer crime. Two of the more reputable studies diverge widely in their estimates.

The Stanford Research Institute (SRI) study was funded by grants from the National Science Foundation, the Atomic Energy Commission, and various private organizations. Data for the study on computer abuse were gathered over a four-year period, and results were published beginning in 1973.[3] By far the most ambitious study of its kind ever undertaken, the study initially identified 381 computer abuse cases from official sources, magazines, and newspapers. Admittedly, the accuracy of some stories printed in the popular media is questionable, both with regard to the exact nature of the crime and the amount of money involved. Despite these drawbacks, the study has generally been well received, since imperfect information may be better than none. Even fictitious accounts have some value as they allow us to study the feasibility of such a crime.

Some of the conclusions drawn by the study illustrate the potential dangers of computer abuse. It found that the average loss per crime was $450,000 and speculated that the crimes reported represented only a fraction of the actual total. However, this finding should be balanced by the fact that the number of cases compiled by the Stanford Research Institute was an insignificant percentage of the 100,000 computers in use at the time.

The Government Accounting Office (GAO) made a search among ten federal agencies for incidents of computer crime.[4] Its findings were much more modest than those of the SRI study. The search generated a list of 69 cases. Nine of these were privacy invasion cases that involved no monetary loss. The average dollar loss for those crimes that did involve money was $44,110, a far cry from the SRI's $450,000 figure. In addition, the GAO report stated that 50 of the crimes were committed by technically naive users.

The large differences in the findings of these two reports does not necessarily reflect biases on the part of those collecting data but rather reflects the more fundamental problems of defining computer abuse, estimating unknown incidents, and allowing for inconsistency in media coverage. Simply stated, there is a great deal of mystery surrounding computer crime that defies quantification with any degree of accuracy. What is certain, however, is that computer crime is real, it involves potentially large sums of money, and it is not likely to decrease as use of computers continues to spread rapidly.

The Effect of Media Publicity

Part of the difficulty in assessing the extent and impact of computer crime is due to its widespread and occasionally inaccurate exposure in the media. Newspapers and magazines have focused on incidents of computer abuse for several reasons. The average person (and perhaps the average newspaper reporter) understands little about the complexities of modern computers and therefore is easily intimidated by them. This lack of understanding is often reflected in media reports, which at times tend to exaggerate the severity of a crime. For instance, a story that appeared in the *San Francisco Chronicle* reported how one computer technician was able to gain access to a computer, resulting in unauthorized use of computer time worth "possibly millions." When the case was brought to court, expert testimony put the value of stolen services at $2,000. Of course, the exaggerated version makes the headlines, while the true version often does not get publicized at all.

As we noted, computer abuse also tends to be glamorized. Often the perpetrator is portrayed as an eccentric genius engaged in a Robin Hood–type operation, stealing from a large, impersonal machine, the epitome of the "establishment." Such a point of view, by fostering sympathy for the lone bandit, leads the public to ignore the high cost to society that such crimes exact.

The Vulnerability of the Computer

There are several factors that make a computer an attractive target for criminals. Among them are the speed with which the computer does its work, making many small thefts of a few cents potentially profitable; the invisibility of records stored in a computer's memory; and the use of programmed processing controls, which can be manipulated or bypassed altogether.

Often overlooked is the fact that although the computer is itself a complex machine, many of the processes used with it are relatively simple. There are five key areas in the operation of a computer, all of which are subject to abuse:

a. Input operations may be manipulated to avoid legitimate charges to a user or to cause the computer to print a check in payment for nonexistent services. Fictitious accounts, and even whole companies, have been created in this way.

b. A program controls the computer's operations and if tampered with can benefit the criminal at the expense of the entity that owns the computer. Also, programs themselves are valuable items that are subject to theft.

c. The central processing unit may be exposed to vandalism or destruction. A user's exclusive reliance on it for vital functions makes it a prime target for vandalism or ransom demands.

d. Output, though the least likely target for criminal attack, can still present serious criminal problems. Valuable data, such as mailing lists, can be stolen. The computer output, such as checks, is usually the goal of the criminal who manipulates the system.

e. The communication process is vital to all information flowing in and out of the computer. This data can be intercepted from the lines of communication through wiretapping or the communication facilities themselves can be destroyed.

Types of Computer Crimes

The variety of computer crimes is quite extensive and can be classified into four broad categories: (a) sabotage, (b) theft of services, (c) property crimes, and (d) financial crimes. This section examines each of these categories and gives examples drawn from actual crimes.

Sabotage

This type of computer crime is usually, though not exclusively, directed against computer hardware. Sabotage of computers often resembles traditional sabotage because a computer facility's unique capabilities would not typically be used to carry out the destruction, although sabotage may require some sophistication if computer-assisted security systems must be thwarted or the system is manipulated to do harm to itself.

Computers are targets of sabotage and vandalism especially during times of political activism. Dissident political groups during the 1960s, for instance, conducted assaults on computer installations, often causing extensive damage. Other forms of physical violence have included shooting a computer with a revolver and flooding the computer room. One fired employee simply walked through the data storage area with an electromagnet, thereby erasing valuable company records. A computer's power source can also be the target of a saboteur.

Obviously, these acts of violence do not require any special expertise on the part of the criminal. Sabotage may, however, be conducted by dissatisfied former employees who may put to use some of their knowledge of company operations to gain access to and destroy both hardware and software.

Though computer sabotage is not the type of computer crime that people see as threatening in the same way as if the secrets of the computer were manipulated by a misguided genius, its potential threat should not be taken lightly. The degree of sophistication in a computer crime does not necessarily correlate with the cost of rectifying the damage.

Theft of Services

Computer services may be abused in a variety of ways, depending upon the individual system. Some examples of theft of computer services have involved politicians using a city's computer to conduct campaign mailings or employees conducting unauthorized free-lance services on a company computer after working hours.

Time-sharing systems have been exposed to great amounts of abuse due to inadequate or nonexistent security precautions. It is much easier to gain unauthorized access to a time-sharing system than to a closed system. Though most require passwords to gain access, such a system is only as good as the common sense and caution of its users. A time-sharing system that does not require regular changing of access codes is inviting the theft of valuable computer time. The amazing lack of care exercised by supposedly sophisticated users in this regard made national headlines . . . when it was discovered that a group of high school computer buffs in Milwaukee had accessed numerous information systems, including those of banks, hospitals, and even the defense research center in Los Alamos, New Mexico. The students reportedly gained access by using each system's password, some of which had not been changed for years and many of which were obtained from public sources.

Wiretapping is another technique used to gain unauthorized access to a time-sharing system. By "piggybacking" onto a legitimate user's line, one can have free use of the user's privileges whenever the line is not being used by the authorized party.

One of the prime examples of computer services theft took place at the

University of Alberta. In 1976, a student at the university undertook an independent study under the supervision of a professor to investigate the security of the university's computer system, a time-sharing system with more than 5,000 users, some as far away as England. After discovering several gaps in the system's security, he was able to develop a program that reduced the possibility for unauthorized use as well as for other tampering. He brought this program to the attention of the computer center, which took no action on the student's recommendations because it was assumed that planned changes in the system would remove security shortcomings. However, the changes were not implemented for another nine months, and during this period, the program was leaked to several students on campus. "Code Green," as the program was nicknamed, was eventually invoked several thousand times.

The university attempted to crack down on the unauthorized users and revoked several students' access privileges. Among these students were two who had been able to manipulate the program to get the computer to display the complete listing of all user passwords, including those at the highest privilege levels. In essence, this gave them unlimited access to the computer's files and programs. These students retaliated against the university administration by occasionally rendering the system inoperable, as well as less harmful acts such as periodically inserting an obscenity into the payroll file. With an unlimited supply of IDs, they were able to escape detection, compiling a library of the computer's programs and even monitoring the implementation of the new security system. The desperate university computer personnel focused exclusively on this situation, keeping a detailed log of all terminal dialogues. This effort led them to a terminal in the geology department one evening, and the students were apprehended.

Though an extreme example, the situation at the University of Alberta shows the extent to which the theft of computer services can be committed in the absence of adequate security measures. Perhaps a more difficult problem exists in dealing with the theft of computer services by employees who are authorized to use the computer for employment purposes. Recent court cases have dealt with the issue of whether an employee's unauthorized use of the employer's computer for personal use constitutes a crime, with varying results.

In *United States* v. *Sampson,*[5] for example, an employee of a computer service company under contract with NASA was charged with theft of a "thing of value" belonging to the United States[6] after he was discovered using the company computer for his own personal gain. The federal court held that computer time did qualify as a thing of value within the scope of the relevant federal criminal statute.

In *People* v. *Weg,*[7] a computer programmer employed by a board of education was charged with a misdemeanor of theft of services for allegedly using the board's computer without permission for various personal projects, including calculating a racehorse handicapping system and tracing the genealogy of horses that he owned. The statute provided that a defendant had committed a "theft of services" when

> obtaining or having control over labor in the employ of another person, or of business, commercial, or industrial equipment or facilities of another person, knowing that he is not entitled to the use thereof, and with intent to derive a commercial or other substantial benefit for himself or a third person, he uses or diverts to the use of himself or a third person such labor, equipment, or facilities.

The judge dismissed the case, finding that since the computer was owned by the public school board, no "business, commercial, or industrial equipment or facilities of another person" were involved; that is, a school board was not considered to be a business. In addition, the judge ruled that the charges failed to include any factual allegation that the defendent intended to derive a commercial benefit from the services.

Property Crimes

The most obvious computer crime that comes to mind in crimes of property is the theft of computer equipment itself. This has been more common with the increasing miniaturization of computer components and the advent of home computers. Such crimes, like acts of vandalism, are easily absorbed into traditional concepts of crime and present no unique legal problems. . . .

Computer crimes of property theft frequently involve merchandise of a company whose orders are processed by computers. These crimes are usually committed by internal personnel who have a thorough knowledge of the operation. By manipulating records, dummy accounts can be created causing orders to be shipped to an accomplice outside the organization. Similarly, one can cause checks to be paid out for receipt of nonexistent merchandise.

Theft of property need not be limited to actual merchandise but may also extend to software. Those with access to a system's program library can easily obtain copies for their own use or, more frequently, for resale to a competitor. Technical security measures in a computer installation are of little use when dishonest personnel take advantage of their positions of responsibility.

Commission of property theft is by no means limited to those within the company structure, however. A computer service having specialized programs but poor security may open itself up to unauthorized access by a competitor. All that is necessary is that the outsider gain access to proper codes. This can be done in a number of ways, including clandestine observation of a legitimate user logging on from a remote terminal or use of a remote minicomputer to test for possible access codes.

Financial Crimes

Although not the most common, financial computer crimes are perhaps the most serious in terms of monetary loss. With the projected increasing dependence on electronic fund transfers, implications for the future are indeed ominous.

A common method of committing a financial computer crime involves checks. These mass-produced negotiable instruments can be manipulated in a number of ways. An employee familiar with a firm's operations can cause multiple checks to be made out to the same person. Checks can also be rerouted to a false address or to an accomplice. Such crimes do not seem so incredible when one realizes the scope of *unintentional* mistakes that have been made with computerized checks. For example, the Social Security Administration once accidentally sent out 100,000 checks to the wrong addresses while the system's files were being consolidated.

A form of a financial computer crime that has captured the attention of many authors, but has probably been used much less frequently than one would expect from media discussion, is known as the round-off fraud. In this crime, the thief, perhaps a bank employee, collects the fractions of cents in customers' accounts that are created when the applicable interest rates are applied. These fractions are then stored in an account created by the thief. The theory is that fractions of cents collected from thousands of accounts on a regular basis will yield a substantial amount of money. It has been suggested that in reality, however, round-off schemes may not yield enough money to make all the manipulations and risks worthwhile.[8]

Another type of financial crime involves juggling confidential information within a computer, both personal and corporate. Once appropriate access is gained to records, the ability to alter them can be highly marketable. At least one group operating in California engaged in the business of creating favorable credit histories to clients seeking loans.

By far the most massive fraud of this nature that ever occurred was the Equity Funding fraud, involving $2 billion over a period of ten years. This fraud, too complex to explain briefly in any detail, occurred in three distinct stages. Much of the criminal activity did not involve the company's computer per se, but there is no doubt that the speed of its data-processing facilities made possible the theft of this exorbitant amount of money.

The Equity Funding Corporation of America (EFCA) had four main activities: the sale of investment programs to the public, the financing of its operations, the purchase of mutual fund shares, and the issuing of insurance. The initial phase of the fraud began in 1964 and involved inflating the company's reported earnings, which made its funded life insurance plan more attractive to investors. An individual bought mutual fund shares, then borrowed on them to pay life insurance premiums over a ten-year period. The hope was that the income generated by the mutual funds would cover the cost of borrowing the money and pay for part of the insurance premium. This phase of the fraud emphasized sales appeal at the expense of profitability, and required only manual entries into company books. It involved about $85 million.

The second phase was known as the foreign phase. It consisted of borrowing funds from foreign subsidiaries without recording the borrowing as liabilities on the company records. Between 1968 and 1970 this scheme allowed the apparently fast-growing organization to acquire several banks, insurance companies, and other financial institutions.

Equity Funding had thus grown from a marketing organization to an insurance conglomerate, bringing into play the third phase of the fraud, carried out with the assistance of a computer. To generate short-term cash flow, Equity Funding sold many of its insurance policies to a co-insurer, Pennsylvania Life Insurance Company. By this time, the company was losing substantial amounts of money due to the necessity of servicing fraudulent policies already in existence, so it created wholly fictitious insurance policies and resold them immediately to co-insurers. To maintain profitability, Equity Funding would have had to sell vast amounts of new insurance, which it failed to do.

The company used its computer to create the new policies, mainly in the form of mass-marketing policies not using individual billing. Eventually, 64,000 fraudulent policies were issued. The fact that many of Equity Funding's subsidiaries were audited by unconnected auditing firms made it possible to shift assets from company to company as the need arose. For instance, an asset on the books of one company would not appear as a liability on another.

At the same time, the computer was programmed to fabricate the appropriate number of deaths, cancellations, and lapses that were to be expected from actual policies. Whenever individual audits were requested, the company would claim that the file was temporarily in service, instruct a programmer to prepare a false file overnight, and have it delivered the next morning. Obviously, the computer's speed and great capacity were central to the success of this phase of the fraud.

The fraud, long suspected by some, was finally exposed in 1973 by a surprise audit by examiners sent by the Illinois Insurance Commission.

The company officers were caught off guard and were unable to manipulate financial records quickly enough to perpetuate their massive fraud.

Twenty-two people were convicted of federal crimes in this $2 billion fraud and at least 50 civil suits were filed in connection with the ongoing crime. The computer, though instrumental only in the final stages of the fraud, was the means by which Equity Funding was able to obtain most of its illicit funds. Obviously, a fraud of this proportion is not the result of a breach in the security system of a computer but rather was made possible by deliberate corporate policy.[9] The only bodies capable of exposing fraud of this degree are external entities, such as private securities investigators or governmental regulatory agencies, and the process took ten years in this case.

Another celebrated computer crime was that perpetrated by Stanley Mark Rifkin, a computer consultant retained by the Seattle Pacific Bank. Rifkin was able to penetrate the bank's computerized system and transferred $10.2 million to a numbered Swiss bank account. In this case, the media worked both for and against the criminal. Although somewhat glamorizing the criminal as a clever loner swindling a corporate giant, the substantial publicity given to his subsequent attempt to steal an additional $50 million while his case was awaiting trial was also widely reported in the press, quite likely contributing to the judge's uncharacteristically stiff sentence at the end of his trial: eight years in a federal penitentiary.[10]

Finally, one of the more ingenious financial crimes perpetrated through the use of a computer occurred in 1977 at Florida's Flagler Dog Track. The dog-racing odds were figured by computer, and often the races were conducted so quickly that the odds would not be figured completely until after the race was over. A conspiracy was developed whereby an operator of the computer received the race results from an accomplice observing the race. He then stopped the computer program in progress, deducted a number of losers and added a corresponding number to the pool of winners in computer storage. The program was restarted and shortly finished its run. False winning tickets were then printed, also by computer, and were cashed in the next day. Since winners were paid from a pool formed by the losers' money, there was no way to detect the loss. Rather, each winner's share was somewhat less than it should have been.

The examples cited in this section represent some main areas in which computer crime can occur. It is by no means comprehensive, because the possibilities are nearly limitless for one with computer expertise and a fertile imagination. It should also be noted, however, that most of these crimes, many involving extremely large sums of money, could have been prevented or detected earlier through adequate security measures.

Notes

1. See, e.g., Kling, "Computer Abuse and Computer Crime as Organizational Activities," 2 Comp. L. J. 403 (1980); D. Parker, S. Nycum, & S. Aura, "Computer Abuse" (Stan. Research Inst. Rep. 1973).
2. See, e.g., Taber, *A Survey of Computer Crime Studies,* 2 Comp. L.J. 275 (1980).
3. D. Parker, S. Nycum, & S. Aura, "Computer Abuse" (Stan. Research Inst. Rep. 1973). *See* Parker, "Computer Abuse Research Update," 2 Comp. L.J. 329, 351 (1980) for a bibliography of the SRI project for 1975–1980. See also D. Parker, *Crime by Computer* (1976).
4. General Accounting Office, *Computer Related Crimes in Federal Programs* (1976), reprinted in Problems Associated with Computer Technology in Federal Programs and Private Industry, Computer Abuses, Sen. Comm. on Gov't Operations, 94th Cong. 2d Sess. 71–91 (Comm. Print 1976).
5. 6 CLSR 879 (N.D. Cal. 1978).
6. 18 USC 641 (1976).
7. 113 Misc.2d 1017, 450 N.Y.S.2d 957 (N.Y. City Crim. Ct. 1982). See also "Using Computer Time No Crime, Judge Says," 68 ABA J. 671 (1982).
8. Taber, supra note 2.
9. *See* Kling, "Computer Abuse and Computer Crime as Organizational Activities," 2 Comp. L.J. 403 (1980).
10. *See* Becker, "Rifkin, A Documentary History," Comp. L.J. 471 (1980).

Beyond Legal Mandates:
Moral Responsibility

The Association for Computing Machinery's code of ethics, reproduced in reading 32, tries to cope with some of the problems raised elsewhere in this book. Like codes of ethics for other professions, it provides some general rules for practitioners to consider when faced with moral questions raised by the diverse applications and growing technological sophistication of computers.

David Parnas (reading 33) applies a professional code to help examine a compelling and topical computer-related moral dilemma: whether to participate in efforts to create the software needed for space-based defensive systems against nuclear attack (e.g., Star Wars). As a member of a military advisory panel, Parnas concluded that such a system was infeasible and undesirable because it would fail in its mission and waste enormous amounts of money.

Walter Morrow (reading 34) does not disagree with Parnas's central contention that the production of error-free software is impossible. Rather, Morrow argues that a less-than-perfect SDI system "could so diminish the success of an attack that the Soviets would be discouraged for fear of a counterstrike." [This lukewarm endorsement is the strongest analytical defense of SDI that the editors could locate.]

Code of Professional Conduct

Association for Computing Machinery

BYLAW 19. ACM CODE OF PROFESSIONAL CONDUCT

PREAMBLE

Recognition of professional status by the public depends not only on skill and dedication but also on adherence to a recognized code of Professional Conduct. The following Code sets forth the general principles (Canons), professional ideals (Ethical Considerations), and mandatory rules (Disciplinary Rules) applicable to each ACM Member.

The verbs "shall" (imperative) and "should" (encouragement) are used purposefully in the Code. The Canons and Ethical Considerations are not, however, binding rules. Each Disciplinary Rule is binding on each individual Member of ACM. Failure to observe the Disciplinary Rules subjects the Member to admonition, suspension, or expulsion from the Association as provided by the Procedures for the Enforcement of the ACM Code of Professional Conduct, which are specified in the ACM Policy and Procedures Guidelines. The term "member(s)" is used in the Code. The Disciplinary Rules of the Code apply, however, only to the classes of membership specified in Article 3, Section 4, of the Constitution of the ACM.

CANON 1

An ACM member shall act at all times with integrity.

Ethical Considerations

EC1.1. An ACM member shall properly qualify himself when expressing an opinion outside his areas of competence. A member is encouraged to express his opinion on subjects within his area of competence.

EC1.2. An ACM member shall preface any partisan statements about information processing by indicating clearly on whose behalf they are made.

EC1.3. An ACM member shall act faithfully on behalf on his employers or clients.

From Bylaw 19. Reprinted from Bylaws of the Association for Computing Machinery.

Disciplinary Rules

DR1.1.1. An ACM member shall not intentionally misrepresent his qualifications or credentials to present or prospective employers or clients.

DR1.1.2. An ACM member shall not make deliberately false or deceptive statements as to the present or expected state of affairs in any aspect of the capability, delivery, or use of information processing systems.

DR1.2.1. An ACM member shall not intentionally conceal or misrepresent on whose behalf any partisan statements are made.

DR1.3.1. An ACM member acting or employed as a consultant shall, prior to accepting information from a prospective client, inform the client of all factors of which the member is aware which may affect the proper performance of the task.

DR1.3.2. An ACM member shall disclose any interest of which he is aware which does or may conflict with his duty to a present or prospective employer or client.

DR1.3.3. An ACM member shall not use any confidential information from any employer or client, past or present, without prior permission.

CANON 2

An ACM member should strive to increase his competence and the competence and prestige of the profession.

Ethical Considerations

EC2.1. An ACM member is encouraged to extend public knowledge, understanding, and appreciation of information processing, and to oppose any false or deceptive statements relating to information processing of which he is aware.

EC2.2. An ACM member shall not use his professional credentials to misrepresent his competence.

EC2.3. An ACM member shall undertake only those professional assignments and commitments for which he is qualified.

EC2.4. An ACM member shall strive to design and develop systems that adequately perform the intended functions and that satisfy his employer's or client's operational needs.

EC2.5. An ACM member should maintain and increase his competence through a program of continuing education encompassing the techniques, technical standards, and practices in his fields of professional activity.

EC2.6. An ACM member should provide opportunity and encouragement for professional development and advancement of both professionals and those aspiring to become professionals.

Disciplinary Rules

DR2.2.1. An ACM member shall not use his professional credentials to misrepresent his competence.

DR2.3.1. An ACM member shall not undertake professional assignments without adequate preparation in the circumstances.

DR2.3.2. An ACM member shall not undertake professional assignments for which he knows or should know he is not competent or cannot become adequately competent without acquiring the assistance of a professional who is competent to perform the assignment.

DR2.4.1. An ACM member shall not represent that a product of his work will perform its function adequately and will meet the receiver's operational needs when he knows or should know that the product is deficient.

CANON 3

An ACM member shall accept responsibility for his work.

Ethical Considerations

EC3.1. An ACM member shall accept only those assignments for which there is reasonable expectancy of meeting requirements or specifications, and shall perform his assignments in a professional manner.

Disciplinary Rules

DR3.1.1. An ACM member shall not neglect any professional assignment which has been accepted.

DR3.1.2. An ACM member shall keep his employer or client properly informed on the progress of his assignments.

DR3.1.3. An ACM member shall not attempt to exonerate himself from, or to limit his liability to clients for his personal malpractice.

DR3.1.4. An ACM member shall indicate to his employer or client the consequences to be expected if his professional judgment is overruled.

CANON 4

An ACM member shall act with professional responsibility.

Ethical Considerations

EC4.1. An ACM member shall not use his membership in ACM improperly for professional advantage or to misrepresent the authority of his statements.

EC4.2. An ACM member shall conduct professional activities on a high plane.

EC4.3. An ACM member is encouraged to uphold and improve the professional standards of the Association through participation in their formulation, establishment, and enforcement.

Disciplinary Rules

DR4.1.1. An ACM member shall not speak on behalf of the Association or any of its subgroups without proper authority.

DR4.1.2. An ACM member shall not knowingly misrepresent the policies and views of the Association or any of its subgroups.

DR4.1.3. An ACM member shall preface partisan statements about information processing by indicating clearly on whose behalf they are made.

DR4.2.1. An ACM member shall not maliciously injure the professional reputation of any other person.

DR4.2.2. An ACM member shall not use the services of or his membership in the Association to gain unfair advantage.

DR4.2.3. An ACM member shall take care that credit for work is given to whom credit is properly due.

CANON 5

An ACM member should use his special knowledge and skills for the advancement of human welfare.

Ethical Considerations

EC5.1. An ACM member should consider the health, privacy, and general welfare of the public in the performance of his work.

EC5.2. An ACM member, whenever dealing with data concerning individuals, shall always consider the principle of the individual's privacy and seek the following:

—To minimize the data collected.

—To limit authorized access to the data.

—To provide proper security for the data.

—To determine the required retention period of the data.

—To ensure proper disposal of the data.

Disciplinary Rules

DR5.2.1. An ACM member shall express his professional opinion to his employers or clients regarding any adverse consequences to the public which might result from work proposed to him.

33

Professional Responsibility to Blow the Whistle on SDI

David Lorge Parnas

In May of 1985 I was asked by the Strategic Defense Initiative Organization, the group within the Office of the U.S. Secretary of Defense that is responsible for the "Star Wars" program, to serve on a $1000/day advisory panel, the SDIO Panel on Computing in Support of Battle Management. The panel was to make recommendations about a research and development program to solve the computational problems inherent in space-based defense systems.

. . . I consider the use of nuclear weapons as a deterrent to be dangerous and immoral. If there is a way to make nuclear weapons "impotent and obsolete" and end the fear of such weapons, there is nothing I would rather work on. However, two months later I had resigned from the panel. I have since become an active opponent of the SDI. This article explains why I am opposed to the program.

My View of Professional Responsibility

My decision to resign from the panel was consistent with long-held views about the individual responsibility of a professional. I believe this responsibility goes beyond an obligation to satisfy the short-term demands of the immediate employer.

As a professional:

- I am responsible for my own actions and cannot rely on any external authority to make my decisions for me.
- I cannot ignore ethical and moral issues. I must devote some of my energy to deciding whether the task that I have been given is of benefit to society.
- I must make sure that I am solving the real problem, not simply providing short-term satisfaction to my supervisor.

From David Lorge Parnas, "SDI: A Violation of Professional Responsibility," *Abacus*, Vol. 4, No. 2, Winter 1987, pp. 46–52. Copyright © 1987 Springer-Verlag New York, Inc.

Some have held that a professional is a "team player" and should never "blow the whistle" on his colleagues and employer. I disagree. As the Challenger incident demonstrates, such action is sometimes necessary. One's obligations as a professional precede other obligations. One must not enter into contracts that conflict with one's professional obligations.

My Views on Defense Work

Many opponents of the SDI oppose all military development. I am not one of them. I have been a consultant to the Department of Defense and other components of the defense industry since 1971. I am considered an expert on the organization of large software systems, and I lead the U.S. Navy's Software Cost Reduction Project at the Naval Research Laboratory. Although I have friends who argue that "people of conscience" should not work on weapons, I consider it vital that people with a strong sense of social responsibility continue to work within the military/industrial complex. I do not want to see that power completely in the hands of people who are *not* conscious of social responsibility.

My own views on military work are close to those of Albert Einstein. Einstein, who called himself a militant pacifist, at one time held the view that scientists should refuse to contribute to arms development. Later in his life he concluded that to hold to a "no arms" policy would be to place the world at the mercy of its worst enemies. Each country has a right to be protected from those who use force, or the threat of force, to impose their will on others. Force can morally be used only against those persons who are themselves using force. Weapons development should be limited to weapons that are suitable for that use. Neither the present arms spiral nor nuclear weapons are consistent with Einstein's principles. One of our greatest scientists, he knew that international security requires progress in political education, not weapons technology.

SDI Background

The Strategic Defense Initiative, popularly known as "Star Wars," was initiated by a 1983 presidential speech calling on scientists to free us from the fear of nuclear weapons. President Reagan directed the Pentagon to search for a way to make nuclear strategic missiles impotent and obsolete. In response, the SDIO has embarked upon a project to develop a network of satellites carrying sensors, weapons, and computers to detect intercontinental ballistic missiles (ICBMs) and intercept them before they can do much damage. In addition to sponsoring work on the basic technologies of sensors and weapons, SDI has funded a number of Phase

I "architecture studies," each of which proposes a basic design for the system. The best of these have been selected, and the contractors are now proceeding to "Phase II," a more detailed design.

My Early Doubts

As a scientist, I wondered whether technology offered us a way to meet these goals. My own research has centered on computer software, and I have used military software in some of my research. My experience with computer-controlled weapon systems led me to wonder whether any such system could meet the requirements set forth by President Reagan.

I also had doubts about a possible conflict of interest. I have a project within the U.S. Navy that could benefit from SDI funding. I suggested to the panel organizer that this conflict might disqualify me. He assured me that if I did not have such a conflict, they would not want me on the panel. He pointed out that the other panelists—employees of defense contractors and university professors dependent on DoD [Department of Defense] funds for their research—had similar conflicts. Readers should think about such conflicts the next time they hear of a panel of "distinguished experts."

My Work for the Panel

The first meeting increased my doubts. In spite of the high rate of pay, the meeting was poorly prepared; presentations were at a disturbingly unprofessional level. Technical terms were used without definition; numbers were used without supporting evidence. The participants appeared predisposed to discuss many of the interesting but tractable technical problems in space-based missile defense, while ignoring the basic problems and "big picture." Everyone seemed to have a pet project of their own that they thought should be funded.

At the end of the meeting we were asked to prepare position papers describing research problems that must be solved in order to build an effective and trustworthy shield against nuclear missiles. I spent the weeks after the meeting writing up those problems and trying to convince myself that SDIO-supported research could solve them. I failed. I could not convince myself that it would be possible to build a system that we could trust, nor that it would be useful to build a system we did not trust.

Why Trustworthiness Is Essential

If the U.S. does not trust SDI, it will not abandon deterrence and nuclear missles. Even so, the U.S.S.R. could not assume that SDI would be completely ineffective. Seeing both a "shield" and missiles, it would feel impelled to improve its offensive forces in an effort to compensate for SDI. The U.S., not trusting its defense, would feel a need to build still more nuclear missiles to compensate for the increased Soviet strength. The arms race would speed up. Further, because NATO would be wasting an immense amount of effort on a system it couldn't trust, we would see a weakening of our relative strength. Instead of a safer world . . . we would have a far more dangerous situation. Thus, the issue of our trust in the system is critical. Unless the shield is trustworthy, it will not benefit any country.

The Role of Computers

SDI discussions often ignore computers, focusing on new developments in sensors and weapons. However, the sensors will produce vast amounts of raw data that computers must process and analyze. Computers must detect missile firings, determine the source of the attack, and compute the attacking trajectories. Computers must discriminate between threatening warheads and mere decoys designed to confuse our defensive system. Computers will aim and fire the weapons. All the weapons and sensors will be useless if the computers do not function properly. Software is the glue that holds such systems together. If the software is not trustworthy, the system is not trustworthy.

The Limits of Software Technology

Computer specialists know that software is always the most troublesome component in systems that depend on computer control. Traditional engineering products can be verified by a combination of mathematical analysis, case analysis, and prolonged testing of the complete product under realistic operating conditions. Without such validation, we cannot trust the product. None of these validation methods works well for software. Mathematical proofs verify only abstractions of small programs in restricted languages. Testing and case analysis sufficient to ensure trustworthiness take too much time. As E. W. Dijkstra has said, "Testing can show the presence of bugs, never their absence."

The lack of validation methods explains why we cannot expect a real program to work properly the first time it is really used. This is confirmed by practical experience. We can build adequately reliable software systems, but they become reliable only after extensive use in the field. Although responsible developers perform many tests, including simulations, before releasing their software, serious problems always remain when the first customers use the product. The test designers overlook the same problems the software designers overlook. No experienced person trusts a software system before it has seen extensive use under actual operating conditions.

Why Software for SDI Is Especially Difficult

SDI is far more difficult than any software system we have ever attempted. Some of the reasons are listed here; a more complete discussion can be found in an article published in *American Scientist* (see reference list).

SDI software must be based on assumptions about target and decoy characteristics, and those characteristics are controlled by the attacker. We cannot rely upon our information about them. The dependence of any program on local assumptions is a rich source of effective countermeasures. Espionage could render the whole multibillion-dollar system worthless without our knowledge. It could show an attacker how to exploit the inevitable differences between the computer model on which the program is based and the real world.

The techniques used to provide high reliability in other systems are hard to apply for SDI. In space, the redundancy required for high reliability is unusually expensive. The dependence of SDI on communicating computers in satellites makes it unusually vulnerable. High reliability can be achieved only if the failures of individual components are statistically independent; for a system subject to coordinated attacks, that is not the case.

Overloading the system will always be a potent countermeasure, because any computer system will have a limited capacity, and even crude decoys would consume computer capacity. An overloaded system must either ignore some of the objects it should track, or fail completely. For SDI, either option is catastrophic.

Satellites will be in fixed orbits that will not allow the same one to track a missile from its launch and to destroy it. The responsibility for tracking a missile will transfer from one satellite to another. Because of noise caused by the battle and enemy interference, a satellite will require data from other satellites to assist in tracking and discrimination. The result is a distributed real-time data base. For the shield to be effective, the data will have to be kept up-to-date and consistent in real time. This means that satellite clocks will have to be accurately synchronized. None of this can be done when the network's components and communication links are unreliable, and unreliability must be expected during a real battle in which an enemy would attack the network. Damaged stations are likely to inject inaccurate or false data into the data base.

Realistic testing of the integrated hardware and software is impossible.

Thorough testing would require "practice" nuclear wars, including attacks that partially damage the satellites. Our experience tells us that many potential problems would not be revealed by lesser measures such as component testing, simulations, or small-scale field tests.

Unlike other weapon systems, this one will offer us no opportunity to modify the software during or after its first battle. It must work the first time.

These properties are inherent in the problem, not some particular system design. As we will see below, they cannot be evaded by proposing a new system structure.

My Decision to Act

After reaching the conclusions described above, I solicited comments from other scientists and found none that disagreed with my technical conclusions. Instead, they told me that the program should be continued, not because it would free us from the fear of nuclear weapons, but because the research money would advance the state of computer science! I disagree with that statement, but I also consider it irrelevant. Taking money allocated for developing a shield against nuclear missiles, while knowing that such a shield was impossible, seemed like fraud to me. I did not want to participate, and submitted my resignation. Feeling that it would be unprofessional to resign without explanation, I submitted position papers to support my letter. I sent copies to a number of government officials and friends, but did not send them to the press until after they had been sent to reporters by others. They have since been widely published.

SDIO's Reaction

The SDIO's response to my resignation transformed my stand on SDI from a passive refusal to participate to an active opposition. Neither the SDIO nor the other panelists reacted with a serious and scientific discussion of the technical problems that I raised.

The first reaction came from one of the panel organizers. He asked me to reconsider, but not because he disagreed with my technical conclusions. He accepted my view that an effective shield was unlikely, but argued that the money was going to be spent and I should help to see it well spent. There was no further reaction from the SDIO until a *New York Times* reporter made some inquiries. Then, the only reaction I received was a telephone call demanding to know who had sent the material to the *Times*.

After the story broke, the statements made to the press seemed, to me, designed to mislead rather than inform the public. Examples are given below. When I observed that the SDIO was engaged in "damage control," rather than a serious consideration of my arguments, I felt that I should inform the public and its representatives of my own view. I want the public to understand that no trustworthy shield will result from the SDIO-sponsored work. I want them to understand that technology offers no magic that will eliminate the fear of nuclear weapons. I consider this part of my personal professional responsibility as a scientist and an educator.

Critical Issues

Democracy can only work if the public is accurately informed. Again, most of the statements made by SDIO supporters seem designed to mislead the public. For example, one SDIO scientist told the public that there could be 100,000 errors in the software and it would still work properly. Strictly speaking, this statement is true. If one picks one's errors very carefully, they won't matter much. However, a single error caused the complete failure of a Venus probe many years ago. I find it hard to believe that the SDIO spokesman was not aware of that.

Another panelist repeatedly told the press that there was no fundamental law of computer science that said the problem could not be solved. Again, strictly speaking, the statement is true, but it does not counter my arguments. I did not say that a correct program was impossible; I said that it was impossible that we could trust the program. It is not impossible that such a program would work the first time it was used; it is also not impossible that 10,000 monkeys would reproduce the works of Shakespeare if allowed to type for five years. Both are highly unlikely. However, we could tell when the monkeys had succeeded; there is no way that we could verify that the SDI software was adequate.

Another form of disinformation was the statement that I—and other SDI critics—were demanding perfection. Nowhere have I demanded perfection. To trust the software we merely need to know that the software is free of catastrophic flaws, flaws that could cause massive failure or that could be exploited by a sophisticated enemy. That is certainly easier to achieve than perfection, but there is no way to know when we have achieved it.

A common characteristic of all these statements is that they argue with statements other than the ones I published in my papers. In fact, in some cases SDIO officials dispute statements made by earlier panels or by other SDIO officials, rather than debating the points I made.

The "90 Percent" Distraction

One of the most prevalent arguments in support of SDI suggests that if there are three layers, each 90 percent effective, the overall "leakage" would be less than 1 percent because the effectiveness multiplies. This argument is accepted by many people who do not have scientific training. However,

- There is no basis for the 90 percent figure; an SDI official told me it was picked for purpose of illustration.

- The argument assumes that the performance of each layer is independent of the others, when it is clear that there are actually many links.

- It is not valid to rate the effectiveness of such systems by a single "percentage." Such statistics are only useful for describing a random process. Any space battle would be a battle between two skilled opponents. A simple percentage figure is no more valid for such systems than it is as a way of rating chess players. The performance of defensive systems depends on the opponent's tactics. Many defensive systems have been completely defeated by a sophisticated opponent who found an effective countermeasure.

The "Loose Coordination" Distraction

The most sophisticated response was made by the remaining members of SDIO's Panel on Computing in Support of Battle Management, which named itself the Eastport group, in December 1985. This group of SDI proponents wrote that the system structures proposed by the best Phase I contractors, those being elaborated in Phase II, would not work because the software could not be built or tested. They said that these "architectures" called for excessively tight coordination between the "battle stations"—excessive communication—and they proposed that new Phase I studies be started. However, they disputed my conclusions, arguing that the software difficulties could be overcome using "loose coordination."

The Eastport Report neither defines its terms nor describes the structure that it had in mind. Parts of the report imply that "loose coordination" can be achieved by reducing the communication between the stations. Later sections of the report discuss the need for extensive communication in the battle-station network, contradicting some statements in the earlier section. However, the essence of their argument is that SDI could be trustworthy if each battle station functioned autonomously, without relying on help from others.

Three claims can be made for such a design:

- It decomposes an excessively large program to a set of smaller ones, each one of which can be built and tested.
- Because the battle stations would be autonomous, a failure of some would allow the others to continue to function.
- Because of the independence, one could infer the behavior of the whole system from tests on individual battle stations.

The Eastport group's argument is based on four unstated assumptions:

1. Battle stations do not need data from other satellites to perform their basic functions.
2. An individual battle station is a small software project that will not run into the software difficulties described above.
3. The only interaction between the stations is by explicit communication. This assumption is needed to conclude that test results about a single station allow one to infer the behavior of the complete system.
4. A collection of communicating systems differs in fundamental ways from a single system.

All of these assumptions are false!

1. The data from other satellites is essential for accurate tracking, and for discriminating between warheads and decoys in the presence of noise.
2. Each battle station has to perform all of the functions of the whole system. The original arguments apply to it. Each one is unlikely to work, impossible to test in actual operating conditions, and consequently impossible to trust. Far easier projects have failed.
3. Battle stations interact through weapons and sensors as well as through their shared targets. The weapons might affect the data produced by the sensors. For example, destruction of a single warhead or decoy might produce noise that makes tracking of other objects impossible. If we got a single station working perfectly in isolation, it might fail completely when operating near others. The failure of one station might cause others to fail because of overload. Only a real battle would give us confidence that such interactions would not occur.
4. A collection of communicating programs is mathematically equivalent to a single program. In practice, distribution makes the problem harder, not easier.

Restricting the communication between the satellites does not solve the problem. There is still no way to know the effectiveness of the system, and it would not be trusted. Further, the restrictions on communication are likely to reduce the effectiveness of the system. I assume that this is why none of the Phase I contractors chose such an approach.

The first claim in the list is appealing, and reminiscent of arguments made in the 1960s and 1970s about modular programming. Unfortunately, experience has shown that modular programming is an effective technique for making errors easier to correct, not for eliminating errors. Modular programming does not solve the problems described earlier in this paper. None of my arguments was based on an assumption of tight coupling; some of the arguments do assume that there will be data passed from one satellite to another. The Eastport Report, like earlier reports, supports that assumption.

The Eastport group is correct when it says that designs calling for extensive data communication between the battle stations are unlikely to work. However, the Phase I contractors were also right when they assumed that without such communication the system could not be effective.

Redefining the Problem

The issue of SDI software was debated in March 1986 at an IEEE [Institute of Electrical and Electronics Engineers] computer conference. While two of us argued that SDI could not be trusted, the two SDI supporters argued that it did not matter. Rather than argue the computer-science issues, they tried to use strategic arguments to say that a shield need not be considered trustworthy. One of them argued, most eloquently, that the president's "impotent and obsolete" terminology was technical nonsense. He suggested that we ignore what "the president's speechwriters" had to say and look at what was actually feasible. Others argue[d] that increased uncertainty is a good thing—quite a contrast to President Reagan's promise of increased security.

In fact, the ultimate response of the computer scientists working on SDI is to redefine the problem in such a way that there is a trivial solution and improvement is always possible. Such a problem is the ideal project for government sponsorship. The contractor can always show both progress and the need for further work. Contracts will be renewed indefinitely!

.   .   .   .   .

Broader Questions

Is SDIO-Sponsored Work of Good Quality?

Although the Eastport panel were unequivocally supportive of continuing SDI, their criticisms of the Phase I studies were quite harsh. They assert[ed] that those studies, costing a million dollars each, overlooked elementary problems that were discussed in earlier studies. If the Eastport group is correct, the SDIO contractors and the SDIO evaluators must be considered incompetent. If the Eastport group's criticisms were unjustified, or if their alternative is unworkable, *their* competence must be questioned.

Although I do not have access to much of the SDIO-sponsored work in my field, I have had a chance to study some of it. What I have seen makes big promises, but is of low quality. Because it has bypassed the usual scientific review processes, it overstates its accomplishments and makes no real scientific contribution.

Do Those Who Take SDIO Funds Really Disagree with Me?

I have discussed my views with many who work on SDIO-funded projects. Few of them disagree with my technical conclusions. In fact, since the story became public, two SDIO contractors and two DoD agencies have sought my advice. My position on this subject has not made them doubt my competence.

Those who accept SDIO money give a variety of excuses. "The money is going to be spent anyway; shouldn't we use it well?" "We can use the money to solve other problems." "The money will be good for computer science."

I have also discussed the problem with scientists at the Los Alamos and Sandia National Laboratories. Here, too, I found no substantive disagreement with my analysis. Instead, I was told that the project offered lots of challenging problems for physicists.

In November 1985, I read in *Der Speigel* an interview with a leading German supporter of Star Wars. He made it clear that he thought of SDI as a way of injecting funds into high technology and not as a military project. He even said that he would probably be opposed to participation in any deployment should the project come to fruition.

The Blind Led by Those with Their Eyes Shut

My years as a consultant in the defense field have shown me that unprofessional behavior is common. When consulting, I often find people doing something foolish. Knowing that the person involved is quite competent, I may say something like, "You know that's not the right way to do that." "Of course," is the response, "but this is what the customer asked for." "Is your customer a computer scientist? Does he know what he is asking?" ask I. "No" is the simple reply. "Why don't you tell him?" elicits the response: "At XYZ Corporation, we don't tell our customers that what they want is wrong. We get contracts."

That may be a businesslike attitude, but it is not a professional one. It misleads the government into wasting taxpayers' money.

The Role of Academic Institutions

Traditionally, universities provide tenure and academic freedom so that faculty members can speak out on issues such as these. Many have done just that. Unfortunately, at U.S. universities there are institutional pressures in favor of accepting research funds from any source. A researcher's ability to attract funds is taken as a measure of his ability.

The president of a major university in the U.S. recently explained his acceptance of a DoD institute on campus by saying, "As a practical matter, it is important to realize that the Department of Defense is a major administrator of research funds. In fact, the department has more research funds at its disposal than any other organization in the country. . . . Increases in research funding in significant amounts can be received only on the basis of defense-related appropriations."

Should We Pursue SDI for Other Reasons?

I consider such rationalizations to be both unprofessional and dangerous. SDI endangers the safety of the world. By working on SDI, these scientists allow themselves to be counted among those who believe that the program can succeed. If they are truly professionals, they must make it very clear that an effective shield is unlikely, and a trustworthy one impossible. The issue of more money for high technology should be debated without the smoke screen of SDI. I can think of no research that is so important as to justify pretending that an ABM system can bring security to populations. Good research stands on its own merits; poor research must masquerade as something else.

I believe in research; I believe that technology can improve our world in many ways. I also agree with Professor Janusz Makowski of the Technion Institute, who wrote in the *Jerusalem Post,* "Overfunded research is like heroin, it leads to addiction, weakens the mind, and leads to prostitution." Many research fields in the U.S. are now clearly overfunded, largely because of DoD agencies. I believe we are witnessing the proof of Professor Makowski's statement.

My Advice on Participation in Defense Projects

I believe that it's quite appropriate for professionals to devote their energies to making the people of their land more secure. In contrast, it is not professional to accept employment doing "military" things that do not advance the legitimate defense interests of that country. If the project would not be effective, or if, in one's opinion, it goes beyond the legitimate defense needs of the country, a professional should not participate. Too many do not ask such questions. They ask only how they can get another contract.

It is a truism that if each of us lives as though what we do does matter, the world will be a far better place than it is now. The cause of many serious problems in our world is that many of us act as if our actions do not matter. Our streets are littered, our environment polluted, and children neglected because we underestimate our individual responsibility.

The arguments given to me for continuation of the SDI program are examples of such thinking. "The government has decided; we cannot change it." "The money will be spent; all you can do is make good use of it." "The system will be built; you cannot change that." "Your resignation will not stop the program."

It is true, my decision not to toss trash on the ground will not eliminate litter. However, if we are to eliminate litter, I must decide not to toss trash on the ground. We all make a difference.

Similarly, my decision not to participate in SDI will not stop this misguided program. However, if everyone who knows that the program will not lead to a trustworthy shield against nuclear weapons refuses to participate, there will be no program. Every individual's decision is important.

It is not necessary for computer scientists to take a political position; they need only be true to their professional responsibilities. If the public were aware of the technical facts, if they knew how unlikely it is that such a shield would be effective, public support would evaporate. We do not need to tell the public not to build SDI. We only need to help them understand why it will never be an effective and trustworthy shield.

References

Einstein, Albert, and Sigmund Freud. 1972. *Warum Krieg?* Zürich: Diogenes Verlag.

Parnas, D. L. 1985. Software aspects of strategic defense systems. *American*

Scientist, September–October: 432–40. Also published in German in Kursbuch 83, *Krieg und Frieden—Streit um SDI,* by Kursbuch/ Rotbuch Verlag, March 1986; and in Dutch in *Informatie,* Nr. 3, March 1986: 175–86.

Parnas, D. L. On the criteria to be used in decomposing systems into modules. *Communications of the ACM* 15, 12: 1053–58.

Eastport Group. Summer study 1985. A Report to the Director—Strategic Defense Initiative Organization, December 1985.

Wer kuscht, hat keine Chance. *Der Spiegel,* Nr. 47, 18 November 1985.

34

SDI Research Is Critical

Walter E. Morrow, Jr.

The U.S. Congress and White House are facing critical decisions about the Strategic Defense Initiative (SDI), the research program that focuses on developing a defense against ballistic nuclear missiles. At issue is the type of experiments allowed under the U.S.-Soviet Anti-Ballistic Missile Treaty, whether to deploy the system early, and the proper funding level for the research program.

Successful defense of ballistic missiles through SDI is by no means certain, but the danger of nuclear weapons to humanity warrants exploring all possibilities. Congress should therefore maintain a strong research program. This will help determine whether some form of early deployment would be viable. And SDI field experimentation that is compatible with the ABM Treaty is critical in the face of the large Soviet missile-defense effort. A capable Soviet missile defense in the absence of a comparable U.S. system would gravely threaten U.S. and NATO security.

SDI proposes to change the U.S. approach to national security from complete reliance on the threat of retaliation to a partial or complete reliance on defense. While the 1970s missile-defense systems protected limited areas, such as missile fields or the national capital, SDI has been

From *Technology Review,* July 1987, pp. 24–25, 77. Reprinted with permission from *Technology Review,* copyright 1987.

proposed as a defense of the entire U.S. land mass and population. As the program has evolved, the defense of U.S. allies and military forces has also been considered.

These proposed changes in U.S. strategy are based on changes in technical approach. The focus has shifted from stopping ballistic missiles close to their targets to destroying them at any point in their trajectory, including during the boost phase. This "multilayer" approach offers the promise of intercepting much higher percentages of attacking missiles than the "single-layer" systems of the 1970s.

SDI would also entail a transition from ground-based defensive missiles with nuclear warheads to space-based and ground-based non-nuclear weapons. SDI would employ kinetic-energy weapons (which destroy targets by physical impact), lasers, or particle beams. In contrast to nuclear weapons, these weapons would have to hit their targets directly and would therefore require far more precise guidance systems.

To accomplish this, several new types of sensors based in aircraft and in space have been proposed to augment older ground-based radars, which bounce waves off targets. Included would be various "passive" optical sensors that "look" at the electromagnetic waves naturally emitted by targets, and "active" sensors such as high-resolution optical radars. By observing projectiles' perturbations, these sensors could discriminate between warheads and decoys traveling through space.

Viable or Not?

The question of whether SDI will be viable has proved intensely controversial. SDI opponents claim that intercontinental ballistic missiles (ICBMs) and their warheads could be protected by shielding that reflects lasers and particle beams. These critics also claim that the missiles could disperse decoys or accelerate before releasing their warheads into space, and that space-based SDI components would be vulnerable to enemy attack. Opponents have raised further concerns about the reliability of the required software. Most important, they claim that the Soviets could counter an SDI system by increasing their offensive forces for less than it would cost to improve their defenses.

Proponents of SDI claim that ICBM shielding would be ineffective, especially against kinetic weapons, and that laser and particle-beam weapons would be able to destroy ICBMs even if they accelerated quickly. They say that decoys could be detected by advanced sensors, and that SDI's space-based components could be defended. And they believe that technological advances will enable defenses to be less expensive than offensive forces.

Even the Soviet Union has conflicting views on the question of SDI's viability. A study by the Soviet Academy of Sciences concludes that SDI is technically infeasible, while Soviet political attempts to hold back SDI research suggest that the Soviets believe that the United States might succeed in developing a defensive system that would work to their disadvantage.

I believe that this divergence of opinion is largely due to a lack of scientific and engineering knowledge of the proposed technology.

Requirements for a Successful System

Perhaps the best criteria for judging the potential of an SDI system have been developed by arms negotiator Paul H. Nitze, who has been involved in national-security issues since the mid-1940s. According to Nitze, the system must be "effective" and not easily vulnerable to attack, and it must cost less than the offensive forces it defends against.

"Effective" could mean that under heavy ICBM attack, no nuclear weapons would explode on U.S. soil. Or it could mean that a less-than-perfect SDI system could so diminish the success of an attack that the Soviets would be discouraged for fear of a counterstrike.

I would accept the second interpretation as sufficient. In my opinion, perfection is not needed to provide an effective counter to offensive forces. The ability to destroy 90 percent of attacking missiles would probably deter an enemy. Our current offensive systems are imperfect, after all, but they have provided protection for a number of decades.

The issue of the defensive system's vulnerability is essentially a matter of cost. SDI's total cost has to include that of adequately protecting the system. If this total is less than the cost of an attack (including the price of destroying the SDI system), and both the United States and Soviet Union can develop an SDI-like system for roughly the same price, the two governments will be inclined to put their money into non-nuclear defense rather than nuclear offense.

By what margin should the defense cost less than the offense? A significant amount, perhaps one-half to one-third as much. In other words, the United States could contemplate spending $100 billion on SDI, since the cost of the Soviet Union's land- and sea-based missile systems is estimated at several hundred billion dollars.

What then are the chances that SDI could contribute to national security? Even though many prominent scientists and engineers have doubts, I think the chances are good. In the past eminent "experts" have often made enormous errors in judging weapons systems. Consider how wrong

Adm. William Daniel Leahy was in 1945 when, advising President Truman on the atomic bomb, he said, "The bomb will never go off, and I speak as an expert on explosives." However, the time and effort required for technical developments have often been greater than anticipated. For example, in the decade after the transistor's invention in 1947, successful applications were sparse. In the succeeding decade, however, the inventors' wildest dreams were exceeded.

Today one can estimate with reasonable confidence the range of SDI's potential effectiveness. Although a perfect defense against a massive ICBM attack is most unlikely, I agree with many experts that defensive systems can provide limited protection for our land-based intercontinental ballistic missiles. Enough forces would survive to pose a counterthreat even if a defensive system allowed 50 percent of the missiles to be destroyed.

Between these extremes lies a region of uncertainty. Given our present knowledge, no one can judge how effective a system can be built. The answer will depend on how technology advances. While there are probably many infeasible combinations of technology, it takes only one set of workable measures to produce an effective SDI system.

If a viable system can be developed in the next 5 to 10 years, it will most likely employ kinetic weapons. This technology has been demonstrated, and shielding against it is impossible. Sensors for early systems will probably involve infrared optics that sense heat from incoming ICBM rocket engines and warheads. Such sensors have also been demonstrated in a variety of tests.

Substantive engineering issues that remain unresolved include the basing choices for these weapons and sensors and the design options that would enable enough components to survive an attack. Space-based kinetic weapons would be effective because they could attack ICBMs in the boost phase, when all the warheads and any decoys would still be attached to the rocket. However, a large fraction of the kinetic weapons would be orbiting on the opposite side of the earth during an attack and could not be employed. Moreover, it might be difficult to enable kinetic weapons, which would be stationed in low-earth orbit, to survive an attack.

Ground-based weapons could be better protected from attack, and all could participate in a defense. But they would have to attack ICBMs during the mid-course or terminal portions of their flights, after the warheads had dispersed. The ground-based weapons would have the problem of distinguishing which warheads were real and which were decoys.

An initial defensive system's performance might be upgraded later using advanced laser weapons and better sensors. Orbiting laser weapons could attack at the speed of light and therefore operate at higher altitudes than kinetic weapons. That would increase the fraction of weapons that could be engaged in a defense, and might improve their ability to survive attack. The addition of laser radar sensors could prove significant in distinguishing decoys from warheads during the ICBMs' mid-course flight.

In the past, building more offensive weapons was cheaper than defending against them. The situation is likely to change with continued advances in defense-associated technologies such as sensors, information processing, space launch capability, and laser weapons. Technologies associated with offensive forces have advanced less dramatically. I believe that defense could become less expensive than offense if we do not demand unrealistically effective defense.

Our national security and our ability to continue living in a democracy require that we persist in studying new strategic weapon systems, both offensive and defensive. At a minimum we must invest in the research that is necessary to understand the potential of systems for strategic defense.

In the near term, most if not all of the important research issues, including field experimentation associated with near-term technology, can be explored within the terms of the ABM Treaty as it is traditionally interpreted. Consider, for example the primary issue of testing kinetic weapons' ability to observe targets. Data relating to the detection, tracking, and discrimination of targets by ground- and space-based weapons can be gathered by equipment that is not suitable for operational use. Therefore treaty questions are not raised.

We are not alone in pursuing such knowledge. The Soviet Union has an extensive research program in this area, about which we know relatively little. In addition, it has experience in deploying, operating, and upgrading its extensive terminal defensive system, which is built around Moscow and permitted by the ABM Treaty.

Both the United States and the Soviet Union are interested in defending against nuclear missiles. If both countries can meet the Nitze criteria in developing SDI-like defenses, I believe it would be to our mutual interest to deploy them.